Fundamentals of Charged Particle Transport in Gases and Condensed Matter

Monograph Series in Physical Sciences

Recent books in the series:

Exchange Bias: From Thin Film to Nanogranular and Bulk Systems
Surender Kumar Sharma

Fundamentals of Charged Particle Transport in Gases and Condensed Matter
Robert Robson, Ronald White, and Malte Hildebrandt

Fundamentals of Charged Particle Transport in Gases and Condensed Matter

By
Robert E. Robson, Ronald D. White and
Malte Hildebrandt

CRC Press
Taylor & Francis Group
Boca Raton London New York

CRC Press is an imprint of the
Taylor & Francis Group, an **informa** business

Cover Image: Simulations of energy deposition of positrons in liquid water, often used in modeling as a surrogate for human tissue. Points of higher (lower) energy deposition are indicated by blue (red) spheres, while trajectories of positrons between collisions, represented by black lines, are biased towards the direction of an applied electric field. Eventually the positrons slow down sufficiently to annihilate with the electrons of the medium, producing two back-to-back gamma rays, as in PET (positron emission tomography) investigations. (Courtesy of Wade Tattersall)

CRC Press
Taylor & Francis Group
6000 Broken Sound Parkway NW, Suite 300
Boca Raton, FL 33487-2742

Printed on acid-free paper

International Standard Book Number-13: 978-1-4987-3636-7 (Hardback)

Library of Congress Cataloging-in-Publication Data

Names: Robson, R. (Robert), 1946- author. | White, Ronald, author. |
Hildebrandt, Malte, author.
Title: Fundamentals of charged particle transport in gases and condensed
matter / Robert Robson, Ronald White, Malte Hildebrandt.
Description: Boca Raton, FL : CRC Press, Taylor & Francis Group, [2017] |
Series: Monograph series in physical sciences
Identifiers: LCCN 2017011666| ISBN 9781498736367 (hardback ; alk. paper) |
ISBN 149873636X (hardback ; alk. paper)
Subjects: LCSH: Kinetic theory of gases. | Transport theory. | Fluid
dynamics. | Condensed matter.
Classification: LCC QC175.13 .R63 2017 | DDC 533/.7–dc23
LC record available at https://lccn.loc.gov/2017011666

Visit the Taylor & Francis Web site at
http://www.taylorandfrancis.com

and the CRC Press Web site at
http://www.crcpress.com

This book is dedicated to Carola,
Marcella and Isabelle,

and

to the memory of Bhala Paranjape,
Edward A. Mason, Kurt Suchy
and Peter Nicoletopoulos

Contents

Monograph Series in Physical Sciences

This monograph series brings together focused books for researchers and professionals in the physical sciences. They are designed to offer expert summaries of cutting edge topics at a level accessible to non-specialists. As such, authors are encouraged to include sufficient background information and an overview of fundamental concepts, together with presentation of state of the art theory, methods, and applications. Theory and experiment are both covered. This approach makes these titles suitable for some specialty courses at the graduate level as well. Subject matter addressed by this series includes condensed matter physics, quantum sciences, atomic, molecular, and plasma physics, energy science, nanoscience, spectroscopy, mathematical physics, geophysics, environmental physics, and other areas.

Proposals for new volumes in the series may be directed to Lu Han, senior publishing editor at CRC Press/Taylor & Francis Group (lu.han@taylorandfrancis.com).

Preface

The foundations of modern transport theory were laid 150 years ago in a seminal paper presented to the Royal Society of London by J. Clerk Maxwell. He formulated the equations of change for the physical properties of a gas, represented as moments or averages over a velocity distribution function and paid particular attention to the influence of collisions. Six years later, Ludwig Boltzmann, undoubtedly influenced by Maxwell's results, presented a kinetic equation to the German Physical Society in Berlin, whose solution furnished the required distribution function. In spite of early criticism and subsequent intense scrutiny, Boltzmann's equation has withstood the test of time and has gone on to become a mainstay in the field of non-equilibrium statistical mechanics, in general, and charged particle transport, in particular, the subject of this book. The key to the success and longevity of Boltzmann's equation is not only its ability to furnish accurate theoretical values of experimentally measured quantities, but also its remarkable flexibility and adaptability to systems and physics that Boltzmann could not possibly have foreseen. Thus, there are generalizations of the kinetic equation to condensed matter, as discussed in this book, and to quantum and relativistic systems, discussed elsewhere. In addition, there are many adaptations and applications of Boltzmann's equation to traditional and contemporary areas of basic physics research and technology. To take just one example of cutting edge science: laser acceleration of particles to very high energies over distances several orders of magnitude smaller than conventional accelerators has been modelled through methods which are similar, at least in principle, to the ideas of Boltzmann and Maxwell. It would take several volumes to do justice to all of the fields on which the Boltzmann equation has had an impact and any single exposition, like the present, is necessarily circumscribed. Nevertheless, the scope of this book is broad and, moreover, the treatment is unique in that we provide a unified approach to the transport theory of particles of various types (electrons, ions, atoms, positrons, and muons) in various media (gases, soft-condensed matter, and amorphous materials). The applications are many and diverse, ranging from traditional drift tube experiments, positron emission tomography, and muon-catalyzed fusion, through to recent developments in materials physics.

One of the problems in writing a book such as this has been to overcome the perception that transport theory, beyond the simplistic mean free path arguments of some undergraduate books and courses, is somehow excessively difficult. On the one hand, it is true that a rigorous solution of the Boltzmann kinetic equation in phase space requires sophisticated mathematics and numerical procedures, and even the senior author of a well-known, formidable treatise on kinetic theory is reputed to have compared the exercise to "chewing glass." On the other hand, the original approach of Maxwell, using moment or "fluid" equations in configuration space, provides a complementary, semi-quantitative picture from which it is possible to obtain physical understanding while maintaining rigour. Both methods are employed in this book to provide a comprehensive treatment of charged particle transport phenomena.

The material has formed the basis of lecture courses given over the past 10 years in Australia and the United States at the senior undergraduate and graduate student level.

We thank Professor Michael Morrison of the University of Oklahoma; Professor Zoran Petrovic of the Institute of Physics, Belgrade; Professor Toshiaki Makabe of Keio University; and Dr. Bernhard Schmidt, originally at the University of Heidelberg and nowadays at DESY, Hamburg, for stimulating discussions and encouragement over many years. The dedication and contributions of the past and current staff, postdoctoral researchers, and post-graduate students at James Cook University cannot be understated. Particular thanks go to Kevin Ness, Bo Li, Sasa Dujko, Daniel Cocks, Gregory Boyle, Bronson Philippa, Wade Tattersall, Peter Stokes, Madalyn Casey, and Nathan Garland. The support of the Alexander von Humboldt Foundation, the Paul Scherrer Institut, the Australian Research Council, and James Cook and Griffith Universities is gratefully acknowledged.

The authors thank the publishers of "Introductory Transport Theory for Charged Particles in Gases," by R.E. Robson, Copyright 2006, World Scientific Publishing Company Pty. Ltd, for granting us permission to adapt and reproduce parts of this publication in the present book.

About the Authors

Robert Robson obtained his PhD in theoretical physics in 1972 at the Australian National University, Canberra. After a postdoctoral fellowship at the University of Alberta, Canada, he went on to lecture and research in physics and meteorology in Australia, the United States, Japan, and Europe. He was an Alexander von Humboldt Fellow at the University of Düsseldorf, Germany and held the Hitachi Chair of Electrical Engineering at Keio University, Japan. His most recent research has been on soft-condensed matter and amorphous semiconductors at James Cook University, Australia, and modelling relativistic electron beams in plasma-based accelerators at Deutsches Elektronen-Synchrotron (DESY), Hamburg. He is a distinguished member of the Australian Association of von Humboldt Fellows, and a Fellow of the American Physical Society and the Royal Meteorological Society.

Ronald White obtained his PhD in theoretical physics in 1997 with a study of electron transport in gases relevant to plasma processing studies at James Cook University. After research appointments in Australia, Japan, and the United States, he returned to James Cook University where he took up a lectureship in 2002 in the Mathematics Department. He was promoted to full Professor in 2015 and is currently the Head of Physical Sciences at James Cook University, which encompasses the Mathematics, Physics, and Chemistry Departments. His research interests are focused on the non-equilibrium transport of charged particles in gases, soft-condensed matter, liquids, and organic matter.

Malte Hildebrandt studied physics at the University of Heidelberg, Germany, and worked for his diploma thesis on electron swarm experiments. In 1999, he completed his PhD on the development of particle detectors for high energy physics experiments. He went on to a postdoctoral position at the University of Zürich, Switzerland, and moved later to the Paul Scherrer Institut, Switzerland. Since 2009, he has been Head of the Detector Group of the Laboratory for Particle Physics at the Paul Scherrer Institut. His work focuses on particle detectors, in particular, gaseous detectors, for charged particles and neutrons.

Glossary of Symbols and Acronyms

Symbol	Meaning
$\mathbf{a}$	external force per unit mass
α	scaling factor for velocity $\sqrt{\frac{m}{k_B T}}$
b	impact parameter
χ	scattering angle in centre of mass
D	diffusion tensor (starred quantities are "flux" while non-starred are "bulk")
ϵ	energy
ε	spatially uniform energy
E_e or E_{eff}	equivalent of effective electric field
$f(\mathbf{v})$, $f_0(\mathbf{v}_0)$	particle and neutral velocity distribution functions
$\phi(\tau)$	relaxation function for de-trapping
$\phi_m^{(\nu,l)}(\mathbf{v})$	Burnett function
$\mathbf{g}$, $\mathbf{G}$	relative and centre-of-mass velocities
γ	gradient energy parameter
Γ	particle flux
I, U	electric current and applied voltage
J, $J^\dagger$	collision operator and its adjoint
$\mathbf{J}_q$	heat flux vector
K, $\mathcal{K}$	mobility and reduced mobility coefficients
K_j	spectral wave number
λ_D	Debye length
m, m_0	particle and neutral molecular masses
n, n_0	particle and neutral number densities
N	total particle number
ν_m, ν_e	momentum and energy-transfer collision frequencies
$\vec{\nu}_i$, $\overleftarrow{\nu}_i$	inelastic and superelastic collision frequencies
ν_I	ionization collision frequency
ν_*	reactive loss collision frequency
$\widetilde{\nu}_m$	structure-modified collision frequency
ω, Ω_L	angular frequency of applied electric field, gyrofrequency of magnetic field
Ω	inelastic collision transfer term
P	pressure tensor
$\sigma(g, \chi)$	differential cross section

Symbol	Meaning
$\sigma^{(l)}, \sigma_m$	lth partial and momentum-transfer cross sections
$S(K, \Omega)$	structure function
T, T_0, T_b	particle, neutral, and basis temperatures
$\mathbf{v}, \mathbf{v}_0$	velocities of particles and neutrals
$\langle\rangle, \langle\rangle_0$	averages over particle and neutral velocities
$\langle\mathbf{v}\rangle$	average particle velocity
$\mathbf{v}_d$, $\mathbf{v}_d^*$	bulk and flux drift velocities
$\hat{\mathbf{v}}$	unit vector in direction of $\mathbf{v}$
$\langle\mathbf{vv}\rangle$	second rank tensor with components $\left\langle v_i v_j \right\rangle$
$V(r)$	interaction potential
$w(\alpha, v)$	Maxwellian distribution function
$Y_m^{(l)}(\hat{\mathbf{v}})$	spherical harmonic
Z	plasma dispersion function
BGK	Bhatnagar–Gross–Krook
μCF	muon-catalyzed fusion
PET	positron emission tomography
MTT	momentum transfer theory
GER	generalized Einstein relation

CHAPTER 1

Introduction

1.1 Boltzmann's Equation

1.1.1 A little history

In 1872, Ludwig Boltzmann proposed a kinetic equation of the form

$$\left(\frac{\partial}{\partial t}+L\right)f=\left(\frac{\partial f}{\partial t}\right)_{\text{col}} \tag{1.1}$$

for the velocity distribution function f of a low density gas, where L is a linear "streaming" operator in phase space, and $\left(\frac{\partial f}{\partial t}\right)_{\text{col}}$ accounts for binary, elastic collisions between the constituent atoms [1]. The expression for the latter was formulated on the basis of an Ansatz (or hypothesis), which effectively introduces an arrow of time into the evolution of the system, leading to the *H*-theorem and establishing a connection with the second law of thermodynamics. Although Boltzmann suffered criticism from his contemporaries, and the Ansatz has been the subject of considerable critical scrutiny since then, no satisfactory alternative has emerged, and the Boltzmann equation, modified by Wang Chang et al. to include inelastic collisions [2,3] remains to this day the preferred means of investigating gases in a non-equilibrium state.

Boltzmann's equation and the distribution function f play the same role in *kinetic theory* as do Schrödinger's equation and the wave function ψ in quantum mechanics. Once f is obtained from solution of Equation 1.1 all quantities of physical interest can be obtained as appropriate velocity "moments," similar to expectation values formed with $|\psi|^2$ in quantum physics (see Appendix A).

The centenary of Boltzmann's work was marked by a special publication [4] of both a biographical and scientific nature, which illustrated the extent of the influence that this remarkable equation has had on many areas of physics, involving both gases and condensed matter. Indeed, Boltzmann's contributions to the wider field of statistical mechanics are profound and are remembered in a special way (see Figure 1.1).

1.1.2 From the "golden" era of gas discharges to modern times

The emergence of Boltzmann's equation in the latter part of the nineteenth century coincided with an era of great interest in electrical discharges

Figure 1.1 The equation $S = k \log W$ linking entropy S with the number of microstates W of a system appears on Boltzmann's memorial headstone in Vienna.

in gases, though mutual recognition took some time. These investigations were motivated by the earlier observation of striations (alternating light and dark bands in the discharge) by Abria [5] (and more recently [6]), and culminated in the seminal drift tube experiments around the turn of the century and in the early 1900s. For example, Kaufmann and Thomson independently determined the elementary charge-to-mass ratio, e/m, which in turn led to Thomson's discovery of the electron, while the seminal experiment of Franck and Hertz confirmed Bohr's predictions of the quantized nature of atoms. As a result, there has been tremendous progress in science and technology, and it is not surprising that in the first three decades of the twentieth century, the field produced more than its fair share of Nobel laureates. Historical surveys of the "golden era" of drift tube experiments have been given by a number of authors, including Brown [7], Müller [8], Loeb [9], and Huxley and Crompton [10].

Investigations of gaseous discharges also spawned the field of plasma physics, with applications ranging from hot, fusion plasmas ($T \sim 10^6 K$ or more), with the promise of virtually limitless clean energy, to low temperature ($T \sim 10^4 K$) plasmas, of such importance in the microchip fabrication industry [11–13] and finally through to low density, low energy "swarms" of electrons and ions in gases [14], with applications in such diverse areas as fundamental atomic and molecular physics [15] and gaseous radiation detectors [16]. In the course of time, Equation 1.1 has come to be regarded as *de rigueur* for analyzing experiments involving charged particles in gases and condensed matter [17], along with applications of both a technological and scientific nature.

1.1.3 Transport processes: Traditional and modern descriptions

In general, non-equilibrium systems are characterized by non-uniformity and gradients in properties which result in an irreversible flow or "flux" of these properties in such a direction as to restore uniformity and equilibrium. Such *transport processes* are *traditionally* represented by well-known empirical linear flux-gradient relations, such as Fourier's law of heat conduction, and Fick's law of diffusion of matter, in which the constants of proportionality define *transport coefficients*, namely, the thermal conductivity and diffusion coefficient tensor, respectively. These coefficients can be calculated theoretically from approximate solution of the Boltzmann's equation, through linearizing in temperature and density gradient, respectively. However, one should be cautious in applying these traditional ideas to interpret drift tube experiments, for two reasons:

- Experiments are traditionally analyzed using the *diffusion equation*, which represents overall particle balance in the bulk of the system, and the coefficients in the diffusion equation differ from those defined by Fick's law when particles are created or lost, for example, by ionization and attachment, respectively. In these circumstances, experiments do not measure the traditional transport coefficients.
- Flux-gradient relations and the diffusion equation are valid only for systems which have attained a state called the *hydrodynamic regime*. Some systems never get to that state and are intrinsically non-hydrodynamic, for example, the steady state Townsend and Franck-Hertz experiments. Neither Fick's law nor the diffusion equation are physically tenable in these cases, and neither is description in terms of transport coefficients (however defined) possible. Measurable properties can be calculated theoretically only by solving Boltzmann's equation without approximation.

1.1.4 Theme of this book

In essence, Boltzmann's equation takes us from the laws of physics governing behaviour on the microscopic (atomic) scale, collisions in particular, to the level of macroscopically measurable quantities. The microscopic–macroscopic connection is the theme of our discussion, and explaining just *how* the connection is made provides the substance of this book. Put succinctly, the program is to solve Equation 1.1 for f, and then form velocity averages to find the macroscopic quantities of interest, for example, electric currents, or total particle number, which are measured in experiment.

1.2 Solving Boltzmann's Equation

1.2.1 The path to solution

- *Chapman–Enskog method:* The Chapman–Enskog method [18] is a perturbative procedure which was developed about 100 years ago to solve Boltzmann's equation for systems close to equilibrium. It was applied to gaseous ions in the 1950s by Kihara [19] and Mason and Schamp [20] but, by virtue of the limitations of the procedure, results could be obtained for only the weak field regime. Given that the systems of interest are often driven far from equilibrium by strong fields, this procedure is inadequate for most purposes.
- *Light particles, Lorentz approximation:* It was recognized early on that $\left(\frac{\partial f}{\partial t}\right)_{\text{col}}$ could be approximated in differential form for electrons undergoing elastic collisions in gases [18,21]. This simplification, together with an assumption of near-isotropy of f in velocity space, originally attributed to Lorentz [22], enables Boltzmann's equation to be solved, sometimes analytically, without any restriction on the magnitude of the field. These ideas underpin the field of gaseous electronics [23], which has maintained a distinct identity over many decades.
- *Light particles in liquids and soft matter:* Cohen and Lekner [24] modified $\left(\frac{\partial f}{\partial t}\right)_{\text{col}}$ to account for coherent scattering of electrons in liquids and, as for gaseous media, f was also assumed to be nearly isotropic in velocity space. Nevertheless, Cohen and Lekner's results have become well established in the literature and provide the basis for more sophisticated transport analysis of both electrons and positrons in liquids and soft-condensed matter.
- *Light particles, inelastic processes:* In many cases of interest, electrons also undergo inelastic collisions with the molecules of the medium, and consequently $\left(\frac{\partial f}{\partial t}\right)_{\text{col}}$ no longer assumes a simplified

differential form. The Lorentz approximation is also questionable if inelastic processes are significant and, all in all, solution of Boltzmann's equation becomes more difficult. In fact, the degree of difficulty is on a par with heavier ions, for which there is significant anisotropy in velocity space even if inelastic processes are absent. This points towards the need for a general procedure for solving Boltzmann's equation for particles of all masses and types.

- *Wannier's theory:* In the 1950s, Wannier [25] solved Boltzmann's equation for dilute ions in gases in the strong field regime, though specifically for special models of interaction. He also formulated a relationship between the mean ion energy and average velocity, and sowed the seeds of an idea for a semi-quantitative alternative to rigorous numerical solution of Boltzmann's equation, which is nowadays called "momentum-transfer theory."
- *The Viehland–Mason solution for ions:* Around the time of the Boltzmann centenary in 1972, computing power had reached a level where rigorous numerical solution of the Boltzmann equation for ions had become possible for realistic forms of interaction, and without resorting to any perturbation method. In a series of papers commencing in 1975, Viehland, Mason, and collaborators developed a general method of solution of Boltzmann's equation for dilute ions in gases in electric fields of arbitrary strength [26–28]. The modern era of charged particle kinetic theory can be traced from this time.
- *Towards a unified kinetic theory:* Lin et al. [29] combined the essentials of the Viehland–Mason approach with Kumar's tensor formalism, adapted from atomic and nuclear physics [30], to develop a rigorous solution of Boltzmann's equation, modified to include inelastic collisions for light particles, avoiding the traditional *a priori* assumption of near-isotropy of f in velocity space. The method has been refined over the years, and nowadays provides the basis of a comprehensive kinetic theory of charged particles, ions, electrons, positrons, muons, and so on, in both gases and condensed matter. The reader can find a number of reviews and books detailing developments from the immediate post-Viehland–Mason era to more modern times [31–35].
- *Charge carriers in semiconductors:* The kinetic theory of free charge carriers (electrons and holes) scattered by phonons (lattice vibrations) in crystalline semiconductors was developed in parallel to gases [17]. It is sometimes remarked that there exists a one-to-one correspondence with scattering of charged particles from molecules and atoms in gases, even though the collision term $\left(\frac{\partial f}{\partial t}\right)_{\text{col}}$ in the kinetic equation (still referred to as

"Boltzmann's equation") is different. The role of transport theory in understanding experiments related to the development of solid-state devices including the transistor has a long history [36]. On the other hand, charge carriers are said to exhibit anomalous behaviour in disordered, non-crystalline amorphous media, such as organic semiconductors, due to trapping effects. These materials are being intensely investigated [37] and it appears that yet another technological revolution is underway [38,39]. The kinetic theory associated with these processes is, however, a "work in progress," with only simple forms of $\left(\frac{\partial f}{\partial t}\right)_{\text{col}}$ having so far been employed [40,41].

1.2.2 A complementary approach: Fluid modelling

After solving the Boltzmann equation as described above, quantities of physical interest are formed by taking appropriate velocity averages of f. An alternative approximate, more computationally economical and physically appealing alternative is to find the averages directly by solving approximate moment or fluid equations in configuration space. These equations can be formed either by taking velocity moments of Boltzmann's equation, or from first principles, as Maxwell [42] did 6 years before Boltzmann formulated his kinetic equation. In fact, the roles can be completely reversed, as we show in this book, and Boltzmann's equation can be obtained (and later solved) using the moment equation method.

Maxwell paid particular attention to the collision terms in the moment equations and showed that they could be evaluated exactly for a particular model, in which the interaction varied inversely as the fifth power of the distance. The Maxwell model, which corresponds to a point-charge, induced dipole interaction, is particularly suitable as a first approximation when discussing charged particles in gases. It provides the basis for "momentum transfer" theory [33], which has proved particularly successful in semi-quantitative fluid modelling of charged particle transport phenomena [43].

1.3 Experiment and Simulation

1.3.1 An idealized apparatus

Although this book focuses on theory, we touch briefly on experiments [10,15,34,36,44,45] though it is not possible to discuss technical details. We instead focus on principles of operation, following the style of Kumar [14], using as an example the idealized experimental arrangement shown in Figure 1.2.

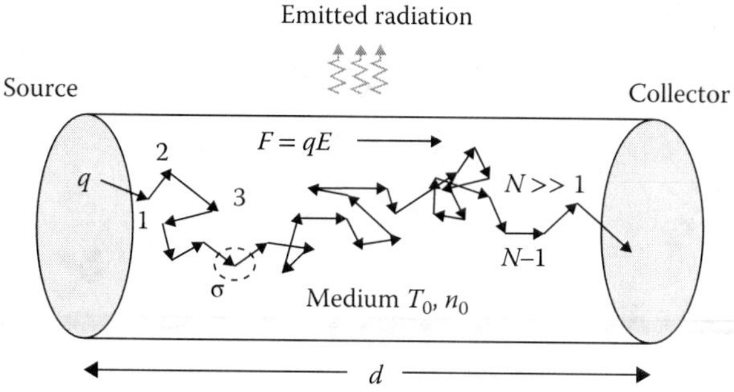

Figure 1.2 A schematic representation of an experiment in which particles of charge q are emitted by the source electrode and travel through a medium of known properties to a collecting electrode a distance d away under the influence of an electric field. Collisions are represented by the vertices of the trajectory and are characterized by appropriate scattering cross sections σ.

Particles of charge q emitted by a source electrode are forced by a uniform electric field E to move a distance d through a chamber containing a medium of known properties (gas or condensed matter) to a collecting electrode. Particle number density n is assumed sufficiently small so that mutual interactions are negligible in comparison with interactions of particles with the constituents of the medium. Such collisions are assumed to be local, that is, to take place in a region small compared with any macroscopic dimension, effectively at a point, and are represented by the vertices in the particle trajectory shown in the figure.

The source may operate in a pulsed or continuous mode. In some experiments, particles incident on the collecting electrode form the current measured in an external circuit. In the Franck–Hertz experiment [46], there is a modulating grid in front of the collecting electrode. In the Cavalleri experiment [10], it is the total number of particles within the chamber that is determined as a function of time. In yet other experiments, the radiation emitted by atoms and molecules returning to a lower energy level after excitation in a collision may be used as a diagnostic tool, as in the photon flux technique [47].

For a *gaseous* medium, particles may be considered to collide with individual atoms and molecules and the various processes (elastic, inelastic, ionizing, reacting, etc.) are characterized by a corresponding binary scattering cross section σ. Collisions take place on a time scale small compared with any relevant macroscopic scale, and to all intents and purposes are instantaneous. In the *time-of-flight experiment*, an initial sharp pulse of particles injected at the source spreads at a constant rate about its centre-of-mass, which moves with constant velocity v_d through the medium, as shown in Figure 1.3. Although the pulse spreads in the course of time,

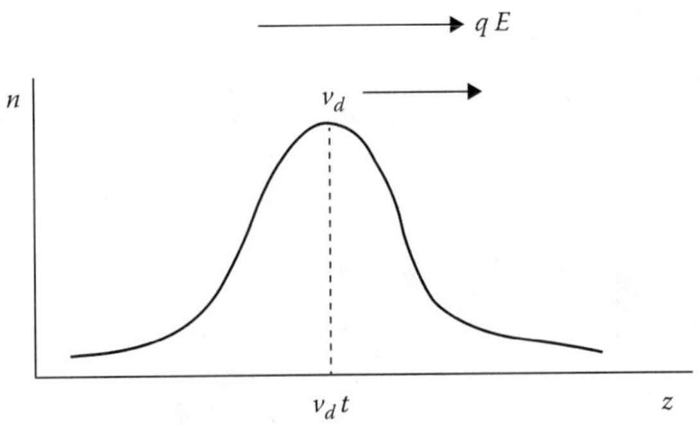

Figure 1.3 The number density n of charged particles as a function of distance z from the source at a time t after injection into the medium, initially as a sharp pulse. After a sufficient number of collisions, the pulse has spread out and its centre-of-mass travels with constant velocity v_d, determined by field, the scattering cross sections, and the properties of the medium. The width of the pulse increases with time t in proportion to the (longitudinal) diffusion coefficient.

it still retains its identity, and its two main properties (centre-of-mass velocity and width) are readily measurable. The same properties may be calculated from solution of Boltzmann's equation. Naturally, the theoretical values should be calculated to at least the same accuracy as the experimentally measured counterparts. Typically, the accuracy in swarm drift tube experiments is 0.1%–1.0% for the drift velocity v_d [10].

The picture is similar for charge carriers scattered from phonons in a *crystalline semiconductor,* and there too the time-of-flight experiment is the canonical experiment.

For a *soft matter* medium with short-range order, particles are scattered simultaneously (diffracted) by many constituent molecules. Nevertheless, the picture of local, instantaneous interactions at the vertices of Figure 1.2 prevails, and a pulse in a time-of-flight experiment in this medium generally maintains its distinct identity. There are cases, however (e.g., electrons in neon), where this is not the case [48], where electrons can get caught and released from "bubble" states.

For *amorphous materials,* such as organic semiconductors, this is generally not the case, where particles are trapped for finite times in localized states. The picture shown in Figure 1.2 still holds, but vertices now represent "collisions" (trapping/de-trappings) lasting finite times, rather than taking place instantaneously. Particles may be trapped for significant times over the entire length of the chamber; consequently, the particle density profile in a time-of-flight experiment is qualitatively quite different. In particular, there is no distinct travelling pulse in a time-of-flight experiment [37].

1.4 About this Book

In this book, we focus on non-relativistic, low density charged particles which interact predominantly with the background medium, and neglect mutual Coulomb interactions and self-consistent fields. The main aims are to:

- Formulate kinetic and fluid equations for charged particles in gases, soft-condensed matter, and amorphous materials, allowing for coherent scattering and/or trapping in localized states where necessary,
- Outline the basic techniques for solving the kinetic equation and for calculating transport properties,
- Understand the link between the microscopic processes and the macroscopic transport properties,
- Apply the theory to traditional and new areas of science, technology, and medical diagnostic techniques.

While rigour is a watchword, we use short arguments and simplified mathematics wherever possible to elucidate the physics.

The structure is as follows:

Part I: Fundamentals of kinetic theory, derivation of Boltzmann's related kinetic equations, as well as calculation of classical cross sections.

Part II: Simplified treatment of transport processes through a fluid equation analysis, in which Boltzmann's kinetic equation in phase space is replaced by a set of approximate "moment" equations in configuration space.

Part III: Procedures and techniques for solution of Boltzmann's equation.

Part IV: Applications include boundary effects and diffusion cooling, Franck–Hertz experiment, anomalous transport in amorphous semiconductors, calculation of positron range in positron emission tomography (PET), muon-catalyzed fusion, and gaseous radiation detectors.

Part V: Gives a series of appendices providing extra information, miscellaneous proofs and values of numerical constants, together with a set of exercises aimed at reinforcing the material in the text, and a comprehensive list of references to books and original papers.

Additional General Reading Materials

- A good introductory text on statistical mechanics: D.V. Schroeder, "Thermal Physics" (Addison-Wesley, Longman, 2000).

- A good introductory background to kinetic theory can be found in the following article: E.D.G. Cohen, *Amer. J. Phys.* 61:524, 1993 (Sections I and II A,B,C only).
- A widely used text for graduate level statistical mechanics: K. Huang, "Statistical Mechanics" 2nd Edition (Wiley, 1987), especially Chapters 3–5.
- A favourite classical mechanics text: H. Goldstein, "Classical Mechanics", 2nd Edition (Addison-Wesley, 1980).
- Graduate level texts dealing with charged particles in gases:
  - R.E. Robson, "Introductory Transport Theory for Charged Particles in Gases" (World Scientific Singapore, 2006).
  - M. Charlton and J.W. Humberston, "Positron Physics" (Cambridge University Press, 2001).
  - E.H. Holt and R.E. Haskell, "Plasma Dynamics" (Macmillan, 1965).
  - E.W. McDaniel, "Collision Phenomena in Ionized Gases" (Wiley, New York, 1964).
  - M.A. Uman, "Introduction to Plasma Physics" (McGraw-Hill, 1964).
  - D.C. Montgomery and D.A. Tidman, "Plasma Kinetic Theory" (McGraw-Hill, 1964).
- Books dealing with transport processes in semiconductors and solid-state devices include:
  - H. Haug and A. Jauho, "Quantum Kinetics in Transport and Optics of Semiconductors" (Springer, Berlin, 2008).
  - S.M. Sze and K.K. Ng, "Physics of Semiconductor Devices" 3rd Edition (Wiley, New York, 2007).
  - K. Seeger, "Semiconductor Physics" (Springer, Berlin, 1989).
  - E. Conwell, "High field transport in semiconductors," Suppl. No. 9 to "Solid State Physics," editors H. Ehrennreich, F. Seitz and D. Turnbull (Academic Press, New York, 1967).
  - C. Kittel, "Elementary solid state physics" 8th Edition, (Wiley, New York, 2005).
- A good introduction to charge carriers in amorphous materials is given by R. Zallen, "The Physics of Amorphous Solids" (Wiley, New York, 1983).
- Although not directly related to the theme of this book, the monograph by M.M.R. Williams "Mathematical Methods in Particle Transport Theory" (Butterworths, London, 1971), contains much useful information, along with important theorems of a general nature and details of mathematical techniques.

- Advanced general kinetic theory references:
  - R. L. Liboff, "Kinetic Theory," 2nd edition (Wiley, New York, 1998), Chapters 3 and 4.
  - A.R. Hochstim and G. Massell, "Kinetic Processes in Gases and Plasmas" (Academic Press, New York, 1969).
- A useful reference on thermodynamics and its relation to Boltzmann's equation: S.R. de Groot and P. Mazur, "Non-equilibrium Thermodynamics" (North Holland, Amsterdam, 1969).

Part I

Kinetic Theory Foundations

CHAPTER 2

Basic Theoretical Concepts: Phase and Configuration Space

2.1 Preliminaries

2.1.1 Configuration and velocity space

Consider a non-degenerate, non-relativistic system comprised of N particles each of mass m confined in a volume V, for which there are N "representative points" in *configuration space* (see Figure 2.1).

If there are d^3n such points in a small volume d^3r centred on $\mathbf{r} = (x, y, z)$, the local *number density* is defined to be $n(\mathbf{r}, t) = d^3n/d^3r$. It follows that integration over the entire volume gives the normalization condition:

$$N = \int_V d^3n = \int_V n(r, t)d^3r. \tag{2.1}$$

A quantity of some significance is the average velocity $\langle \mathbf{v} \rangle(\mathbf{r}, t)$, which is the average over all particle velocities $\mathbf{v}$ within d^3r. The *particle flux* is the number of particles crossing unit area in unit time normal to the direction of the flow:

$$\Gamma(\mathbf{r}, t) = n(\mathbf{r}, t) \langle \mathbf{v} \rangle. \tag{2.2}$$

The *equation of continuity* space represents conservation of particle number in configuration space:

$$\frac{\partial n}{\partial t} + \nabla \cdot \Gamma = 0. \tag{2.3}$$

If particle number is not conserved and there are $S(\mathbf{r}, t)$ particles produced per unit volume per unit time, the equation of continuity becomes

$$\frac{\partial n}{\partial t} + \nabla \cdot \Gamma = S(\mathbf{r}, t). \tag{2.4}$$

Such sources may be internal, caused by "reactive" (non-conservative) collisions within the medium, for example, by electron impact ionization, or external through, for example, an ultraviolet ionizing radiation source. It is also possible to have particle loss mechanisms (e.g., electron attachment), in which case S is negative. These internal sources/sinks can influence the transport properties of charged particles in matter in

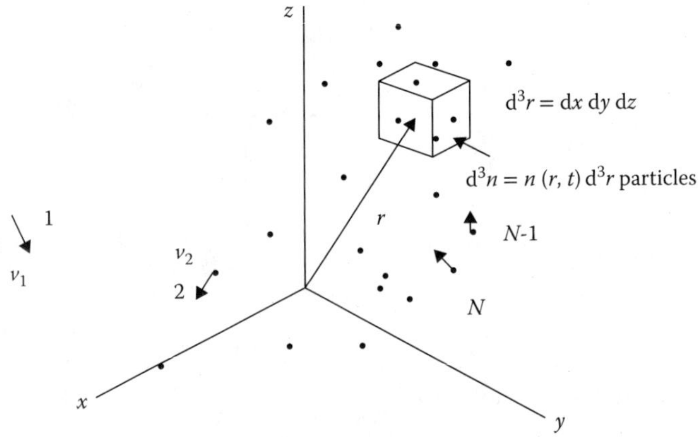

Figure 2.1 Configuration space, in which a system comprised of N particles has a corresponding number of representative points. The number of points per unit volume $n(\mathbf{r}, t)$ is called the *number density*.

fundamental ways. Note, however, that neither of these forms of the equation of continuity is of much practical use until Γ and S are further specified. Equation 2.4 can be derived from either first principles or as a velocity "moment" of the kinetic equation, as discussed in Section 2.4.2.

An *ideal gas* is defined to be one in which there are no forces acting between particles. The particles we consider definitely do interact! Furthermore, the systems dealt with are almost always *not* in equilibrium and cannot be studied using the methods of equilibrium statistical mechanics [49].

Let us return to the infinitesimal volume element d^3r of configuration space. The d^3n particles lying within have a whole range of possible velocities $\mathbf{v}$, which may be represented by points in a *velocity space* (see Figure 2.2)

2.1.2 Distribution function and averaging

The *phase space distribution function* $f(\mathbf{r}, \mathbf{v}, t)$ is defined such that $f(\mathbf{r}, \mathbf{v}, t)\, d^3r\, d^3v \equiv$ number of particles with velocities in the region d^3v of $\mathbf{v}$ and positions within d^3r of $\mathbf{r}$. If we integrate over all $\mathbf{v}$, while holding position $\mathbf{r}$ fixed, we must regain $d^3n = \left\{ \int f(\mathbf{r}, \mathbf{v}, t) d^3v \right\} d^3r$. After dividing through by d^3r, there follows the normalization condition on f:

$$n(\mathbf{r}, t) = d^3n/d^3r = \int f(\mathbf{r}, \mathbf{v}, t)\, d^3v. \tag{2.5}$$

Here and elsewhere, it is implicit by virtue of the non-relativistic assumption that velocity integrals are over the infinite range $(-\infty, \infty)$.

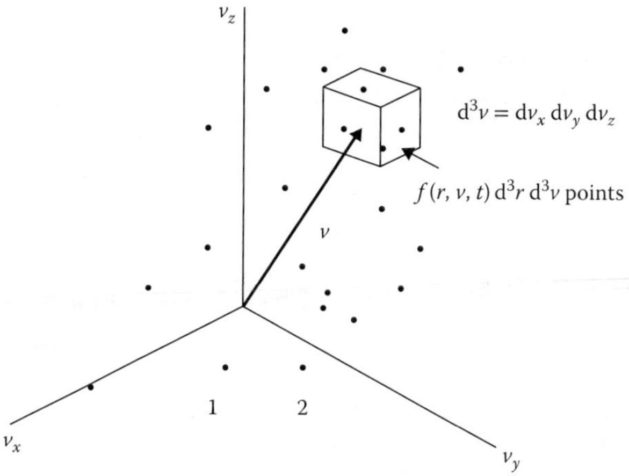

Figure 2.2 A *velocity space*, as shown above, exists in each infinitesimal volume of the configuration space. We denote the number of representative points in a velocity space element d^3v embedded within d^3r by $f(\mathbf{r}, \mathbf{v}, t)\, d^3r\, d^3v$. Thus, $f(\mathbf{r}, \mathbf{v}, t)$ is effectively the density of points in the *phase space* comprised of configuration and velocity spaces.

Averages over velocity space at fixed position $\mathbf{r}$ are defined for arbitrary functions $\phi(\mathbf{v})$ by

$$\langle \phi(\mathbf{v}) \rangle = \frac{1}{n(\mathbf{r}, t)} \int \phi(\mathbf{v})\, f(\mathbf{r}, \mathbf{v}, t)\, d^3v. \tag{2.6}$$

For example, setting $\phi(\mathbf{v}) = \mathbf{v}$ gives the local average velocity $\langle \mathbf{v} \rangle(\mathbf{r}, t)$ referred to previously. Other quantities of physical interest are the mean energy $\epsilon(\mathbf{r}, t) = \langle \frac{1}{2} m v^2 \rangle$, found by setting $\phi(\mathbf{v}) = \frac{1}{2} m v^2$, and temperature T, defined by

$$\frac{3k_B T}{2} = \frac{1}{2} m \langle V^2 \rangle, \tag{2.7}$$

where $k = 1.38 \times 10^{-23}$ J/ deg is *Boltzmann's constant* and

$$\mathbf{V} = \mathbf{v} - \langle \mathbf{v} \rangle$$

is called the *peculiar velocity*. Note that averages of functions of velocity only may in general be space–time dependent.

While velocity space averages like Equation 2.6 are our main focus, averages of functions $\Phi(\mathbf{r}, \mathbf{v})$ over all or part of phase space,

$$\langle \Phi(\mathbf{r}, \mathbf{v}) \rangle = \frac{1}{N} \int_V d^3r \int d^3v\, \Phi(\mathbf{r}, \mathbf{v})\, f(\mathbf{r}, \mathbf{v}, t), \tag{2.8}$$

where V is the volume of the region under consideration, can also be of interest.

2.1.3 Polar coordinates and symmetries

An important point concerns symmetries; particularly, the fact that a symmetry in velocity space does not necessarily translate to symmetry in configuration space. A proper treatment of diffusion of electrons in gases was delayed for many years because of just such a misunderstanding (see Chapter 7).

To deal with symmetry properties, it is usually better to work with spherical or cylindrical coordinate systems, rather than Cartesian coordinates. The situation in velocity space for spherical velocity coordinates $v = (v, \theta, \varphi)$ is shown in Figure 2.3. Note that an element of solid angle in **v**-space is $d\Omega_{\mathbf{v}} = \sin\theta \, d\theta \, d\varphi$, while a volume element is expressible in terms of spherical polar coordinates as $d^3v = v^2 dv \, d\Omega_v = v^2 dv \, \sin\theta \, d\theta \, d\varphi$.

As an example, consider a spatially uniform system of electrons in methane gas, with an electric field directed in the negative z-direction, so that $-\mathbf{E}$ defines the v_z-axis. Clearly, the distribution function depends upon v and θ, but not φ, that is, $f = f(v, \theta)$ since the system is rotationally invariant about the v_z axis. The picture in velocity space (see Figure 2.4, [50]) is the same everywhere in configuration space in this special case. Note that in Figure 2.4, we have used electron energy $\epsilon = \frac{1}{2}mv^2$ rather than the speed v as independent coordinate.

The diagram is in the nature of a polar plot, showing lines of constant $f = f(\epsilon, \theta)$ drawn in much the same way as isobars (lines of constant pressure) on a weather chart. Spheres of constant ϵ are also shown for comparison. Note the distortion and displacement of the $f-$ contours reflect the preferential flow of electrons in the field direction (actually in the direction of $-\mathbf{E}$, since we are dealing with negatively charged particles). If the field were zero and the electrons were in thermal

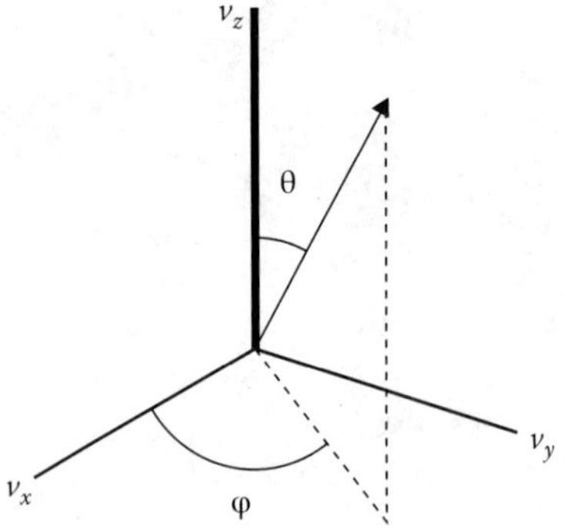

Figure 2.3 Polar coordinates (v, θ, φ) are often used in velocity space. Elements of solid angle and volume are then $d^2\Omega_{\mathbf{v}} = \sin\theta \, d\theta \, d\varphi$ and $d^3v = v^2 dv \, d^2\Omega_{\mathbf{v}}$, respectively.

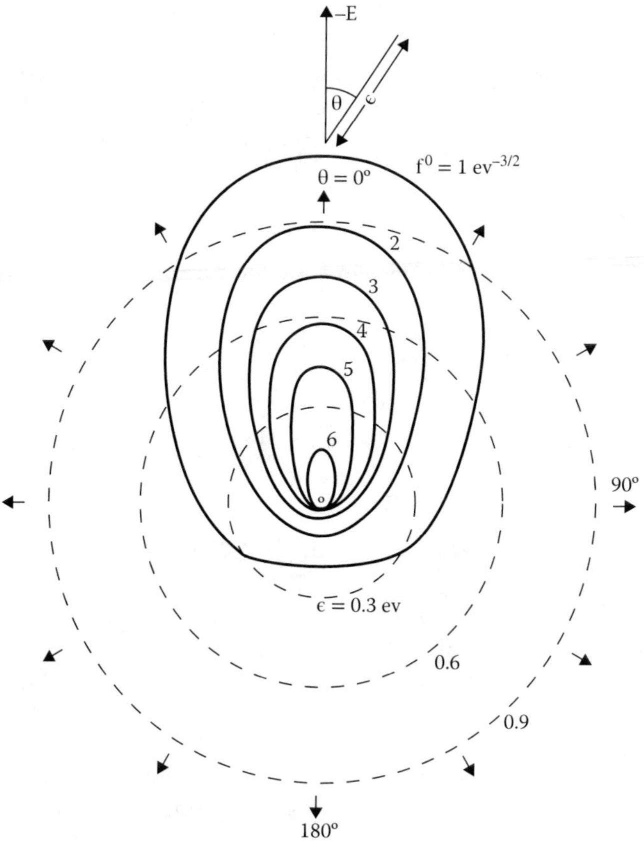

Figure 2.4 Representation of the velocity distribution function for electrons in methane gas under spatially uniform conditions. Polar angles θ are measured with respect to the $-E$ direction, which defines an axis of rotational symmetry (no φ− dependence). The electron velocity distribution function is denoted here by $f(\epsilon, \theta)$. Contours (solid lines) correspond to lines of constant f, while dashed lines represent spheres of constant energy ϵ. The reduced field is $E/n_0 = 5\,\text{Td}$, where 1 *townsend* (Td) $= 10^{-21}\,\text{Vm}^2$, and the mean energy is 0.6 eV. (From K. F. Ness, Kinetic theory of charged particle swarms with applications to electrons, PhD Thesis, James Cook University, 1986). See also [51] for further examples.

equilibrium with the methane gas molecules, both would have an equilibrium (Maxwellian) distribution of velocities, depending upon v only, with corresponding spherical f−contours.

Consider next a cylindrical tube containing an ionized gas, of the sort found in plasma processing devices, and take cylindrical spatial coordinates $\mathbf{r} = (r_\perp, \varphi_r, z)$ with the z-axis defining the axis of the cylinder. Correspondingly, we have cylindrical velocity coordinates $\mathbf{v} = (v_\perp, \varphi, v_z)$. Suppose that properties can vary in the radial direction $r_\perp$ due, for example, to charged particle flow along a radially directed field or gradient but that the system is invariant with respect to the rotations about the z-axis. Thus, looking along any radial vector, we expect to see the same physical properties, independent of the value of the spatial azimuthal coordinate φ_r. in particular, the distribution of velocities should look the same in all radial directions. But does this mean that $f(r, v, t)$ should be independent of φ_r or φ? Certainly not! In fact, it can be readily shown that the dependence upon coordinates must be of the form $f(r_\perp, \varphi_r, z; v_\perp, \varphi, v_z; t) = f(r_\perp, z; v_\perp, \varphi - \varphi_r, z; t)$ [52]. This is just one example where properties in configuration space and velocity space may be quite different.

The above also illustrates the strategy that we adopt in calculations: identify any symmetries at the outset, thereby deducing the maximum number of independent quantities to be found, and then solve the kinetic equations (refer Chapter 12 for more details).

2.2 Phase Space and Kinetic Equation

2.2.1 Trajectories in phase space

We now combine configuration and velocity space into one six-dimensional phase space (μ-space), and represent each of the N constituent particles by a point corresponding to its velocity and position. (Traditionally, phase space is defined by spatial and *momentum* coordinates, $\mathbf{r}$ and $\mathbf{p}$, respectively, but for non-relativistic situations, where the particle mass is close to the rest mass m, $\mathbf{r}$, and velocity $\mathbf{v} = \mathbf{p}/m$ offer an equivalent description.) A "volume" element in phase space is denoted by $\mathrm{d}^6 n = \mathrm{d}^3 r\ \mathrm{d}^3 v$ and the number of points in this volume is, according to the definition of the distribution function, equal to $f(\mathbf{r}, \mathbf{v}, t)\ \mathrm{d}^3 r\ \mathrm{d}^3 v$. Thus, $f(\mathbf{r}, \mathbf{v}, t)$ is effectively the density of points in μ-space and, like its counterpart in three-dimensional configuration space, $n(\mathbf{r}, t)$, satisfies a balance equation. Points move in μ-space for three reasons:

1. Position coordinates $\mathbf{r}$ change smoothly by virtue of the fact that the particles have velocity $\mathbf{v}$, that is, $\frac{\mathrm{d}\mathbf{r}}{\mathrm{d}t} = \mathbf{v}$.
2. Velocities change smoothly due to the action of a field of force,* $\mathbf{F}$, that is, $\frac{\mathrm{d}\mathbf{v}}{\mathrm{d}t} = \mathbf{F}/m$.
3. Collisions between particles scatter points in and out of phase space regions. In a dilute *gas* or *crystalline semiconductor*, particle velocities change *abruptly* due to *binary collisions*, which may be considered to be both *local* and *instantaneous*. That is, collisions take place in a spatial region and on a time scale, which are

* In general, the force field has both external and internal (self-consistent) contributions.

very small compared with any relevant macroscopic scales (see Figure 2.5).

In *soft-condensed matter*, particles may be coherently scattered from many molecules simultaneously. While collisions are strictly speaking not binary, they may nevertheless still be regarded as local and instantaneous. On the other hand, in *amorphous (and other disordered) materials*, particles can be trapped in localised states and then de-trapped after a finite time back into the conduction band. Collisions cannot be treated as instantaneous, though they are still essentially *local.*

In all cases, the motion in μ-space can be represented as in Figure 2.6 by a smooth trajectory, characterized by a six-dimensional "velocity" $v^{(6)} = (\dot{\mathbf{r}},\dot{\mathbf{v}}) = (\mathbf{v}, \mathbf{F}/m)$, interspersed with abrupt, vertical changes in the velocity coordinate.

2.2.2 Kinetic equation in phase space

The *kinetic equation* represents the way in which a "fluid" of representative points behaves in phase space and is essentially an extension of the equation of continuity (Equation 2.4) in configuration space to six-dimensional phase space. Thus, we postulate

$$\frac{\partial f}{\partial t} + \nabla^{(6)} \cdot \Gamma^{(6)} = S^{(6)}, \tag{2.9}$$

where the representative point "flux" is defined by $\Gamma^{(6)} = f\, v^{(6)}$, and $\nabla^{(6)} = (\nabla, \frac{\partial}{\partial \mathbf{v}})$. The scattering in/out of a cell in phase space by collisions

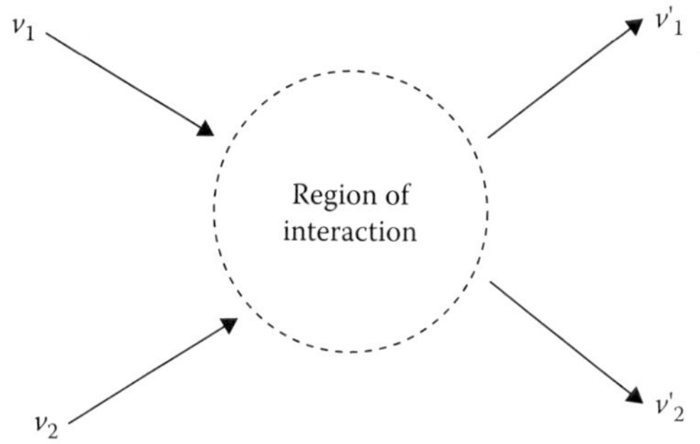

Figure 2.5 A basic assumption of kinetic theory is that collisions take place in a region very small compared with any relevant macroscopic spatial dimension, effectively at a point. In gases and crystalline semiconductors, particles interact with the medium through two-body (binary) collisions, but in soft-condensed matter, the interaction involves many-body collisions. Except for amorphous media, where trapping occurs, the duration of a collision may be assumed to be very small (effectively instantaneous) compared with any relevant macroscopic time scale.

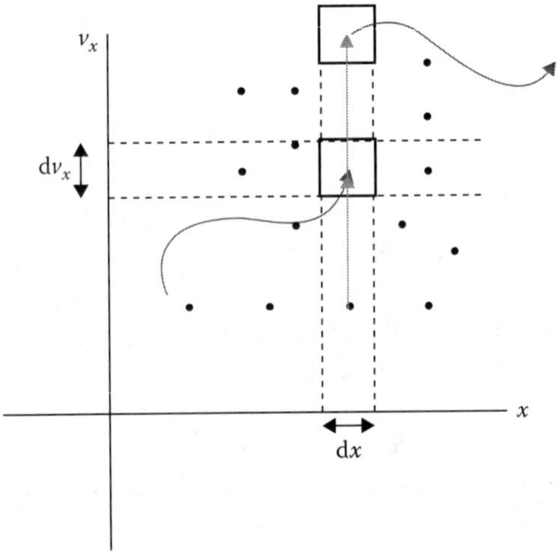

Figure 2.6 The trajectory of a point in phase space varies smoothly (curved arrows) due to the intrinsic velocity and action of external forces, but collisions induce abrupt changes (vertical arrows) into and out of a phase space element. For particles in gases and soft-condensed matter, collisions are effectively instantaneous, but for amorphous materials, scattering back into the element is delayed by trapping. In all cases, collisions take place effectively at a point, and the corresponding transitions in phase space are vertical.

corresponds to a local source/sink term in phase space and has been written as $S^{(6)}$. Writing the latter as $\left(\frac{\partial f}{\partial t}\right)_{\text{col}}$ and substitution for the six-dimensional quantities in the left hand side, then gives

$$\frac{\partial f}{\partial t} + \nabla \cdot (f\ \mathbf{v}) + \frac{\partial}{\partial \mathbf{v}} \cdot (f\ \mathbf{F}/m) = \left(\frac{\partial f}{\partial t}\right)_{\text{col}}.$$

Since $\mathbf{r}$ and $\mathbf{v}$ are independent coordinates and, if the force term is either independent of $\mathbf{v}$, or of the form $\mathbf{F} = q(\mathbf{E} + \mathbf{v} \times \mathbf{B})$, then $\frac{\partial}{\partial \mathbf{v}} \cdot \mathbf{F} = 0$. Thus, it follows that

$$\frac{\partial f}{\partial t} + \mathbf{v} \cdot \nabla f + a \cdot \frac{\partial f}{\partial \mathbf{v}} = \left(\frac{\partial f}{\partial t}\right)_{\text{col}}, \tag{2.10}$$

where

$$\mathbf{a} \equiv \mathbf{F}/m$$

is the acceleration suffered by a particle due to the force field $\mathbf{F}$, which may be of external or internal origin, or a combination of both.

Equation 2.10 is the generic form of the *kinetic equation* which forms the basis for the (non-relativistic) investigations carried out in this book.

Relativistic systems require modification of the left hand side and are investigated in a phase space spanned by $(\mathbf{r}, \mathbf{p})$ rather than $(\mathbf{r}, \mathbf{v})$ [53].

2.2.3 Equilibrium

A system is said to be in thermodynamic equilibrium if it is in thermal, mechanical and chemical equilibrium. In the context of the present discussion, equilibrium pertains when the properties of the system are spatially uniform, and constant in time, with no force field. The left hand side of Equation 2.10 then vanishes, $\frac{\partial f}{\partial t} + \mathbf{v}\cdot\nabla f + \mathbf{a}\cdot\frac{\partial f}{\partial \mathbf{v}} = 0$, and hence

$$\left(\frac{\partial f}{\partial t}\right)_{\text{col}} = 0$$

The solution of this equation (see Chapter 3) for a gas of atoms of mass m_0 at temperature T_0 is the *Maxwellian* velocity distribution function,

$$f_0(\mathbf{v}) = n\left(\frac{\alpha^2}{2\pi}\right)^{\frac{3}{2}} \exp\left(-\frac{1}{2}\alpha^2 v^2\right), \tag{2.11}$$

where $\alpha \equiv \sqrt{\frac{m_0}{k_B T_0}}$. For a molecular gas this generalizes to the Maxwell–Boltzmann distribution over velocities and internal states (see Chapter 5).

2.3 Kinetic Equations for a Mixture

2.3.1 The general kinetic equation

Equation 2.10 can be generalized in a straightforward way to a mixture. Thus for a system of S species, specified by an index $s = 1, 2, ..., S$ there are S coupled kinetic equations,

$$\frac{\partial f_s}{\partial t} + \mathbf{v}_s\cdot\nabla f_s + \mathbf{a}_s\cdot\frac{\partial f_s}{\partial \mathbf{v}_s} = \sum_{s'=1}^{S}\left(\frac{\partial f}{\partial t}\right)_{\text{col}}^{(s,s')} \qquad (s = 1, 2, ..., S) \tag{2.12}$$

for each of the distribution functions $f_s\ (\mathbf{r}, \mathbf{v}, t)$. Here $\left(\frac{\partial f}{\partial t}\right)_{\text{col}}^{(s,s')}$ is the rate of change of f_s due to collisions between particles of species s and s', and $\mathbf{a}_s = \mathbf{F}_s/m_s$ is the acceleration suffered by a particle of species s due to the force field. Equation 2.12 provides the basis for the study of non-relativistic charged particles in gases and condensed matter.

2.3.2 Dilute particles in a neutral medium

For a two-component system ($S = 2$) consisting of dilute particles (species $s = 1$) in a neutral medium (species $s = 2$), there are, in principle, two kinetic equations, one for each of the respective distribution functions f_1 and f_2. However, if the charged particles are of sufficiently low density, then mutual collisions between them are infrequent compared with their interaction with the molecules of the background medium, then $\left|\left(\frac{\partial f}{\partial t}\right)_{\text{col}}^{(1,1)}\right| \ll \left|\left(\frac{\partial f}{\partial t}\right)_{\text{col}}^{(1,2)}\right|$ and, to a good approximation, the kinetic equation for f_1 simplifies to

$$\frac{\partial f_1}{\partial t} + \mathbf{v}_1 \cdot \nabla f_1 + \mathbf{a}_1 \cdot \frac{\partial f_1}{\partial \mathbf{v}_1} = \left(\frac{\partial f}{\partial t}\right)_{\text{col}}^{(1,2)} + \left(\frac{\partial f}{\partial t}\right)_{\text{col}}^{(1,1)} \approx \left(\frac{\partial f}{\partial t}\right)_{\text{col}}^{(1,2)}. \tag{2.13}$$

Furthermore, the particles, being relatively few in number, perturb the background medium only slightly and, if initially in equilibrium, it remains so. Thus, to a good approximation $f_2 \approx f_0$, the Maxwellian equation (Equation 2.11) and no second kinetic equation is required. Equation 2.13 is the prototype kinetic equation for most of the systems discussed in this book.

2.3.3 Locality, instantaneity, and linearity

For particles in gaseous and soft-condensed matter media, $\left(\frac{\partial f}{\partial t}\right)_{\text{col}}^{(1,2)}$ represents collisions which are *instantaneous* in time, and *local* in space (Figure 2.5). Such collisions change the velocity coordinate of $f_1(\mathbf{r}, \mathbf{v}, t)$, but not $\mathbf{r}$ or t. On the other hand, in *amorphous materials,* where trapping and de-trapping occur, collisions are local, but cannot be regarded as instantaneous. Consequently, $\left(\frac{\partial f}{\partial t}\right)_{\text{col}}^{(1,2)}$ involves changes in both the velocity and time coordinates of $f_1(\mathbf{r}, \mathbf{v}, t)$.

In all cases, $\left(\frac{\partial f}{\partial t}\right)_{\text{col}}^{(1,2)}$ is *linear* in f_1, while $\left(\frac{\partial f}{\partial t}\right)_{\text{col}}^{(1,1)}$ is non-linear. For dilute particles, the latter may be neglected, as in Equation 2.13. Furthermore, if the force field can be prescribed externally, the kinetic equation itself is also linear, making for a considerably simplified problem.

2.4 Moment Equations

2.4.1 The general moment equation

Equations for velocity "moments" can be formulated by multiplying the kinetic equation by an arbitrary function $\phi(\mathbf{v})$ of velocity and integrating over all $\mathbf{v}$. Thus from Equation 2.10, we obtain

$$\frac{\partial n\langle\phi\rangle}{\partial t} + \nabla \cdot n\langle \mathbf{v}\phi(\mathbf{v})\rangle - n\left\langle \mathbf{a} \cdot \frac{\partial \phi}{\partial \mathbf{v}} \right\rangle = \int \mathrm{d}^3 v \, \phi(\mathbf{v}) \left(\frac{\partial f}{\partial t}\right)_{\text{col}}, \tag{2.14}$$

where averages are defined by Equation 2.6 and superscripts have been dropped for simplicity. This procedure can be thought of as a projection of a six-dimensional phase space onto three-dimensional configuration space. Equation 2.14 provides the basis of the so-called "fluid" method, as discussed in Chapters 7–10.

2.4.2 Equation of continuity

Setting $\phi = 1$ in Equation 2.14 yields

$$\frac{\partial n}{\partial t} + \nabla \cdot \Gamma = S(\mathbf{r}, t), \tag{2.15}$$

where particle density is given by Equation 2.5 and the particle flux follows from Equation 2.2:

$$\Gamma = n\langle \mathbf{v} \rangle = \int \mathrm{d}^3 v \, \mathbf{v} \, f(\mathbf{r}, \mathbf{v}, t).$$

The source term

$$S \equiv \int \mathrm{d}^3 v \left(\frac{\partial f}{\partial t} \right)_{\mathrm{col}} \tag{2.16}$$

represents the rate of creation of particles through collisions. In general $S \neq 0$, but if collisions are instantaneous, and if particle number is conserved, that is, the number and type of particles remain unchanged, as in Figure 2.5, then $S = 0$. In spite of transitions into and out of the phase space cells, the *total* number of points in a vertical column in Figure 2.6 remains constant. The losses and gains cancel out overall, and hence at any time and position the integral on the right hand side of Equation 2.16 vanishes. This is the case for conservative collisions in gaseous, crystalline, and soft-condensed matter media.

2.5 Concluding Remarks

This chapter gives a broad overview of classical, non-relativistic kinetic theory, both in its general aspects and the way in which it is to be applied to charged particles in a background medium. The key quantity is the phase space distribution function $f(\mathbf{r}, \mathbf{v}, t)$, which can be found by solving the kinetic equation incorporating a collision term $\left(\frac{\partial f}{\partial t} \right)_{\mathrm{col}}$ appropriate to the physical system under investigation. Expressions for $\left(\frac{\partial f}{\partial t} \right)_{\mathrm{col}}$ are derived in the following chapters, starting with Boltzmann's famous collision integral for elastic collisions between point particles in gases.

CHAPTER 3

Boltzmann Collision Integral, H-Theorem, and Fokker–Planck Equation

3.1 Classical Collision Dynamics

3.1.1 Conservation laws

We now consider *elastic* binary collisions between two structureless, classical, and spinless point particles. The masses are m_1 and m_2, respectively, while velocities before and after the collision are denoted by $\mathbf{v}_1$, $\mathbf{v}_2$ and $\mathbf{v}'_1$, $\mathbf{v}'_2$, respectively. In what follows, we shall use primes to denote post-collision properties.

Of central importance are the center-of-mass (CM) and relative velocities

$$\mathbf{G} = M_1\mathbf{v}_1 + M_2\mathbf{v}_2$$
$$\mathbf{g} = \mathbf{v}_1 - \mathbf{v}_2,$$

where

$$M_i = m_i/(m_1 + m_2) \quad (i = 1, 2).$$

The inverse relationships are

$$\begin{aligned} \mathbf{v}_1 &= \mathbf{G} + M_2\mathbf{g} \\ \mathbf{v}_2 &= \mathbf{G} - M_1\mathbf{g}. \end{aligned} \tag{3.1}$$

It is left as an exercise to show that momentum and energy conservation require

$$\begin{aligned} \mathbf{G}' &= \mathbf{G} \\ g' &= g. \end{aligned} \tag{3.2}$$

Note that for point particles, "energy" means "kinetic energy" and the "elastic" means that the *total* kinetic energy is conserved, *not* just the kinetic energy of one of the colliding partners. Inelastic collisions may occur when one or both of the colliding partners possesses internal structure, leading to a change of internal energy levels during a collision and non-conservation of total kinetic energy. This is discussed in detail in Chapter 5.

3.1.2 Transformation of coordinates

Next, consider a coordinate system translating with the CM velocity $\mathbf{G}$, in which reference frame the particle velocities before and after collision are given by

$$\begin{aligned}
\tilde{\mathbf{v}}_1 &= \mathbf{v}_1 - \mathbf{G} = M_2\mathbf{g} \\
\tilde{\mathbf{v}}_2 &= \mathbf{v}_2 - \mathbf{G} = -M_1\mathbf{g} \\
\tilde{\mathbf{v}}_1' &= \mathbf{v}_1' - \mathbf{G} = M_2\mathbf{g}' \\
\tilde{\mathbf{v}}_2' &= \mathbf{v}_2' - \mathbf{G} = -M_1\mathbf{g}'.
\end{aligned}$$

In the CM system, only the relative velocities $\mathbf{g}$, $\mathbf{g}'$ need be considered, and it suffices to focus attention on only one of the particles, say "1," because the the collision partner "2" always moves in exactly the opposite direction. Thus, the problem reduces to the simpler, but equivalent one of scattering of one particle by a fixed center of force. The momentum of particle 1 in the CM system is $m_1\tilde{\mathbf{v}}_1 = \mu\mathbf{g}$ where

$$\mu = m_1 M_2 = \frac{m_1 m_2}{m_1 + m_2} \tag{3.3}$$

is the *reduced mass*. Furthermore, since we are dealing with spinless point particles, we can assume that they interact through a *central force* potential $V(r)$, which depends only on the distance r between them. Angular momentum is conserved for central force interactions, and the trajectories lie in a plane.

For future reference, we note that the Jacobian of the transformation $(\mathbf{v}_1, \mathbf{v}_2) \longrightarrow (\mathbf{g}, \mathbf{G})$ is unity, that is,

$$d^3v_1\, d^3v_2 = d^3g\, d^3G. \tag{3.4}$$

3.2 Differential Cross Section

3.2.1 Basic collision parameters

We next consider a beam of particles, each of mass μ, speed g, impinging on a fixed scattering centre, as shown in Figure 3.1.

Two more parameters are needed to specify the collision completely:

- The *impact parameter* b (distance of the initial trajectory from the corresponding trajectory for a head-on collision), which is in one-to-one correspondence with the polar *scattering angle* χ, defined as

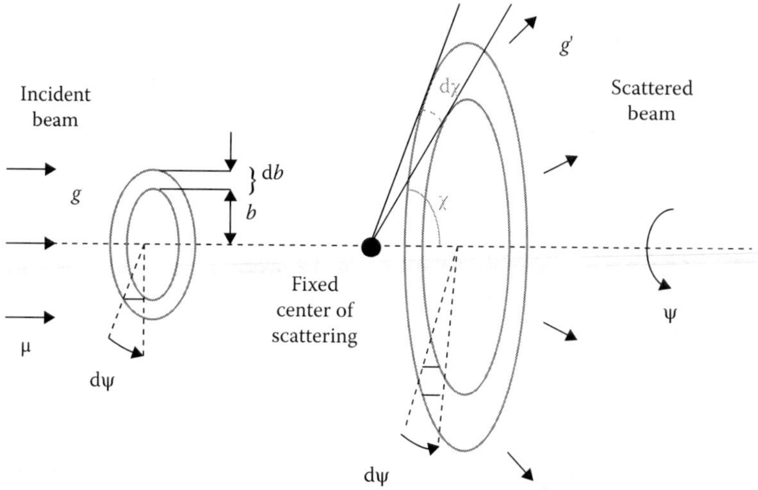

Figure 3.1 A beam of particles each of mass μ, velocity **g**, incident on a fixed centre with impact parameter between b and $b + db$ is scattered at an angle between χ and $\chi + d\chi$. All particles passing through the small annulus on the left hand side pass through the larger one on the right. For central forces, each trajectory lies in a plane, at some fixed value of ψ, and the picture is symmetric under rotations about the scattering axis (dashed line).

the angle between the directions $\hat{\mathbf{g}}$ and $\hat{\mathbf{g}}'$ of the initial and relative velocity vectors respectively, that is,

$$\cos\chi = \hat{\mathbf{g}} \cdot \hat{\mathbf{g}}' \tag{3.5}$$

and

- The azimuthal scattering angle ψ.

In Figure 3.1, we therefore picture a class of collisions, in which all particles have the same initial relative velocity **g**, but different b and ψ. Note that the latter is largely redundant for central force scattering problems, for two reasons:

- There is rotational symmetry about the scattering axis.
- ψ does not vary during a collision, since angular momentum conservation is conserved and every trajectory lies in a plane.

Thus, all particles passing through the shaded portion of the left hand ring of radius b, thickness db, with azimuthal angles lying between ψ and $\psi + d\psi$, are scattered into the solid angle defined by the angles χ and $\chi + d\chi$, ψ and $\psi + d\psi$. It is sometimes convenient to work with the coordinate system shown in Figure 3.2 in which the angles $\Omega_{g'}$ of $\mathbf{g}'$ are (χ, ψ),

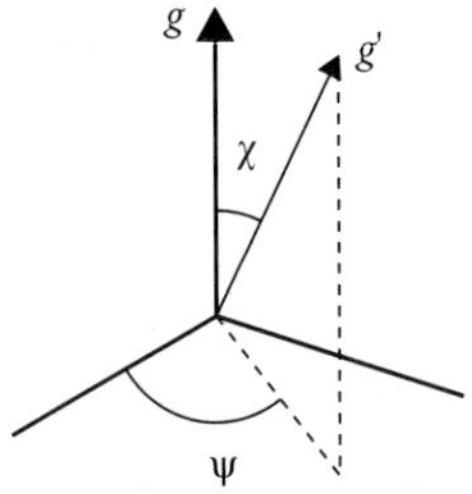

Figure 3.2 In a coordinate frame with **g** directed along the z-axis, the polar angles of $\mathbf{g}'$ are the scattering angles χ, ψ.

and the solid angle of scattering is

$$d^2\Omega_{\mathbf{g}'} = \sin\chi \, d\chi \, d\psi. \tag{3.6}$$

We are now ready to introduce the quantity of main interest in scattering problems, the differential cross section, which we write for the moment in its most general form as $\sigma(\mathbf{g}, \mathbf{g}')$.

Following the usual convention [54], we define the differential cross section by

$$\sigma(\mathbf{g}, \mathbf{g}') \, d^2\Omega_{\mathbf{g}'} = \frac{\text{Number of particles scattered into solid angle } d\Omega_{\mathbf{g}'}}{\text{Incident particle flux}}. \tag{3.7}$$

This definition applies with equal validity to both classical and quantum mechanical descriptions of scattering. There are two general properties of the differential cross that can be usefully discussed at this point.

3.2.2 Symmetries in space and time

3.2.2.1 Time-reversal invariance

Since Newton's equations of motion are invariant under the operation $t \to -t$, then the forward scattering event as shown in Figure 3.1 is indistinguishable physically from the time-reversed situation. Thus, the two events shown in Figure 3.3 are equivalent, with identical cross sections for each process.

Furthermore, the equations describing a scattering event are indistinguishable from those describing its mirror image. Thus, the parity operation applied to the last picture produces the *inverse collision* shown in Figure 3.4.

The associated cross section must also be the same as for the original event shown in Figure 3.3. This leads to the identity

$$\sigma(\mathbf{g}, \mathbf{g}') = \sigma(-\mathbf{g}', -\mathbf{g}) = \sigma(\mathbf{g}', \mathbf{g}), \tag{3.8}$$

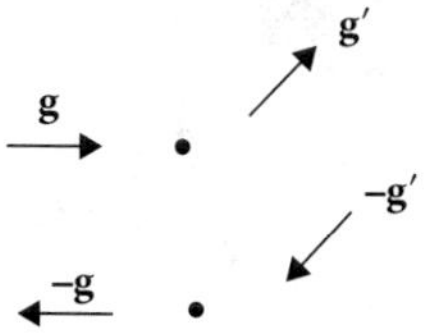

Figure 3.3 A scattering event (top) and its time-reversed counterpart (below).

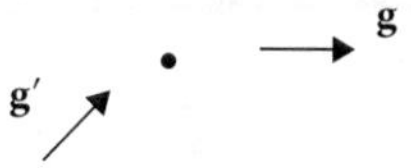

Figure 3.4 The mirror image of the lower picture in Figure 3.3.

that is, the cross section is the same for the direct and *inverse* collisions, the latter being defined as the scattering event for which the roles of the initial and final relative velocities are reversed.

3.2.2.2 Angular dependence for central force problems

Here the equivalence of incident velocities from all directions, plus the symmetry about the scattering axis of Figure 3.1 guarantee that the differential cross section depends on $\mathbf{g}$ and $\mathbf{g}'$ *solely* through their (common) magnitude and the angle χ between them, that is,

$$\sigma(\mathbf{g}, \mathbf{g}') = \sigma(g, \hat{\mathbf{g}} \cdot \hat{\mathbf{g}}') \equiv \sigma(g, \chi). \tag{3.9}$$

This result is true for both elastic and inelastic collisions, and obviously Equation 3.8 is identically satisfied.

There is a further identity which is needed for manipulations of the Boltzmann integral. If we define the differential

$$d^8\vartheta = g\,\sigma(\mathbf{g}, \mathbf{g}')\, d^2\Omega_{\mathbf{g}'} d^3g\, d^3G \tag{3.10}$$

then under the transformation $(\mathbf{v}_1, \mathbf{v}_2) \longrightarrow (\mathbf{v}_1', \mathbf{v}_2')$ this becomes

$$d^8\vartheta' = g'\,\sigma(\mathbf{g}', \mathbf{g})\, d^2\Omega_{\mathbf{g}}\, d^3g'\, d^3G'. \tag{3.11}$$

By virtue of Equations 3.2 and 3.8, and the fact that volume elements in velocity space have the form

$$d^3g = g^2 dg\, d^2\Omega_{\mathbf{g}} \tag{3.12}$$

and similarly for $d\mathbf{g}'$, it can be shown that

$$d^8\vartheta' = d^8\vartheta \tag{3.13}$$

the proof of which is straightforward and is left as an exercise. This result can also be shown to carry over to the more general case where inelastic collisions occur as considered in Chapter 5.

3.2.3 Partial cross sections

In transport theory, the differential cross section generally enters the calculations through integral *partial cross sections*,

$$\begin{aligned}\sigma^{(l)}(g) &\equiv \int P_l(\cos\chi)\,\sigma(\mathbf{g},\mathbf{g}')\,d^2\Omega_{\mathbf{g}'} \\ &= \int_0^{2\pi} d\psi \int_0^{\pi} \sin\chi\,d\chi\,P_l(\cos\chi)\,\sigma(g,\chi) \\ &= 2\pi\int_0^{\pi} \sin\chi\,d\chi\,P_l(\cos\chi)\,\sigma(g,\chi),\end{aligned} \tag{3.14}$$

where $P_l(\cos\chi)$ is a Legendre polynomial of order $l = 0, 1, 2, 3, \ldots$. The following relation then follows from orthogonality of these polynomials:

$$\sigma(g,\chi) = \sum_{l=0}^{\infty} \frac{2l+1}{4\pi}\sigma^{(l)}(g)\,P_l(\cos\chi) \tag{3.15}$$

showing that knowledge of all the partial cross sections is equivalent to knowing the differential cross section itself.

3.2.4 Calculation of cross sections

Let the incident particle flux (i.e., the number of particles crossing unit area normal to the beam direction in unit time) in Figure 3.1 be denoted by I. As observed, all incident particles passing through the left hand ring are scattered through the right hand ring, that is, by the definition Equation 3.7

$$I\,b\,db\,d\psi = I\,\sigma(g,\chi)\,d\Omega_{\mathbf{g}'},$$

which together with Equation 3.6 gives

$$\sigma(g,\chi) = \left|\frac{b}{\sin\chi}\frac{db}{d\chi}\right|, \tag{3.16}$$

where the modulus ensures that σ is positive.

The relationship $b(\chi)$ between the impact parameter and scattering angle required to evaluate the right hand side of Equation 3.16 is provided by solution of the equations of motion for a prescribed interaction potential [54]. In general, this must be done numerically (see Chapter 12), even for simple power law potentials, and the procedure can be quite involved for more realistic forms of interaction.

A classical description usually suffices for heavier charged particles, such as atomic or molecular ions, but it is inadequate for light particles such as electrons. There, one must calculate cross sections from wave shifts found from solutions of Schrödinger's equation. Note, however, that the cross section $\sigma(g, \chi)$ that appears in the the Boltzmann collision integral derived below can be of either classical or quantum origin: classical kinetic theory does not distinguish between the nature of the collision dynamics, since the region of interaction is effectively a point, as discussed in Chapter 2.

Full details of calculation of classical cross sections can be found in Chapter 4.

3.3 Boltzmann Collision Integral

3.3.1 Collision moment

We now wish to calculate the collision terms on the right hand side of the kinetic equation. To be specific, let us consider scattering of particles of gas 1 by particles of gas 2, that is, we calculate the rate of change of $f_1(\mathbf{r}, \mathbf{v}_1, t)$ due to collisions with particles of type 2. The calculation involves a "trick" which we use a number of times in this book, and which has a spin-off to the extent that it supplies an expression for the all important collision "moment" directly.

Thus, we multiply the relevant collision term $\left(\frac{\partial f}{\partial t}\right)_{\mathrm{col}}^{(1,2)}$ by some *arbitrary* function $\phi(\mathbf{v}_1)$ of $\mathbf{v}_1$, and integrate over all allowed velocities, producing the quantity

$$\int \mathrm{d}^3 v_1 \, \phi(\mathbf{v}_1) \left(\frac{\partial f}{\partial t}\right)_{\mathrm{col}}^{(1,2)} \tag{3.17}$$

$$= \text{rate of change of } \phi \text{ per unit time, per unit volume,}$$
$$\text{due to all collisions with particles of gas 2.}$$

We then calculate the right hand side using the definitions and results from the last section. Note that the collisions are assumed to be binary, that is, to take place between only two particles, and to be effectively instantaneous in time and local in space, that is, $\left(\frac{\partial f}{\partial t}\right)_{\mathrm{col}}^{(1,2)}$ at position $\mathbf{r}$, time t, is assumed to be controlled by the values of the distribution functions of species 1 and 2 only *at that same position and at the same time*. The space-time coordinates $\mathbf{r}, t$ do not play any role in the collision dynamics and we suppress any such dependence in the distribution functions by simply writing $f_1(\mathbf{v}_1)$ and $f_2(\mathbf{v}_2)$, respectively.

There are $f_1(\mathbf{v}_1)\,d^3v_1$ particles 1 per unit volume within an infinitesimal velocity range d^3v_1 of $\mathbf{v}_1$ resulting in a flux $I = g\,f_1(\mathbf{v}_1)\,d^3v_1$ impinging on *each* particle 2, which acts as the fixed scattering center in Figure 3.1. Given that there are $f_2(\mathbf{v}_2)\,d^3v_2$ of the latter per unit volume with velocities in the range d^3v_2 of $\mathbf{v}_2$, the definition Equation 3.7 gives

Number of particles 1 with velocities in the range d^3v_1 of $\mathbf{v}_1$
scattered into solid angle $d^2\Omega_{\mathbf{g}'}$, per unit volume, per unit time,
from particles 2, with velocities in the range d^3v_2 of $\mathbf{v}_2$

$$= \sigma(\mathbf{g},\mathbf{g}')\,d^2\Omega_{\mathbf{g}'} \quad g\,f_1(\mathbf{v}_1)\,d^3v_1 \quad f_2(\mathbf{v}_2)\,d^3v_2.$$

For each such collision in the velocity range d^3v_1 of $\mathbf{v}_1$, the value of $\phi(\mathbf{v}_1)$ changes to $\phi(\mathbf{v}_1')$, so that after integrating over all velocities and solid angles, we find that

The change of ϕ per unit time, per unit volume,
due to all collisions with particles of gas 2

$$= \int d^3v_1 \int d^3v_2\, g\,f_1(\mathbf{v}_1)\,f_2(\mathbf{v}_2) \int d^2\Omega_{\mathbf{g}'}\,\sigma(\mathbf{g},\mathbf{g}') \,\left[\phi(\mathbf{v}_1') - \phi(\mathbf{v}_1)\right]. \tag{3.18}$$

Combining Equations 3.18 and 3.17 yields

$$\int d^3v_1\,\phi(\mathbf{v}_1)\left(\frac{\partial f}{\partial t}\right)_{\mathrm{col}}^{(1,2)} \tag{3.19}$$
$$= \int d^3v_1 \int d^3v_2\, g\,f_1(\mathbf{v}_1)\,f_2(\mathbf{v}_2) \int d^2\Omega_{\mathbf{g}'}\,\sigma(\mathbf{g},\mathbf{g}') \,\left[\phi(\mathbf{v}_1') - \phi(\mathbf{v}_1)\right].$$

Note that here and elsewhere, it is implicit that the integrals are over all allowed values of velocity and solid angles.

3.3.2 Fundamental assumptions

There is an important assumption implicit in the above derivation, namely that before a collision, the distribution functions of the respective particles are uncorrelated, the so-called collision hypothesis ("Stosszahlansatz") of Boltzmann. After a collision that will not be the case. This introduces an "arrow of time" and irreversibility into otherwise time-reversible collision dynamics.

Another assumption is that any external field is negligible compared to the force of interaction during a collision. We also emphasize that the duration of a collision is assumed to be negligibly small compared with the time between collisions, and that any spatial variation of the distribution function can be neglected during scattering. In particular, the "beam" in Figure 3.1 is taken to have uniform cross section.

3.3.3 Calculating $\left(\frac{\partial f}{\partial t}\right)_{\text{col}}^{(1,2)}$

Equation 3.19 is of significance in its own right, and indeed, we shall come back to it time and again in this book. However, to complete the task of finding $\left(\frac{\partial f}{\partial t}\right)_{\text{col}}^{(1,2)}$ itself, we need to go a few steps further. First, with the aid of definition (Equation 3.10), we rewrite Equation 3.19 and then make the transformation $(\mathbf{v}_1, \mathbf{v}_2) \longrightarrow (\mathbf{v}_1', \mathbf{v}_2')$ in the first integral on the right hand side:

$$
\begin{aligned}
&\int \mathrm{d}^3 v_1\, \phi(\mathbf{v}_1) \left(\frac{\partial f}{\partial t}\right)_{\text{col}}^{(1,2)} \\
&\quad = \int \mathrm{d}^8\vartheta\, f_1(\mathbf{v}_1) f_2(\mathbf{v}_2)\, \left[\phi(\mathbf{v}_1') - \phi(\mathbf{v}_1)\right] \\
&\quad = \int \mathrm{d}^8\vartheta\, f_1(\mathbf{v}_1) f_2(\mathbf{v}_2)\, \phi(\mathbf{v}_1') - \int \mathrm{d}^8\vartheta\, f_1(\mathbf{v}_1) f_2(\mathbf{v}_2)\, \phi(\mathbf{v}_1) \\
&\quad = \int \mathrm{d}^8\vartheta'\, f_1(\mathbf{v}_1') f_2(\mathbf{v}_2')\, \phi(\mathbf{v}_1) - \int \mathrm{d}^8\vartheta\, f_1(\mathbf{v}_1) f_2(\mathbf{v}_2)\, \phi(\mathbf{v}_1).
\end{aligned}
$$

Then using Equation 3.13 and finally (Equation 3.10) to convert back to explicit velocity integrals, we get

$$
\begin{aligned}
&\int \mathrm{d}^3 v_1\, \phi(\mathbf{v}_1) \left(\frac{\partial f}{\partial t}\right)_{\text{col}}^{(1,2)} \\
&\quad = \int \mathrm{d}^3 v_1\, \phi(\mathbf{v}_1) \int \mathrm{d}^3 v_2 \int \mathrm{d}^2\Omega_{\mathbf{g}'}\, g\, \sigma(\mathbf{g}, \mathbf{g}') \\
&\quad\quad \times \left[f_1(\mathbf{v}_1') f_2(\mathbf{v}_2') - f_1(\mathbf{v}_1) f_2(\mathbf{v}_2)\right].
\end{aligned} \tag{3.20}
$$

Since $\phi(\mathbf{v}_1)$ is an arbitrary function of velocity, the only way this can be satisfied is if

$$
\begin{aligned}
&\left(\frac{\partial f}{\partial t}\right)_{\text{col}}^{(1,2)} \\
&\quad = \int \mathrm{d}^3 v_2 \int \mathrm{d}^2\Omega_{\mathbf{g}'}\, g\, \sigma(\mathbf{g}, \mathbf{g}')\, \left[f_1(\mathbf{v}_1') f_2(\mathbf{v}_2') - f_1(\mathbf{v}_1) f_2(\mathbf{v}_2)\right],
\end{aligned} \tag{3.21}
$$

which is the famous Boltzmann collision term [1].

3.4 Simple Gas

3.4.1 Classical Boltzmann kinetic equation

Note that we have spoken about particles 1 and 2 as if they were a different species, but the above derivation carries over to the case where they are the same, $f = f_2 \equiv f$. The collision term at velocity $\mathbf{v}$ is then (with minor notational changes)

$$\left(\frac{\partial f}{\partial t}\right)_{\text{col}} = \int d^3\bar{v} \int d^2\Omega_{\mathbf{g}'}\, g\, \sigma(\mathbf{g}, \mathbf{g}') \,\left[f(\mathbf{v}')f(\bar{\mathbf{v}}') - f(\mathbf{v})f(\bar{\mathbf{v}})\right]. \tag{3.22}$$

The Boltzmann kinetic equation for a simple single component gas is then

$$\frac{\partial f}{\partial t} + \mathbf{v}\cdot\nabla f + \mathbf{a}\cdot\frac{\partial f}{\partial \mathbf{v}} = \int d^3\bar{v} \int d^2\Omega_{\mathbf{g}'}\, g\, \sigma(\mathbf{g}, \mathbf{g}') \,\left[f(\mathbf{v}')f(\bar{\mathbf{v}}') - f(\mathbf{v})f(\bar{\mathbf{v}})\right], \tag{3.23}$$

where $\bar{\mathbf{v}}$ is the new integration variable, that is, the velocity of the second particle, and $\mathbf{g} = \mathbf{v} - \bar{\mathbf{v}}$, and so on. This equation is clearly non-linear in f.

3.4.2 Summational invariants

The corresponding expression for the collisional rate of change of some arbitrary property $\phi(\mathbf{v})$ is

$$\int d^3v\, \phi(\mathbf{v}) \left(\frac{\partial f}{\partial t}\right)_{\text{col}} = \int d^3v \int d^3\bar{v}\, f(\mathbf{v})f(\bar{\mathbf{v}}) \int d^2\Omega_{\mathbf{g}'}\, g\, \sigma(\mathbf{g}, \mathbf{g}') \,\left[\phi(\mathbf{v}') - \phi(\mathbf{v})\right]. \tag{3.24}$$

Apart from the terms in $\phi(\mathbf{v})$, the right hand side is invariant under interchange of $\mathbf{v}$ and $\bar{\mathbf{v}}$, for even though $\mathbf{g} \to -\mathbf{g}$, the symmetry condition (Equation 3.8) applies. Thus, we may write

$$\int d^3v\, \phi(\mathbf{v}) \left(\frac{\partial f}{\partial t}\right)_{\text{col}} = \frac{1}{2}\int d^3v \int d^3\bar{v}\, f(\mathbf{v})f(\bar{\mathbf{v}}) \int d^2\Omega_{\mathbf{g}'}\, g\, \sigma(\mathbf{g}, \mathbf{g}') \left[\phi(\mathbf{v}') + \phi(\bar{\mathbf{v}}') - \phi(\mathbf{v}) - \phi(\bar{\mathbf{v}})\right]. \tag{3.25}$$

Given that particle number, momentum and energy are conserved in a collision, then

$$m\mathbf{v} + m\bar{\mathbf{v}} = m\mathbf{v}' + m\bar{\mathbf{v}}'$$

$$\frac{1}{2}m\mathbf{v}^2 + \frac{1}{2}m\bar{\mathbf{v}}^2 = \frac{1}{2}m\mathbf{v}'^2 + \frac{1}{2}m\bar{\mathbf{v}}'^2$$

and it is clear that setting $\phi(\mathbf{v}) = 1$, $m\mathbf{v}$ or $\frac{1}{2}m\mathbf{v}^2$ in Equation 3.25 gives zero on the right hand side. If we denote this set of *summational invariants* by $\left\{\phi_i^{(S)}(\mathbf{v})\right\}_{i=1,2,3}$, then it is clear that

$$\int d^3v\, \phi_i^{(S)}(\mathbf{v}) \left(\frac{\partial f}{\partial t}\right)_{\text{col}} = 0 \qquad (i = 1, 2, 3). \tag{3.26}$$

3.4.3 H-theorem, equilibrium, and the Maxwellian distribution

Again since the integrand in Equation 3.25 is invariant under the transformation $(\mathbf{v}, \bar{\mathbf{v}}) \leftrightarrow (\mathbf{v}', \bar{\mathbf{v}}')$, it follows that we can write it as

$$\begin{aligned}
&\int d^3v\, \phi(\mathbf{v}) \left(\frac{\partial f}{\partial t}\right)_{\text{col}} \\
&= \frac{1}{4}\int d^3v \int d^3\bar{v} \int d^2\Omega_{\mathbf{g}'}\, g\, \sigma(\mathbf{g}, \mathbf{g}') \\
&\quad \times [f(\mathbf{v}) f(\bar{\mathbf{v}}) - f(\mathbf{v}') f(\bar{\mathbf{v}}')] \left[\phi(\mathbf{v}') + \phi(\bar{\mathbf{v}}') - \phi(\mathbf{v}) - \phi(\bar{\mathbf{v}})\right].
\end{aligned} \tag{3.27}$$

Boltzmann introduced the quantity

$$H \equiv \int d^3v\, f(\mathbf{v}) \ln f(\mathbf{v}), \tag{3.28}$$

which for a spatially uniform, field-free situation varies in time according to

$$\begin{aligned}
\frac{dH}{dt} &= \int d^3v\, [1 + \ln f(\mathbf{v}, \mathbf{t})] \frac{\partial f(\mathbf{v}, \mathbf{t})}{\partial t} \\
&= \int d^3v\, [1 + \ln f(\mathbf{v}, \mathbf{t})] \left(\frac{\partial f}{\partial t}\right)_{\text{col}\, l} \\
&= \frac{1}{4}\int d^3v \int d^3\bar{v} \int d^2\Omega_{\mathbf{g}'}\, g\, \sigma(\mathbf{g}, \mathbf{g}')\, [f\, \bar{f} - f'\, \bar{f}'] \\
&\qquad \left[\ln f' + \ln \bar{f}' - \ln f - \ln \bar{f}\right] \\
&= -\frac{1}{4}\int d^3v \int d^3\bar{v} \int d^2\Omega_{\mathbf{g}'}\, g\, \sigma(\mathbf{g}, \mathbf{g}')\, [f\, \bar{f} - f'\, \bar{f}']\, \ln\left[\frac{f\bar{f}}{f'\bar{f}'}\right],
\end{aligned} \tag{3.29}$$

where we have set $\phi = 1 + \ln f(\mathbf{v},t)$ in Equation 3.27, and we now make the time-dependence of f explicit, and use abbreviated notation, $f' = f(\mathbf{v}', t)$, and so on. Now if $ab \geq cd$, then $\ln\left[\frac{ab}{cd}\right] \geq 0$, or if $ab \leq cd$, then $\ln\left[\frac{ab}{cd}\right] \leq 0$, and clearly we always have $[ab - cd]\ln\left[\frac{ab}{cd}\right] \geq 0$. Thus

$$\left[f\bar{f} - f'\bar{f}'\right] \ln\left[\frac{f\,\bar{f}}{f'\,\bar{f}'}\right] \geq 0$$

and obviously

$$\frac{\mathrm{d}H}{\mathrm{d}t} \leq 0, \tag{3.30}$$

which is Boltzmann's H-theorem. The quantity H decreases monotonically with time to the *equilibrium state*, where $\frac{\mathrm{d}H}{\mathrm{d}t} = 0$, for which a necessary and sufficient condition is $f\,\bar{f} = f'\,\bar{f}'$, or equivalently,

$$\ln f' + \ln \bar{f}' - \ln f - \ln \bar{f} = 0.$$

This suggests that *in the equilibrium state*, $f = f^{(\text{equil})}$, where $\ln f^{(\text{equil})}$ must be a linear combination of the summational invariants, that is,

$$\ln f^{(\text{equil})} = \sum_{i=1}^{3} c_i\, \phi_i^{(S)}(\mathbf{v}), \tag{3.31}$$

where the c_i are three constants determined by the three conditions of normalization to a particle number density n, with mean velocity $\langle \mathbf{v} \rangle$, and temperature T. After some algebra, it can be shown that

$$f^{(\text{equil})}(\mathbf{v}) = n\left(\frac{m}{2\pi k_B T}\right)^{\frac{3}{2}} \exp\left[-\frac{\frac{1}{2}m(\mathbf{v} - \langle \mathbf{v} \rangle)^2}{k_B T}\right] \tag{3.32}$$

the famous *Maxwellian velocity distribution function* of equilibrium statistical mechanics. When this is substituted into the right hand side of Equation 3.28, it can be shown that the equilibrium value of H is

$$\begin{aligned} H^{(\text{equil})} &= n\left[\ln n + \frac{3}{2}\ln\left(\frac{m}{2\pi k_B T}\right) - \frac{3}{2}\right] \\ &= -\frac{1}{k} S^{(\text{equil})}/V + \text{constant}, \end{aligned}$$

where $S^{(\text{equil})}$ is the entropy of the gas calculated from thermodynamics. It is then usual to generalize this to non-equilibrium situations, and write

$$S/V = -kH + \text{constant}$$

and then deduce that the H-theorem (Equation 3.30) for a gas is just another manifestation of the *Second Law of thermodynamics*, which states that for an isolated system, S increases in the course of time $\left(\frac{\mathrm{d}S}{\mathrm{d}t} \geq 0\right)$ until the equilibrium state of *maximum entropy* is attained.

3.5 Fokker–Planck Kinetic Equation

3.5.1 Small deflection collisions

Although for the greater part of this book, we consider dilute particles interacting predominantly with the molecules of a neutral medium through short-range forces, it is instructive to examine the collision operator describing mutual interaction between charged particles via the Coulomb force. Since this is a long-range force, most collisions involve large impact parameters and small angle deflections. For simplicity, we take a single component system of charged particles, but the results can readily be extended to a number of species, for example, electron–electron, electron–ion, and ion–ion interactions in a plasma.

For small angle deflection collisions, the change in velocity,

$$\Delta\mathbf{v} = \mathbf{v}' - \mathbf{v}$$

is generally small, allowing an approximate differential form of Boltzmann collision integral to be formulated. Starting again with the general expression (Equation 3.24) for collision transfer rate of some arbitrary quantity $\phi(\mathbf{v})$, and making the Taylor expansion

$$\phi(\mathbf{v}') = \phi(\mathbf{v}+\Delta\mathbf{v}) = \phi(\mathbf{v}) + \Delta\mathbf{v}\cdot\frac{\partial\phi(\mathbf{v})}{\partial\mathbf{v}} + \frac{1}{2}\Delta\mathbf{v}\Delta\mathbf{v} : \frac{\partial^2\phi(\mathbf{v})}{\partial\mathbf{v}\,\partial\mathbf{v}} + \ldots,$$

we find

$$\int \mathrm{d}^3v\, \phi(\mathbf{v}) \left(\frac{\partial f}{\partial t}\right)_{\mathrm{col}\,l}$$

$$= \int \mathrm{d}^3v \int \mathrm{d}^3\bar{v}\, f(\mathbf{v}) f(\bar{\mathbf{v}}) \int \mathrm{d}^2\Omega_{\mathbf{g}'}\, g\, \sigma(\mathbf{g},\mathbf{g}')\ \ [\phi(\mathbf{v}') - \phi(\mathbf{v})]$$

$$= \int \mathrm{d}^3v \int \mathrm{d}^3\bar{v}\, f(\mathbf{v}) f(\bar{\mathbf{v}}) \int \mathrm{d}^2\Omega_{\mathbf{g}'}\, g\, \sigma(\mathbf{g},\mathbf{g}')$$

$$\times \left[\Delta\mathbf{v}\cdot\frac{\partial\phi(\mathbf{v})}{\partial\mathbf{v}} + \frac{1}{2}\Delta\mathbf{v}\Delta\mathbf{v} : \frac{\partial^2\phi(\mathbf{v})}{\partial\mathbf{v}\,\partial\mathbf{v}} + \ldots\right].$$

Integration by parts n times for the nth term in the expansion (we restrict the discussion to $n=2$ from now on) then gives

$$\int d^3v\, \phi(\mathbf{v}) \left(\frac{\partial f}{\partial t}\right)_{\text{col}}$$
$$= \int d^3v\, \phi(\mathbf{v}) \left[-\frac{\partial}{\partial \mathbf{v}} \cdot \left(f\langle \Delta\mathbf{v}\rangle\right) + \frac{1}{2}\frac{\partial^2}{\partial \mathbf{v}\,\partial \mathbf{v}} : \left(f\langle \Delta\mathbf{v}\Delta\mathbf{v}\rangle\right)\right],$$

where the average rate of change of $\langle \Delta\mathbf{v}\rangle$ due a succession of small angle Coulomb collisions is given by

$$\langle \Delta\mathbf{v}\rangle = \int d^3\bar{v}\, f(\bar{\mathbf{v}}) \int d^2\Omega_{\mathbf{g}'}\, g\, \sigma(\mathbf{g}, \mathbf{g}')\, \Delta\mathbf{v} \tag{3.33}$$

and similarly for $\langle \Delta\mathbf{v}\Delta\mathbf{v}\rangle$. (N.B. This averaging is different from the simple velocity space average (Equation 2.6), but we use the same notation $\langle\ \rangle$ temporarily in order to be consistent with plasma physics text books.) Since $\phi(\mathbf{v})$ is arbitrary, this can only be satisfied if

$$\left(\frac{\partial f}{\partial t}\right)_{\text{col}} = -\frac{\partial}{\partial \mathbf{v}} \cdot \left(f\langle \Delta\mathbf{v}\rangle\right) + \frac{1}{2}\frac{\partial^2}{\partial \mathbf{v}\,\partial \mathbf{v}} : \left(f\langle \Delta\mathbf{v}\Delta\mathbf{v}\rangle\right), \tag{3.34}$$

which is the Fokker–Planck collision operator. The Fokker–Planck kinetic equation is thus

$$\frac{\partial f}{\partial t} + \mathbf{v}\cdot\nabla f + \mathbf{a}\cdot\frac{\partial f}{\partial \mathbf{v}}$$
$$= -\frac{\partial}{\partial \mathbf{v}} \cdot \left(f\langle \Delta\mathbf{v}\rangle\right) + \frac{1}{2}\frac{\partial^2}{\partial \mathbf{v}\,\partial \mathbf{v}} : \left(f\langle \Delta\mathbf{v}\Delta\mathbf{v}\rangle\right). \tag{3.35}$$

3.5.2 Coulomb scattering

It turns out that while the long-range nature of the Coulomb force allows some simplification as described above, it also introduces a complication to the extent that the integrals over scattering angles in Equation 3.33 diverge when the Rutherford differential cross section is inserted in the right hand side [55]. This arises from the fact that we are trying to model many-body Coulomb collisions in terms of a succession of binary collisions. The usual way around this problem is to arbitrarily "cut off" the integrals at some lower limit of scattering angle χ, corresponding to a maximum impact parameter equal to the Debye screening distance [56,57]. No such divergence problems arise for the short-range interactions characterizing charged particle–neutral interactions.

3.6 Concluding Remarks

In this chapter, we derived the classical Boltzmann kinetic equation for particles undergoing *elastic collisions in a gas*, in which there is no change in the internal structure of the colliding partners. We went on to establish the H-theorem for a single-component gas, and showed how the Fokker–Planck equation follows from the Boltzmann equation in the limit of small angle scattering. The kinetic theory of gases is further developed in Chapter 5, with the inclusion of charge-exchange and inelastic collisions and particle loss/creation processes.

CHAPTER 4

Interaction Potentials and Cross Sections

4.1 Introduction

The cross sections $\sigma(g,\chi)$ appearing in the Boltzmann collision integral can be of either theoretical or experimental origin, and there is nothing to *explicitly* differentiate between classical and quantum scattering. It is only when we wish to *calculate* the cross sections that the difference becomes important.

Generally speaking, quantum effects are evident if the de Broglie wavelength λ of the particle is comparable with or larger than the size of the scattering atom. This is more likely for light particles, such as electrons or positrons, since $\lambda = h/mv$ is inversely proportional to the particle mass m. On the other hand, the de Broglie wavelength of a much heavier ion, moving at the same speed v, may be significantly smaller than the dimensions of the atom. Thus, as a rule, ions can be treated classically at all but the lowest energies, while electrons and positrons must, in general, be treated quantum mechanically. A useful analogy may be made between light versus heavy particle scattering and physical versus geometrical optics, respectively.

This chapter discusses classical scattering of ions from neutral atoms and details are given of how the differential cross section (DCS) $\sigma(g,\chi)$, or equivalently, the set of partial cross sections $\sigma^{(l)}(\epsilon)$, are to be calculated from a knowledge of the interaction potential. In general, even for simple power law potentials, the cross sections must be computed numerically (an exception is the Coulomb potential). For realistic potentials, corresponding to forces which are attractive at long range and repulsive at close range there can be significant computational challenges. The procedures are not usually well documented in text books, and we take the opportunity to outline the details in this chapter.

4.2 Classical Scattering Theory

4.2.1 Differential and partial cross sections

4.2.1.1 Scattering angle and impact parameter

For a given centre-of-mass (CM) energy $\epsilon = \frac{1}{2}\mu g^2$, solution of Newton's equations [54] gives the following general expression for the classical scattering angle corresponding to an interaction potential $V(r)$:

$$\chi = \pi - 2b\int_R^\infty \frac{dr}{r^2}\left[1 - \frac{b^2}{r^2} - \frac{V(r)}{\frac{1}{2}\mu g^2}\right]^{-\frac{1}{2}}, \tag{4.1}$$

where R is the distance of closest approach and is given by the largest positive zero of

$$1 - \frac{b^2}{R^2} - \frac{V(R)}{\frac{1}{2}\mu g^2} = 0. \tag{4.2}$$

Evaluation of the the right hand side of Equation 4.1 gives $\chi(b)$, from which the DCS can be calculated as discussed in Chapter 3:

$$\sigma(g,\chi) = \left|\frac{b}{\sin\chi}\frac{db}{d\chi}\right|. \tag{4.3}$$

4.2.1.2 A special case

In general, a numerical calculation is required, but in a handful of cases, the relationship $\chi(b)$ can be derived analytically from first principles. For example, if the colliding particles behave as rigid elastic spheres, then straightforward geometrical considerations yield

$$b = a\cos(\chi/2)$$

where a is the sum of the radii (see Figure 4.1). Substitution in Equation 4.3 then gives

$$\sigma(g,\chi) = \frac{1}{4}a^2. \tag{4.4}$$

The Coulomb potential is another case where $\sigma(g,\chi)$ can be obtained analytically as in Ref. [54].

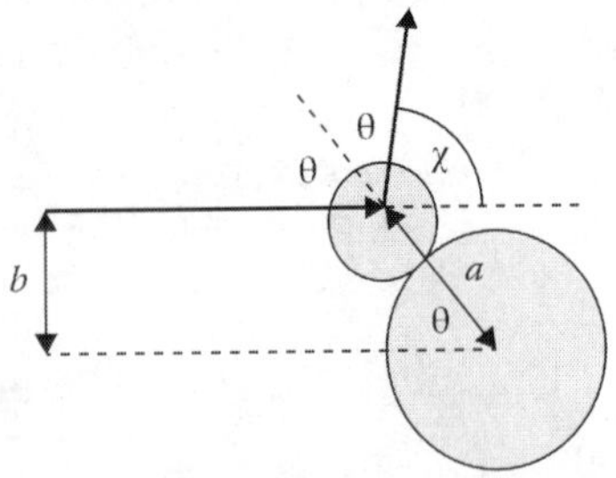

Figure 4.1 For a collision between two rigid spheres whose sum of radii is a, the scattering angle $\chi = \pi - 2\theta$ and impact parameter are related by $b = a\sin\theta = a\cos(\chi/2)$.

4.2.1.3 Partial cross sections and differences

Note that while the DCS appears explicitly in the Boltzmann equation, it is the *partial cross sections*, defined by Equation 3.14, which arise in its solution, or more precisely the *differences*,

$$\Delta\sigma^{(l)} \equiv \sigma^{(0)} - \sigma^{(l)} = 2\pi \int_0^\pi \sigma(g,\chi)[1 - P_l(\cos\chi)] \sin\chi d\chi \quad (l = 1, 2, 3, ...) \tag{4.5}$$

between the partial cross sections $\sigma^{(l)}$ and the total (or integral) cross section

$$\sigma^{(0)} = 2\pi \int_0^\pi \sigma(g,\chi) \sin\chi d\chi. \tag{4.6}$$

For example, the momentum transfer and so-called viscosity cross sections correspond to $l = 1, 2$, respectively,

$$\sigma_m = \Delta\sigma^{(1)} = \sigma^{(0)} - \sigma^{(1)} \tag{4.7}$$

$$\sigma_v = \Delta\sigma^{(2)} = \sigma^{(0)} - \sigma^{(2)}. \tag{4.8}$$

For the rigid sphere model, for example, the orthogonality of the Legendre polynomials implies that $\sigma^{(l)} \equiv 0$, for $l > 0$ and hence

$$\Delta\sigma^{(l)} \equiv \sigma^{(0)} = \pi a^2 \quad (l = 1, 2, 3, ...).$$

While the rigid sphere model proves useful as a reference, the aim is to find cross sections for more realistic forms of interaction.

4.2.2 Inverse power law potentials

4.2.2.1 Dimensional analysis

For inverse power law potentials of the form $V(r) \sim r^{-n}$, we can obtain useful information about $\chi(b)$ directly from Equation 4.1 using dimensional analysis. It is left as an exercise to show that χ depends upon b and energy solely through the combination $b\left(\frac{1}{2}\mu g^2\right)^{\frac{1}{n}}$, or conversely

$$b = \left(\frac{1}{2}\mu g^2\right)^{-\frac{1}{n}} F(\chi) \tag{4.9}$$

where $F(\chi)$ is some function of scattering angle, which we do not need to know for present purposes. Upon substitution into Equation 4.3 there follows:

$$\sigma(g,\chi) = \left(\frac{1}{2}\mu g^2\right)^{-\frac{2}{n}} G(\chi),$$

where $G(\chi) = \frac{1}{2}\left|\frac{F'(\chi)}{\sin\chi}\right|^2$, or in other words,

$$\sigma(g,\chi) \sim g^{-4/n} \tag{4.10}$$

for an inverse *nth* power law potential. The partial cross sections (Equation 4.5) also have this property.

4.2.2.2 Coulomb potential

The Coulomb potential operating between particles of charge q_1 and q_2 corresponds to $n = 1$,

$$V(r) = \xi/r,$$

where $k \equiv q_1 q_2/(4\pi\varepsilon_0)$. In this case, the integral on the right hand side of Equation 4.1 can be evaluated analytically, to give the $\chi(b)$ relation

$$\tan(\chi/2) = \frac{\xi}{b\mu g^2} \tag{4.11}$$

and then substitution in Equation 4.3 yields the well-known Rutherford cross section [54]:

$$\sigma_R(g,\chi) = \left(\frac{\xi}{2\mu g^2}\right)^2 \frac{1}{\sin^4 \chi/2}. \tag{4.12}$$

This has the g^{-4} dependence as expected from the general relation (Equation 4.10). The angular function in this case $G(\chi) \sim \sin^{-4}\chi/2$ has a singularity at $\chi \to 0$, which is due to the long range of the Coulomb force. Consequently, the integrals (Equation 4.5) diverge at the lower limit for $l = 1, 2, \ldots$. The problem arises because the Coulomb interaction actually involves many simultaneous, small angle scattering events, and cannot be adequately treated by binary collision theory. Traditionally, the situation is salvaged *ad hoc* by replacing the lower limit with some small, but finite value χ_{min}, defined by setting the impact parameter equal to the Debye screening distance [56] in Equation 4.11,

$$\lambda_D = \frac{\xi}{\mu g^2} \cot\left(\chi_{min}/2\right).$$

No such difficulty occurs for short-range forces with $n \geq 4$, since the integrals (Equation 4.5) converge for all l. The case of $n = 4$ corresponds to the Maxwell model [42] and occupies a special place in the kinetic theory of charged particle transport.

4.3 Inverse Fourth-Power Law Potential

4.3.1 Polarization potential

Consider a point charge $q = +e$ at a distance r from the CM of an electrically neutral atom of atomic number Z. The electric field $E = \frac{e}{4\pi\epsilon_0 r^2}$ of the ion *polarizes* the atom by repelling the positively charged nucleus and attracting the negatively charged electron cloud, thereby inducing an electric dipole moment whose magnitude is proportional to E, that is,

$$p = \alpha E = \frac{\alpha e}{4\pi\epsilon_0 r^2}, \tag{4.13}$$

where α is called the *polarizability* of the atom. The direction of $\mathbf{p}$ is away from the atom, that is, to the right in Figure 4.2. The field $\mathbf{E}_{\text{dipole}}$ of the polarized atom then exerts an *attractive force* back on the ion. We now calculate this force.

If the distance between the CM of the positive and negative charge respectively in the polarized atom is denoted by $2a$, the dipole moment has magnitude $p = Ze \times 2a$. The electrostatic field due to the dipole is then

$$\begin{aligned} E_{\text{dipole}} &= \frac{-Ze}{4\pi\epsilon_0 (r-a)^2} + \frac{Ze}{4\pi\epsilon_0 (r+a)^2} \\ &= \frac{-Ze\, 4ar}{4\pi\epsilon_0 (r^2 - a^2)^2} \\ &\approx -\frac{p}{2\pi\epsilon_0 r^3}, \end{aligned} \tag{4.14}$$

where terms of order $O(r^{-5})$ have been neglected, it being implicitly assumed that the ion–atom distance is large, in the sense that $r \gg a$. The

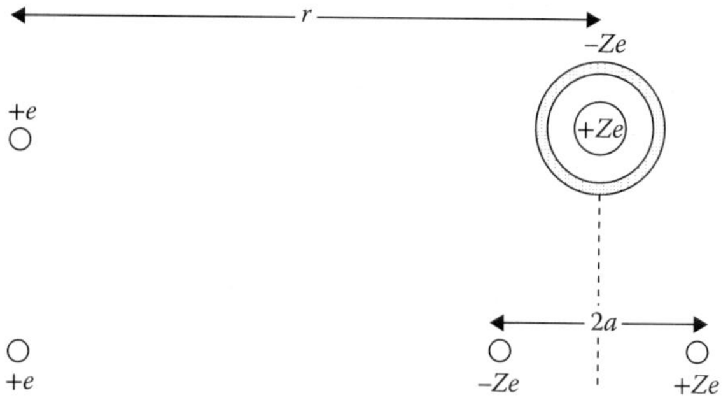

Figure 4.2 A positively charged ion polarizes a nearby atom, which in turn exerts a force on the ion itself which is inversely proportional to the fourth power of r.

force exerted by the atom on the ion is thus

$$F = eE_{\text{dipole}} = -\frac{ep}{2\pi\epsilon_0 r^3} = -\frac{\alpha e^2}{8\pi^2\epsilon_0^2 r^5}, \tag{4.15}$$

which by Newton's third law is also the force exerted by the ion on the atom. The corresponding interaction potential $V(r)$, defined by $F = -\frac{dV}{dr}$, is

$$V(r) = -\frac{\xi}{r^4}, \tag{4.16}$$

where $\xi = \alpha e^2/(\pi^2\epsilon_0^2)$. This is the inverse fourth power law potential discussed in the previous section.

4.3.2 Constant collision frequency

Substituting $n = 4$ in Equation 4.10 gives

$$\sigma(g, \chi) \sim g^{-1} \tag{4.17}$$

and similarly for the partial cross section differences $\Delta\sigma^{(l)}(g) \sim g^{-1}$. It is clear that for this model the momentum transfer collision frequency

$$\nu_m = n_0 g \Delta\sigma^{(1)} = n_0 g[\sigma^{(0)}(g) - \sigma^{(1)}(g)] \tag{4.18}$$

is constant, independent of g (or of energy $\epsilon = \frac{1}{2}\mu g^2$). The constant collision frequency or Maxwell model plays a central rôle in transport theory.

4.4 Realistic Interaction Potentials

4.4.1 The Mason–Schamp potential

Real charged particle–atom interaction potentials are attractive at long range and repulsive at short range, with an intermediate region where the corresponding forces tend to balance, as shown schematically in Figure 4.3. The position and depth of the "potential well" are denoted by r_m and ε_m with typical values being of the order of a few Å and a few tens of meV, respectively. Representation by a power law at long range is soundly physically based, but at short range, it is more empirical in nature.

A potential of the form

$$V(r) = \varepsilon_m \left\{ A\left(\frac{r_m}{r}\right)^{2N} - B\left(\frac{r_m}{r}\right)^6 - C\left(\frac{r_m}{r}\right)^4 \right\} \tag{4.19}$$

can be used to describe ion–atom interactions, where A, B, and C are constants, and $N = 6$ in the potential of Mason and Schamp [20,58]. The

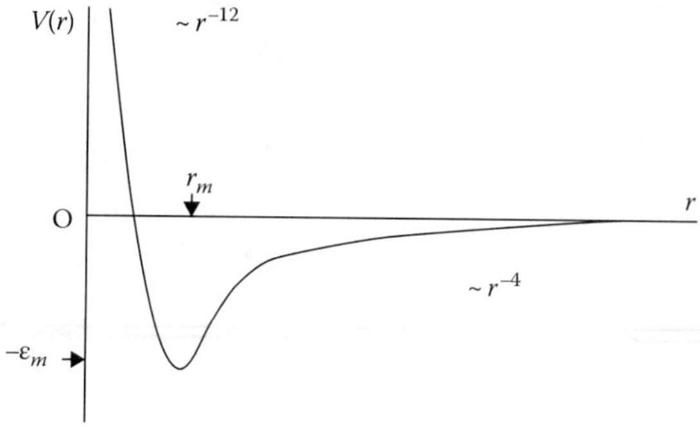

Figure 4.3 Realistic ion–atom interactions are governed at long range by the inverse-fourth power law polarization potential, and at close range by some strongly repulsive core, here represented by an inverse twelfth power law potential.

first and third terms on the right hand side correspond to close range repulsion and polarization potential respectively, while the second term corresponds to point-charge induced quadrupole and London dispersion energy. Two of the constants are fixed by $V(r_m)=\varepsilon_m$ and $V'(r_m)=0$, and a parameter γ is introduced to describe the relative strengths of the attractive terms. Thus, we find

$$\begin{aligned} A &= \frac{2(\gamma+1)}{N-2} \\ B &= 2\gamma \\ C &= \frac{N-2\gamma(N-3)}{N-2}. \end{aligned} \tag{4.20}$$

The calculation of the corresponding cross sections involves a numerical procedure, as discussed in detail in Section 4.5. Tables of cross sections for various values of the parameters N and γ can be found in the appendices of the monograph of Mason and McDaniel [58].

4.4.2 Momentum transfer collision frequency

The detailed procedure for numerical calculations of $\Delta\sigma^{(l)}$ is discussed in the following section. For the present, we focus on the qualitative nature of the momentum transfer collision frequency (Equation 4.18) for a 12-6-4 potential, obtained by setting $N=6$ in Equation 4.19. As shown schematically in Figure 4.4, the polarization limit applies at low g, where the long-range inverse fourth power contribution to the potential dominates, and consequently, ν_m is constant. Conversely, at high energies, the repulsive core r^{-12} of the potential dominates, $\sigma_m \sim g^{-1/3}$ by virtue of

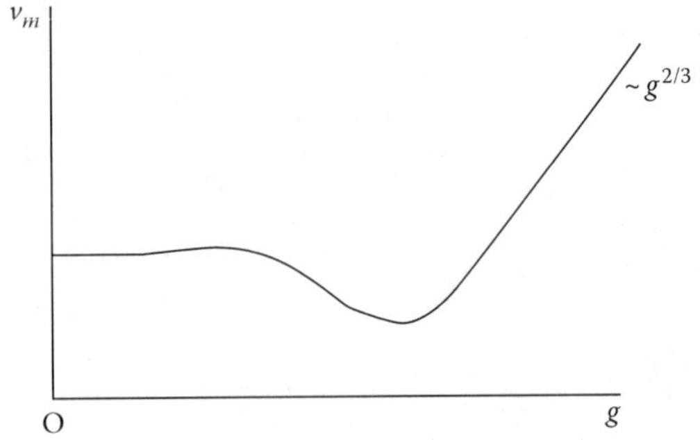

Figure 4.4 Schematic representation of the g-dependence of momentum transfer collision frequency ν_m for a 12-6-4 potential. In the *polarization limit* at low g, ν_m is approximately constant, corresponding to the long-range, inverse fourth-power law part of the potential. At higher g, the repulsive inverse twelfth power law dominates, and ν_m increases appropriately. The "dip" at intermediate g corresponds to an interplay between attractive and repulsive parts of the potential, which acts to minimize scattering.

Equation 4.10 and hence $\nu_m = n_0 g\,\sigma_m \sim g^{2/3}$. At intermediate energies, however, no such simple representation is possible. There is a balance between repulsion and attraction that leads to the phenomenon of "orbiting," which causes a "dip" in ν_m. This in turn is reflected macroscopically as an enhanced "transparency" of the gas to charged particle transport, namely, the "bump" in the mobility versus field curve [58].

The numerical procedure for calculation of cross sections is now given.

4.5 Calculation of Cross Sections for a General Interaction Potential

4.5.1 Transformation of variables

First we use Equation 4.3 to transform the variable of integration in Equation 4.5 from χ to b:

$$\Delta\sigma^{(l)}(\epsilon) = 2\pi\int_0^\infty [1 - P_l(\cos\chi)]\,\sigma(g,\chi)\,\sin\chi\,d\chi = 2\pi\int_0^\infty [1 - P_l(\cos\chi)]\,bdb. \tag{4.21}$$

Subsequently, a further transformation is made of the integration variable from b to the distance of closest approach R using Equation 4.2. This avoids having to find its zeros R for each value of b and ϵ. A similar transformation in Equation 4.1 for the scattering angle yields

$$\chi(R,\epsilon)=\pi-2\left[1-V(R)/\epsilon\right]^{1/2}\int_0^1 \frac{dx}{\sqrt{1-x^2}}\left[1+F(x,R)/\epsilon\right]^{-1/2}, \tag{4.22}$$

where the integration variable is now

$$x=R/r$$

and

$$F(x,R)\equiv\frac{x^2V(R)-V(R/x)}{1-x^2}. \tag{4.23}$$

This function is defined at $x=1$, since

$$\lim_{x\to 1}F(x,R)=-V(R)-\frac{R}{2}V'(R)\equiv G(R). \tag{4.24}$$

While the integrand of Equation 4.22 has a singularity at $x=1$, arising from the term $(1-x^2)^{-1/2}$, this is *removable*, for example, by making the transformation $x=\sin\theta$.

4.5.2 Orbiting, critical energy, and cross sections

While the integrand is well behaved at $x=1$, there is an unremoveable singularity when the term in square brackets in the integrand vanishes, that is, when

$$1+F(x,R)/\epsilon=0 \qquad (0\le x\le 1) \tag{4.25}$$

is satisfied for real, positive values of R. The integral (Equation 4.22) then diverges, the scattering angle $\chi\to-\infty$, and *orbiting* is said to occur. Physically speaking, the particle is trapped indefinitely in the potential well (see page 70 of Ref. [59]). The phenomenon can occur only if the energy ϵ lies below a certain critical value ϵ_c and the impact parameter corresponds to a critical value b_c. Expressions for these important quantities are obtained below.

For energies above the critical value, Equation 4.25 has no real positive solutions; there is no singularity, and R is a continuous function of b. Transformation of the variable of integration in Equation 4.21 from b to R results in the expression

$$\Delta\sigma^{(l)}(\epsilon>\epsilon_c)=2\pi\int_{R_0}^{\infty}\left[1-P_l(\cos\chi)\right]\left[1+G(R)/\epsilon\right]R\,dR, \tag{4.26}$$

where $G(R)$ is defined by Equation 4.24, and R_0 is the turning point for $b=0$, that is, the largest positive zero of

$$1 - V(R)/\epsilon = 0. \tag{4.27}$$

However, when $\epsilon < \epsilon_c$, things are much more complicated due to orbiting when $\chi \to -\infty$ and the integrand oscillates infinitely due to the presence of the term $P_l(\cos\chi)$. An accurate numerical evaluation of $\Delta\sigma^{(l)}(\epsilon > \epsilon_c)$ is difficult in this case.

If orbiting conditions prevail, Equation 4.25 is satisfied by two real, positive values of R, the largest being denoted by R_1 and the smallest by R_2. Intermediate values of R, that is, $R_2 \le R \le R_1$ result in an imaginary and therefore unphysical scattering angle χ. Hence the integral over R in the expression for the partial cross section must be divided into two parts:

$$\Delta\sigma^{(l)}(\epsilon < \epsilon_c) = 2\pi \int_{R_0}^{R_2} [1 - P_l(\cos\chi)] \left[1 + G(R)/\epsilon\right] R\,dR$$

$$+ 2\pi \int_{R_1}^{\infty} [1 - P_l(\cos\chi)] \left[1 + G(R)/\epsilon\right] R\,dR. \tag{4.28}$$

4.5.3 Determination of ϵ_c

The value of R_1 is found by setting $x=1$ in Equation 4.25, which, along with Equation 4.24, requires that R_1 is the largest positive zero of

$$1 + F(1, R)/\epsilon = 1 - \left[V(R) + \frac{R}{2} V'(R)\right] /\epsilon = 0. \tag{4.29}$$

Then R_2 is determined from the condition that Equation 4.25 must be satisfied for $x = \frac{R_2}{R_1}$ and $R = R_2$. Hence R_2 is the smallest positive zero of

$$1 + \left\{ \frac{(R/R_1)^2\, V(R) - V(R_1)}{\epsilon[1 - (R/R_1)^2]} \right\} = 0. \tag{4.30}$$

The critical impact parameter b_c for orbiting is determined by setting $R = R_1$ in Equation 4.2, that is,

$$b_c^2 = R_1^2 \left[1 - V(R_1)/\epsilon\right]. \tag{4.31}$$

While this is not needed explicitly in further discussion, we note that Equation 4.30 for R_2 can be written in the equivalent form

$$1 - \frac{b_c^2}{R^2} - \frac{V(R)}{\epsilon} = 0. \tag{4.32}$$

When $\epsilon = \epsilon_c$, the two zeros converge, $R_1 = R_2 \equiv R_c$. Then Equation 4.29 has a double zero, found by setting the derivative of Equation 4.29 with respect to R equal to zero:

$$3V'(R) + RV''(R) = 0. \tag{4.33}$$

The unique positive zero of this equation is R_c and the critical energy is found by substituting $R = R_c$ in Equation 4.29 and setting $\epsilon = \epsilon_c$:

$$\epsilon_c = V(R_c) + \frac{R_c}{2}V'(R_c). \tag{4.34}$$

We are now in a position to calculate $\Delta\sigma^{(l)}(\epsilon)$ over the entire energy range for any interaction potential $V(r)$. In what follows, we give numerical results initially for a simple power law potential and then for the Mason–Schamp potential.

4.6 Cross Sections for Specific Interaction Potentials

4.6.1 Numerical methods and techniques

4.6.1.1 Numerical quadrature

The integrals over R for the partial cross sections in Equations 4.26 and 4.28 in the respective ranges of energy ϵ can be evaluated efficiently using Gauss–Legendre quadrature, and the accuracy estimated simply by incrementing the order of quadrature and comparing results. A quadrature with 32 nodes, R_i is capable of giving cross sections accurate to a few parts in 10^3 or better at low energies, and about one part in 10^4 at high energies, provided that the scattering angle χ can be found from Equation 4.22 at each node R_i of the integrand to sufficient accuracy, better than $\pm 10^{-4}$. Fast calculation of χ is also essential, since there are many function evaluations for each energy ϵ in the range under consideration.

In general, for realistic potentials, accurate evaluation of χ from Equation 4.22 requires some care, since the integrand is sharply peaked for near-orbiting conditions. The most efficient way of addressing the problem is to use an adaptive quadrature technique, which automatically places the nodes in regions where the integrand is most rapidly varying. The same procedure is also efficient for simple power law potentials, where no orbiting is possible and the integrand in Equation 4.22 is well behaved.

4.6.1.2 Calculation of zeros

In what follows, the zeros of polynomials are found using the Newton–Raphson iteration technique, with initial estimates provided by the interval halving method. This is fast, accurate, and robust, enabling the required zeros to be found to a precision of 1 part in 10^7.

4.6.1.3 Dimensionless parameters

In any numerical computation, it is desirable to work with dimensionless quantities and, to that end, we introduce scaling parameters ϵ_m and r_m for energy and distance, respectively. For realistic potentials as shown in Figure 4.3, these arise in a natural way, as the depth and position of the potential well respectively, but initially they need not be specified as such. The required dimensionless parameters are

$$\varepsilon^* \equiv \epsilon/\epsilon_m;\; b^* = b/r_m;\; \eta = (r_m/R)^2 \tag{4.35}$$

$$\Delta\sigma^{*\,(l)} \equiv \Delta\sigma^{(l)}/\pi r_m^2. \tag{4.36}$$

Other dimensionless variables will also appear without necessarily being explicitly defined, since their meaning should be obvious, for example, the dimensionless critical energy, $\epsilon_c^* = \epsilon_c/\epsilon_m, \eta_1 = (r_m/R_1)^2$.

4.6.2 Power law potentials

To begin the discussion, consider a simple inverse *nth* power law potential,

$$V(r) = k/r^n, \tag{4.37}$$

where k is a constant. It is convenient to consider even power laws, with $n = 2N$, and write the expression in terms of the scaling parameters as

$$V(r) = \epsilon_m \left(\frac{r_m}{r}\right)^{2N}, \tag{4.38}$$

where $k = \epsilon_m r_m^{2N}$. Substituting this into Equations 4.23 and 4.24 gives

$$F(x,R) = \epsilon_m \eta^{2N} \sum_{n=1}^{N-1} x^{2n} \tag{4.39}$$

and

$$G = \epsilon_m (N-1)\eta^{2N} \tag{4.40}$$

respectively. The expression (Equation 4.22) for the scattering angle becomes

$$\chi\left(\eta,\varepsilon^*\right) = \pi - 2\left(1-\xi^N\right)^{1/2}\int_0^1 \frac{dx}{\sqrt{1-x^2}}\left[1+\xi^N \sum_{n=1}^{N-1} x^{2n}\right]^{-1/2}, \tag{4.41}$$

where

$$\xi \equiv \eta^N/\varepsilon^*. \tag{4.42}$$

The integral is readily evaluated since there is no orbiting and the integrand is well behaved. Cross sections can be evaluated over the entire energy range using Equation 4.26, which may also be written in terms of dimensionless variables. The dimensionless cross sections are found to be

$$\Delta\sigma^{*\,(l)} = q_l(N)\left(\varepsilon^*\right)^{-1/N}, \tag{4.43}$$

where the constants $q_l(N)$ are given in Table 4.1 for a range of l and N. Reverting temporarily to dimensional form and the original parameters of the potential (Equation 4.37), this is equivalent to

$$\Delta\sigma^{(l)}(\epsilon) = \pi r_m^2 q_l(N)\left(\varepsilon^*\right)^{-1/N} = \pi q_l(n/2) k^{2/n} \epsilon^{-2/n} \sim g^{-4/n}, \tag{4.44}$$

which is consistent with Equation 4.10 obtained on the basis of a dimensional argument.

4.6.3 Mason–Schamp (12-6-4) potential

For the Mason–Schamp potential (Equation 4.19), the scattering angle may be expressed in terms of dimensionless variables as

$$\chi(\eta,\varepsilon^*) = \pi - 2\left[1 - \left(A\eta^N - B\eta^3 - C\eta^2\right)/\varepsilon^*\right]^{1/2} \times \int_0^1 \frac{dx}{\left(1-x^2\right)^{1/2}}\left[1 + F(x,\eta)/\varepsilon^*\right]^{-1/2},$$

where the constants A, B, C are defined by Equation 4.20, and

$$F(x,\eta) = x^2\left\{\left(\sum_{n=0}^{N-2} x^{2n}\right)A\eta^N - (1+x^2)B\eta^3 - C\eta^2\right\}.$$

The scattering angles are then used to calculate the dimensionless cross sections,

$$\Delta\sigma^{*\,(l)}(\varepsilon^* > \varepsilon_c^*) = \int_0^{\eta_0}\left[1 - P_l(\cos\chi)\right]\left[1 + G(\eta)/\varepsilon^*\right]\frac{d\eta}{\eta^2}$$

Table 4.1 Values of the Constant $q_l(N)$ in Equation 4.43

	N = 2	3	4	6	8
$l = 1$	1.1934	1.1118	1.0462	1.0771	1.0322
2	1.8506	1.5436	1.4018	1.2650	1.1979
3	2.3478	1.8343	1.6072	1.3953	1.2938
4	2.7621	2.0595	1.7601	1.4884	1.3610

and

$$\Delta\sigma^{*(l)}(\varepsilon^* \le \varepsilon_c^*) = \int_0^{\eta_1} [1 - P_l(\cos\chi)]\,[1 + G(\eta)/\varepsilon^*]\,\frac{d\eta}{\eta^2} + \int_{\eta_2}^{\eta_0} [1 - P_l(\cos\chi)]\,[1 + G(\eta)/\varepsilon^*]\,\frac{d\eta}{\eta^2},$$

where

$$G(\eta) = (N-1)A\eta^N - 2B\eta^3 - C\eta^2.$$

The dimensionless form of Equation 4.34 gives the critical energy ε_c^*

$$\begin{aligned} \varepsilon_c^* &= C\eta_c^2 + 2B\eta_c^3 - (N-1)A\eta_c^N \\ &\equiv C(1-2/N)\eta_c^2 + 2B(1-3/N)\eta_c^3 \\ &\equiv \eta_c^2 + 2\gamma(N-3)/N\,(2\eta_c^3 - \eta_c^2), \end{aligned}$$

where by Equation 4.33 η_c is the unique positive root of

$$N(N-)A\eta^{N-2} - 6B\eta - 2C = 0.$$

The dimensionless forms of Equations 4.27, 4.29, 4.31, and 4.32, respectively, show that η_0 is the unique positive root of

$$\varepsilon^* + C\eta^2 + B\eta^3 - A\eta^N = 0,$$

η_1 is the smallest positive root of

$$\varepsilon^* - C\eta^2 - 2B\eta^3 + (N-1)A\eta^N = 0 \qquad (\varepsilon^* < \varepsilon_c^*),$$

$$(b_c^*)^2 = \eta_1^{-1}\left[1 - (A\eta_1^N - B\eta_1^3 - C\eta_1^2)/\varepsilon^*\right]$$

and η_2 is the largest positive root of

$$1 - \eta\,(b_c^*)^2 + (C\eta^2 + B\eta^3 - A\eta^N)/\varepsilon^* = 0, \ (\varepsilon^* < \varepsilon_c^*).$$

For a Mason–Schamp potential with $\gamma = 0.27$, it is found that $\varepsilon_c^* = 0.5337$ and $\eta_c = 0.6949$. Figure 4.5 shows examples of the dimensionless partial cross sections $\Delta\sigma^{*(l)}(\varepsilon^*)$ for $l = 1$ and 4. The observed asymptotic behaviour, $\Delta\sigma^{*(l)} \sim (\varepsilon^*)^{-1/2}$ at low energies and $\sim (\varepsilon^*)^{-1/6}$ at high energies, is consistent with the simple power law result (Equation 4.44), and reflects the long- and short-range nature of the interaction potential, $V(r) \sim r^{-4}$ and $\sim r^{-12}$, respectively. The decrease at intermediate energies arises from orbiting, where there is a balance between attractive and repulsive components of the interaction potential.

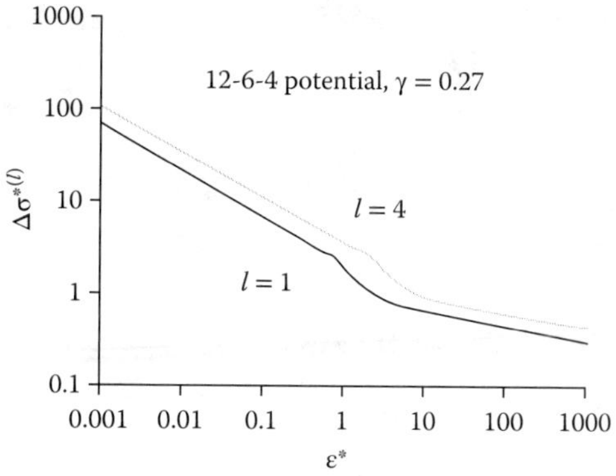

Figure 4.5 Dimensionless partial cross sections $\Delta\sigma^{*\,(l)} = \Delta\sigma^{(l)}/\pi r_m^2$ as a function of dimensionless energy $\varepsilon^* = \epsilon/\epsilon_m$ for the Mason–Schamp potential with $\gamma = 0.27$.

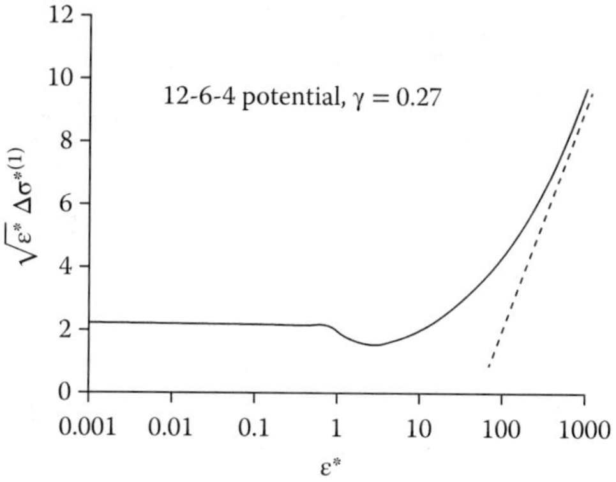

Figure 4.6 Dimensionless momentum transfer collision frequency for the Mason–Schamp potential with $\gamma = 0.27$. The dashed line represents the asymptote at high energies and has slope 1/3.

Cross sections for other values of l can similarly be found, and the corresponding dimensional quantities then follow as

$$\Delta\sigma^{(l)}(\epsilon) = \pi r_m^2 \Delta\sigma^{*(l)}(\epsilon/\epsilon_m)\ (l = 1, 2, 3, 4, \ldots).$$

In practice, where we take as many values of l as are needed to obtain solution of the kinetic equation to the required accuracy.

Figure 4.6 shows a dimensionless momentum transfer collision frequency $\sqrt{\varepsilon^*}\ \Delta\sigma^{*(1)}$ for the 12-6-4 potential with $\gamma = 0.27$. It can be seen that this is approximately constant at low energies, dips to a minimum at intermediate energies, and increases asymptotically as $(\varepsilon^*)^{1/3}$ at high energies, as represented by the dashed line. The dimensional momentum transfer collision frequency,

$$\nu_m(\epsilon) = n_0 \pi r_m^2 \sqrt{\frac{2\epsilon}{\mu}} \Delta\sigma^{*(1)}(\epsilon/\epsilon_m)$$

exhibits the same behaviour with respect to $\epsilon = \frac{1}{2}\mu g^2$ and g (see Figure 4.4).

4.7 Concluding Remarks

This chapter outlines how classical cross sections are found from a given interaction potential, for subsequent use in solution of the Boltzmann kinetic equation for ions in a monatomic gas. Standard quantum mechanics texts generally have an elementary discussion of scattering, see for example Davydov [60]. Cross sections for elastic and inelastic scattering of lighter particle (electrons, positrons) from atoms from molecules must, in general, be calculated quantum mechanically (see e.g. Brunger and Buckman [61] and Morrison [62]). However, for the purposes of transport analysis, it is not necessary to go into these details since it is assumed that the necessary cross sections are given.

CHAPTER 5

Kinetic Equations for Dilute Particles in Gases

5.1 Low Density Charged Particles in Gases

5.1.1 Free diffusion or swarm limit

In this chapter, we develop the kinetic theory of dilute charged particles in gaseous media, commencing with the Boltzmann collision integral for elastic collisions obtained in Chapter 3, and progressively including other types of interaction.

A typical application is the drift tube experiment shown schematically in Figure 5.1. Particles are assumed to be of sufficiently low density so that collisions between them are rare, and the only factors which influence their passage through the gas are interactions with neutrals and the electric field E. They may, in principle, still interact through some self-consistent electric field, but if densities are so low that even this too is negligible,* the "free diffusion" or "swarm" limit pertains. In this case, the average properties of the emerging beam should be independent of particle current, something which can be checked in experiment.

In the interests of simplifying notation, we drop the species index entirely for particle properties and simply write, for example, $n(\mathbf{r}, t)$ and $f(\mathbf{r}, \mathbf{v}, t)$ for particle number density and distribution function, respectively. Subscripts "0" will generally denote a property of the neutral gas, for example, n_0 is the number density of the neutral molecules. Thus, the low density limit means $n/n_0 \ll 1$.

5.1.2 The linear Boltzmann kinetic equation

The distribution function $f(\mathbf{r}, \mathbf{v}, t)$ of dilute charged particles undergoing elastic collisions in a monatomic gas is found by solving the

* The criterion is that the Debye length,

$$\lambda_D \equiv \sqrt{\frac{\varepsilon_0 k_B T}{ne^2}}$$

must be larger than any relevant macroscopic scale, where T is the temperature of the charged particles, $e = 1.6 \times 10^{-19}$C is the magnitude of the electric charge, and $\varepsilon_0 = 8.854 \times 10^{-12}$F/m is the permittivity of free space.

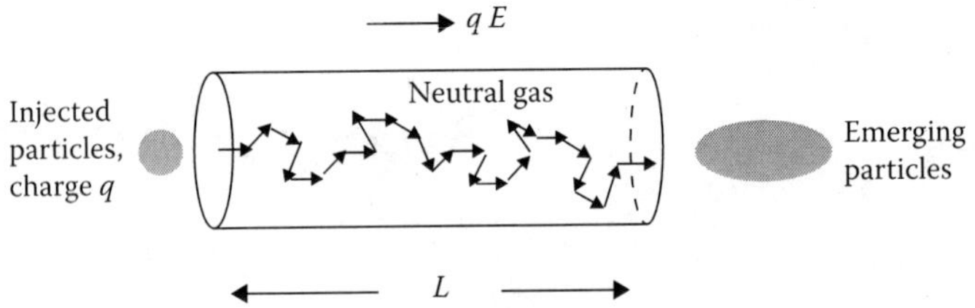

Figure 5.1 Schematic representation of a swarm experiment in which the external field E is carefully controlled and space-charge fields are ensured to be negligible by working at low densities and currents. The experiment essentially measures the characteristics of the emerging particles for a known drift distance L and given gas properties n_0 and T_0.

Boltzmann equation

$$\frac{\partial f}{\partial t} + \mathbf{v} \cdot \nabla f + \mathbf{a} \cdot \frac{\partial f}{\partial \mathbf{v}} = \int \mathrm{d}^3 v_0 \int \mathrm{d}^2 \Omega_{\mathbf{g}'}\, g\sigma(g,\chi) \left[f(\mathbf{r},\mathbf{v}',t) f_0(\mathbf{v}_0') - f(\mathbf{r},\mathbf{v},t) f_0(\mathbf{v}_0) \right] \tag{5.1}$$

in which the non-linear particle-particle collision term has been neglected. The gas can be assumed to remain essentially in undisturbed thermal equilibrium at temperature T_0 and thus f_0 is a stationary Maxwellian (see Equation 3.32):

$$f_0(\mathbf{v}_0) = n_0\, w(\alpha_0, v_0) \equiv n_0 \left(\frac{m_0}{2\pi k_B T_0} \right)^{3/2} \exp\left[-\frac{m v_0^2}{2 k_B T_0} \right]. \tag{5.2}$$

The differential cross section $\sigma(g,\chi)$ characterizing collisions between particles and gas atoms is assumed as given (see Chapter 4).

Equation 5.1 is *linear* in f since we always assume that the external force per unit mass,

$$\mathbf{a} = \frac{q}{m} (\mathbf{E} + \mathbf{v} \times \mathbf{B}) \tag{5.3}$$

derives entirely from *external* electric and magnetic fields.

As we progress through this chapter, we shall introduce contributions from inelastic and reactive processes into the collision term on the right hand side. However, the kinetic equation retains its linearity, as long as mutual interactions between particles can be neglected and any self-consistent fields remain negligible.

5.1.3 Moment equations

In Chapter 3, we foreshadowed the use of moment equations formed by appropriate integration of the kinetic equation with functions $\phi(\mathbf{v})$. For the Boltzmann equation (Equation 5.1), the rate of transfer of $\phi(\mathbf{v})$ in elastic collisions between charged particles and the neutral gas is given by Equation 3.19:

$$\int d^3v\,\phi(\mathbf{v})\left(\frac{\partial f}{\partial t}\right)_{\text{col}}$$
$$= \int d^3v \int d^3v_0\, g f(\mathbf{r},\mathbf{v},t) f_0(\mathbf{v}_0) \int d^2\Omega_{\mathbf{g}'}\, \sigma(g,\chi)\left[\phi(\mathbf{v}') - \phi(\mathbf{v})\right]. \tag{5.4}$$

This too will be modified as other types of interaction are introduced.

5.2 Charge Exchange

5.2.1 Collision model

In many applications of interest, for example, ions in the parent gas, resonant charge exchange is an important process. The corresponding collision integral may be obtained directly from the Boltzmann expression and greatly simplified by means of a model. Thus, we assume that the exchange takes place at long range, with hardly any deflection of the colliding species after the charge exchange takes place, as shown schematically in Figure 5.2.

While in reality, the particles are almost undeflected, it *looks* as if the positive ion with initial velocity $\mathbf{v}$ is scattered backwards with velocity $\mathbf{v}_0$, while the neutral particle of initial velocity $\mathbf{v}_0$ is scattered backwards with velocity $\mathbf{v}$. The relative velocity before the collision is $\mathbf{g} = \mathbf{v} - \mathbf{v}_0$, while after the collision it is $\mathbf{g}' = \mathbf{v}_0 - \mathbf{v} = -\mathbf{g}$, that is, the scattering angle is $\chi = \pi$, just as it

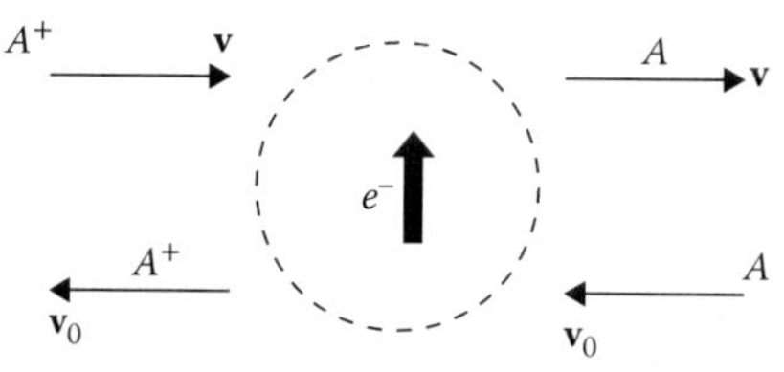

Figure 5.2 Schematic representation of an idealized charge exchange process in which an ionized atom A^+ with velocity $\mathbf{v}$ is neutralized by absorbing an electron released from a neutral atom A with velocity $\mathbf{v}_0$ within the region of interaction, marked by a dashed sphere. The interaction takes place at long range so that the velocities change negligibly. To all intents and purposes, there is a head-on collision, and the relative velocity changes from $\mathbf{g} = \mathbf{v} - \mathbf{v}_0$ to $\mathbf{g}' = \mathbf{v}_0 - \mathbf{v} = -\mathbf{g}$.

is for a head on collision between rigid spheres. Thus, the charge transfer collision integral is

$$\left(\frac{\partial f}{\partial t}\right)_{\text{col}} \approx \int \mathrm{d}^3 v_0 \int \mathrm{d}^2\Omega_{\mathbf{g}'}\, g\sigma(g,\chi)\,[f(\mathbf{r},\mathbf{v}_0,t)f_0(\mathbf{v}) - f(\mathbf{r},\mathbf{v},t)f_0(\mathbf{v}_0)], \tag{5.5}$$

with a differential cross section

$$\sigma(g,\chi) = \frac{\sigma^{(0)}(g)}{2\pi}\,[\delta(\cos\chi - \cos\chi_0)]_{\chi_0\to\pi}$$

and a total cross section $\sigma^{(0)}(g)$. Thus for the ideal resonant charge transfer model,

$$\left(\frac{\partial f}{\partial t}\right)_{\text{col}} \approx \int \mathrm{d}^3 v_0\, g\, \sigma^{(0)}(g)\,[f(\mathbf{r},\mathbf{v}_0,t)f_0(\mathbf{v}) - f(\mathbf{r},\mathbf{v},t)f_0(\mathbf{v}_0)].$$

5.2.2 Polarization potential and Bhatnagar–Gross–Krook equation

At large distances, r, the charge particle–neutral interaction is often well-described by the polarization potential $V(r) \sim r^{-4}$ (see Chapter 4), for which the speed-dependence of the cross section is $\sigma^{(0)}(g) \sim g^{-1}$. Assuming this to be the case, then $g\sigma^{(0)}(g) = \text{constant} = C$, and hence

$$\begin{aligned}\left(\frac{\partial f}{\partial t}\right)_{\text{col}} &\approx C \int \mathrm{d}^3 v_0\, [f(\mathbf{r},\mathbf{v}_0,t)f_0(\mathbf{v}) - f(\mathbf{r},\mathbf{v},t)f_0(\mathbf{v}_0)] \\ &= \nu\,[n(\mathbf{r},t)w(\alpha,v) - f(\mathbf{r},\mathbf{v},t)],\end{aligned} \tag{5.6}$$

where

$$\nu = n_0 C = n_0 g \sigma^{(0)}(g)$$

is the collision frequency for charge transfer, and we have used the normalization condition for the neutrals,

$$\int \mathrm{d}^3 v_0\, f_0(\mathbf{v}_0) = n_0.$$

Equations like Equation 5.6 are well known in plasma and condensed matter kinetic theory, but are generally introduced in an *ad hoc* fashion, without any justification other than a simplification of the mathematics. This equation also goes under different names, for example, a constant relaxation (time between collisions is $\tau = \nu^{-1}$) model, or Bhatnagar–Gross–Krook equation, after the original proponents Bhatnagar et al. [63].

In this book, we too use Equation 5.6 to simplify the analysis (see e.g., Chapter 14), though in the knowledge that it is soundly physically based.

5.3 Collision Term for Extremes of Mass Ratio

5.3.1 Fractional energy exchange

We now consider a binary elastic collision between an ion and a neutral particle of masses m and m_0, with initial velocities $\mathbf{v}$, and $\mathbf{v}_0 = 0$, respectively. The total kinetic energy is conserved, that is,

$$u + u_0 = u' + u_0',$$

where $u = \frac{1}{2}mv^2$, and so on, but we now focus on how much kinetic energy $\Delta u = u - u'$ is transferred from the ion to the neutral in a collision corresponding to center of mass (CM) scattering angle χ.

In this case, where the neutral is initially at rest, the relative and CM velocities before the collision are

$$\mathbf{g} = \mathbf{v} - \mathbf{v}_0 \equiv \mathbf{v}$$
$$\mathbf{G} = M\mathbf{v} + M_0\mathbf{v}_0 \equiv M\mathbf{v},$$

where $M = \frac{m}{m+m_0}$ and $M_0 = \frac{m_0}{m+m_0}$, and the initial kinetic energy of the ion is

$$u = \frac{1}{2}mv^2 = \frac{1}{2}m\left[G^2 + 2M_0\mathbf{g}\cdot\mathbf{G} + M_0^2g^2\right]$$

Since in *any* collision the CM velocity is constant, that is, $\mathbf{G}' = \mathbf{G}$, and for an *elastic* collision $g' = g$, the kinetic energy of the ion after the collision is

$$u' = \frac{1}{2}mv'^2 = \frac{1}{2}m\left[G^2 + 2M_0\mathbf{g}'\cdot\mathbf{G} + M_0^2g^2\right]$$

and hence the energy lost from the ion to the neutral is

$$\begin{aligned}\Delta u = u - u' &= mM_0\,(\mathbf{g} - \mathbf{g}')\cdot\mathbf{G}\\ &= mM_0M(\mathbf{g} - \mathbf{g}')\cdot\mathbf{g}\\ &= mM_0Mg^2(1 - \cos\chi)\\ &= mM_0Mv^2(1 - \cos\chi)\\ &= u2M_0M(1 - \cos\chi).\end{aligned}$$

The *fractional* kinetic energy exchange is therefore

$$\frac{\Delta u}{u} = 2M_0M(1 - \cos\chi) \equiv \frac{2mm_0}{(m+m_0)^2}(1 - \cos\chi). \tag{5.7}$$

The angular term $(1 - \cos\chi)$ lies between 0 and 2 and is of little interest as compared to the mass factor $2mm_0/(m+m_0)^2$. For ions and neutrals of comparable mass, $m \sim m_0$, the energy transfer is generally large. For ions in their parent gas undergoing a head on collision, $\chi = \pi$, $\frac{\Delta u}{u} = 1$. Thus, the ion is brought to rest and the neutral particle takes up all the energy, as in the charge exchange model above. For ions and neutrals of disparate masses, however, the energy exchange is very small and this has important implications, as we shall now see.

5.3.2 Heavy ions and Rayleigh limit

Suppose that the charged particles are very heavy, $m \gg m_0$, so that by Equation 5.7 the fractional energy transfer in a collision $\Delta u/u \sim 2m_0/m$ is very small. Since

$$\Delta \mathbf{v} = \mathbf{v}' - \mathbf{v} = M_0 \Delta \mathbf{g}$$

and by Figure 5.3

$$|\Delta \mathbf{v}| = M_0 \, |\Delta \mathbf{g}| = 2 M_0 g \sin \frac{\chi}{2}, \tag{5.8}$$

it follows that the picture for $M_0 = \frac{m_0}{m + m_0} \ll 1$ is one of heavy particles being buffeted very slightly by collisions with the light background gas.

Expanding the Boltzmann collision term to second order in $\Delta \mathbf{v}$, as in Section 3.5, we again obtain a Fokker–Planck equation,

$$\frac{\partial f}{\partial t} + \mathbf{v} \cdot \nabla f + \mathbf{a} \cdot \frac{\partial f}{\partial \mathbf{v}} = -\frac{\partial}{\partial \mathbf{v}} \cdot \left(f \langle \Delta \mathbf{v} \rangle \right) + \frac{1}{2} \frac{\partial^2}{\partial \mathbf{v} \partial \mathbf{v}} : \left(f \langle \Delta \mathbf{v} \Delta \mathbf{v} \rangle \right), \tag{5.9}$$

with

$$\langle \Delta \mathbf{v} \rangle = \int d^3 v_0 \, f_0(v_0) \int d^2 \Omega_{\mathbf{g}'} \, g \sigma(g, \chi) \Delta \mathbf{v}$$

$$\langle \Delta \mathbf{v} \Delta \mathbf{v} \rangle = \int d^3 v_0 \, f_0(v_0) \int d^2 \Omega_{\mathbf{g}'} \, g \sigma_R(g, \chi) \Delta \mathbf{v} \Delta \mathbf{v}.$$

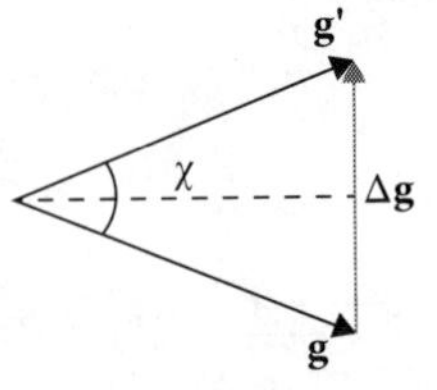

Figure 5.3 In an elastic collision, the relative velocity vector is rotated through an angle χ without changing magnitude, and hence $|\Delta \mathbf{g}| = 2g \sin \frac{\chi}{2}$.

If the thermal motion of the neutral gas particles is negligible, we may make the *cold gas* approximation, $T_0 \to 0$. The neutral velocity distribution function (Equation 5.2) then becomes a delta function,

$$f_0(\mathbf{v}_0) = n_0\, \delta(\mathbf{v}_0) \tag{5.10}$$

and it can be shown that

$$\begin{aligned} \langle \Delta \mathbf{v} \rangle &= -M_0 \mathbf{v} \nu_m \\ \langle \Delta \mathbf{v} \Delta \mathbf{v} \rangle &= M_0^2 \left[\frac{1}{3} v^2 \nu_v \mathbf{I} + (2\nu_m - \nu_v) \mathbf{v}\mathbf{v} \right], \end{aligned} \tag{5.11}$$

where **I** denotes the unit tensor, while

$$\nu_m(v) \equiv n_0 v\, 2\pi \int_0^{\pi} \sigma(v, \chi)[1 - P_1(\cos\chi)] \sin\chi \, d\chi \tag{5.12}$$

and

$$\nu_v(v) \equiv n_0 v\, 2\pi \int_0^{\pi} \sigma(v, \chi)[1 - P_2(\cos\chi)] \sin\chi \, d\chi \tag{5.13}$$

are the collision frequencies for momentum and viscosity transfer, respectively. The kinetic equation for the "Rayleigh gas" is then found by substituting Equation 5.11 into the right hand side of Equation 5.9:

$$\begin{aligned} \frac{\partial f}{\partial t} + \mathbf{v} \cdot \nabla f + \mathbf{a} \cdot \frac{\partial f}{\partial \mathbf{v}} &= M_0 \frac{\partial}{\partial \mathbf{v}} \cdot (\mathbf{v} \nu_m f) \\ &\quad + \frac{M_0^2}{2} \frac{\partial^2}{\partial \mathbf{v} \partial \mathbf{v}} : \left\{ \left[\frac{1}{3} v^2 \nu_v \mathbf{I} + (2\nu_m - \nu_v) \mathbf{v}\mathbf{v} \right] f \right\}. \end{aligned} \tag{5.14}$$

The right hand side differs from the Fokker–Planck equation obtained in Chapter 3 for long-range Coulomb interactions between charged particles in a fully ionized plasma in two important respects:

- It is *linear* in f; and
- The integrals Equations 5.12 and 5.13 can be evaluated without any *ad hoc* "cut-off" since, unlike the Coulomb force, charge–neutral interactions are short range.

5.3.3 Light charged particles and Lorentz gas

5.3.3.1 Two-term approximation

Consider now the case of light particles $\frac{m}{m_0} \ll 1$, for example, light ions, electrons, positrons or muons, in a neutral gas. In this Lorentz gas limit, the fractional energy exchange in an *elastic* collision $\sim \frac{2m}{m_0}$ is small, just as it is for the Rayleigh gas. In contrast, however, M_0 is of the order of unity and the *directional* changes need not be small. Light particles may then be scattered randomly in all directions as they move through gas, but generally with little change $\Delta v = v' - v$ in *speed*. Physically, we therefore expect that the distribution of velocities for light particles undergoing predominantly elastic collisions should be *almost isotropic*, with any external field causing only a slight anisotropy. This is expressed mathematically (suppressing any space-time dependence for the present)

$$f(\mathbf{v}) = f^{(0)}(v) + \mathbf{f}^{(1)} \cdot \hat{\mathbf{v}} + \dots, \tag{5.15}$$

where $f^{(0)}(v)$ is the dominant, isotropic part of the velocity distribution function, $\hat{\mathbf{v}}$ is a unit vector, and $\mathbf{f}^{(1)}$ is a small anisotropic contribution. Equation 5.15 is called the "two-term approximation" of the distribution function.*

If this is first integrated over all angles,

$$\int d^2\Omega_v f(\mathbf{v}) = f^{(0)}(v) \int d^2\Omega_v + \mathbf{f}^{(1)} \cdot \int d^2\Omega_v \hat{\mathbf{v}}, \tag{5.16}$$

where $d^{(2)}\Omega_v = \sin\theta d\theta d\varphi$ is an element of solid angle in $\mathbf{v}$-space, and noting the identities

$$\int d^2\Omega_v = \int_0^{2\pi} d\varphi \int_0^{\pi} \sin\theta \, d\theta = 4\pi; \quad \int d^2\Omega_v \hat{\mathbf{v}} \equiv 0 \tag{5.17}$$

it follows that

$$\int d^2\Omega_v f(\mathbf{v}) = 4\pi f^{(0)}(v).$$

On the other hand, if both sides of Equation 5.15 are first multiplied by $\hat{\mathbf{v}}$ and then integrated over all angles, we get

$$\int d^{(2)}\Omega_v \hat{\mathbf{v}} f(\mathbf{v}) = f^{(0)}(v) \int d^2\Omega_v \hat{\mathbf{v}} + \mathbf{f}^{(1)}(v) \cdot \int d^2\Omega_v \hat{\mathbf{v}}\hat{\mathbf{v}}.$$

* On the other hand, *inelastic* collisions involve exchange of a fixed quantity of energy (the threshold energy of the inelastic process), which may amount to a large fraction of the particle's energy. Consequently, particle velocities may change significantly in magnitude through inelastic scattering, and the distribution function may acquire a distinctly anisotropic component. Additional terms are then generally required in the right hand side of Equation 5.15, making it a "multi-term expansion."

Since the following identity holds,

$$\int d^2\Omega_v\, \hat{\mathbf{v}}\hat{\mathbf{v}} = \frac{4\pi}{3}\mathrm{I}, \tag{5.18}$$

where I is the unit tensor of rank 2, it then follows that

$$\int d^2\Omega_v\, \hat{\mathbf{v}} f(\mathbf{v}) = \frac{4\pi}{3}\mathbf{f}^{(1)}(v). \tag{5.19}$$

Quantities of physical interest then follow, for example,

1. *Number density*

$$n = \int d^3v\, f(\mathbf{v}) = \int_0^\infty v^2 dv \int d^2\Omega_v\, f(\mathbf{v}) = 4\pi \int_0^\infty dv\, v^2 f^{(0)}(v). \tag{5.20}$$

2. *Average velocity*

$$\begin{aligned}\mathbf{v} &= \frac{1}{n}\int d^3v\, \mathbf{v} f(\mathbf{v}) \\ &= \frac{1}{n}\int_0^\infty dv\, v^3 \int d^2\Omega_v\, \hat{\mathbf{v}} f(\mathbf{v}) = \frac{4\pi}{3n}\int_0^\infty dv\, v^3 \mathbf{f}^{(1)}(v).\end{aligned} \tag{5.21}$$

Note that in actual fact, Equation 5.15 shows just the first two terms of an infinite expansion in spherical harmonics, with the higher order contributions assumed negligible. This "two-term" approximation generally works well for electrons in noble gases, where the collisions are primarily elastic. However, it cannot be expected to be always satisfactory for electrons in molecular gases, where inelastic collisions become important and anisotropy in velocity space may be pronounced (see e.g., Figure 2.4) [64]. It is left as an exercise to argue physically why this might be so, and also to prove the identities of Equations 5.17 and 5.18. Since the "two-term" approximation has been widely (and often indiscriminately) used in electron transport theory for many decades, it will therefore be the focus of some attention in this book.

5.3.3.2 Spherical components of the collision terms

As for the respective collision terms, we first note that for central force interactions, collisions do not change the tensor properties of f and therefore we can write, just as in Equation 5.15

$$\left(\frac{\partial f}{\partial t}\right)_{\text{col}} = \left(\frac{\partial f^{(0)}}{\partial t}\right)_{\text{col}} + \hat{\mathbf{v}}\cdot\left(\frac{\partial \mathbf{f}^{(1)}}{\partial t}\right)_{\text{col}} + \dots. \tag{5.22}$$

Next apply the general expression (Equation 5.4) for some arbitrary property ϕ which we take to be a function of *speed* only, that is, $\phi = \phi(v)$:

$$\int d^3v\, \phi(v) \left(\frac{\partial f}{\partial t}\right)_{\text{col}} = \int d^3v \int d^3v_0\, g f(\mathbf{v}) f_0(\mathbf{v}_0) \int d^2\Omega_{\mathbf{g}'} \sigma(g,\chi) \left[\phi(v') - \phi(v)\right]. \tag{5.23}$$

In the left hand side, the integral over angles can be done immediately, using Equation 5.22:

$$\int d^3v\, \phi(v) \left(\frac{\partial f}{\partial t}\right)_{\text{col}} = \int_0^\infty v^2\, dv\, \phi(v) \int d^2\Omega_{\mathbf{v}} \left(\frac{\partial f}{\partial t}\right)_{\text{col}} = 4\pi \int_0^\infty v^2\, dv\, \phi(v) \left(\frac{\partial f^{(0)}}{\partial t}\right)_{\text{col}}.$$

Furthermore, to simplify matters, we take a cold gas as in Equation 5.10, and thus Equation 5.23 becomes

$$\begin{aligned} 4\pi \int_0^\infty dv\, v^2\, \phi(v) \left(\frac{\partial f^{(0)}}{\partial t}\right)_{\text{col}} &= n_0 \int d^3v\, v f(\mathbf{v}) \int d^2\Omega_{\mathbf{v}'}\, \sigma(v,\chi) \left[\phi(v') - \phi(v)\right] \\ &\approx n_0 \int d^3v\, v f(\mathbf{v}) \int d^2\Omega_{\mathbf{v}'}\, \sigma(v,\chi)(v' - v)\frac{\partial \phi}{\partial v}. \end{aligned}$$

Now by Equation 5.7 it follows that

$$\frac{v' - v}{v} = -\frac{\Delta u}{2u} \approx -\frac{m}{m_0}(1 - \cos\chi)$$

and hence

$$\begin{aligned} 4\pi \int_0^\infty v^2\, dv\, \phi(v) \left(\frac{\partial f^{(0)}}{\partial t}\right)_{\text{col}} &= -n_0 \frac{m}{m_0} \int d^3v\, v f(\mathbf{v}) \frac{\partial \phi}{\partial v} \int d^2\Omega_{\mathbf{v}'}\, v\sigma(v,\chi)(1 - \cos\chi) \\ &= -\frac{m}{m_0} \int d^3v\, v\, \nu_m(v)\, f(\mathbf{v})\, \frac{\partial \phi}{\partial v} \end{aligned}$$

$$= -\frac{m}{m_0}\int_0^\infty dv\, \frac{\partial\phi}{\partial v} v^3 \nu_m(v) \int d^2\Omega_{\mathbf{v}}\, f(\mathbf{v})$$

$$= -\frac{4\pi m}{m_0}\int_0^\infty dv\, \frac{\partial\phi}{\partial v} v^3 \nu_m(v)\, f^{(0)}(v)$$

$$= \frac{4\pi m}{m_0}\int_0^\infty dv\, \phi(v)\, \frac{\partial}{\partial v}\left[v^3 \nu_m(v)\, f^{(0)}(v)\right].$$

The momentum transfer collision frequency is defined by Equation 5.12. Since $\phi(v)$ is arbitrary, the only way this can be satisfied is if

$$\left(\frac{\partial f^{(0)}}{\partial t}\right)_{\text{col}} = \frac{m}{m_0}\frac{1}{v^2}\frac{\partial}{\partial v}\left[v^3 \nu_m(v)\, f^{(0)}(v)\right]. \tag{5.24}$$

The derivation including thermal motion of the neutral gas is slightly more complicated, and we leave the details to Chapter 6, where we address the problem from a different viewpoint for (soft-)condensed systems. The expression can be found in standard texts such as Chapman and Cowling [18]:

$$\left(\frac{\partial f^{(0)}}{\partial t}\right)_{\text{col}} = \frac{m}{m_0}\frac{1}{v^2}\frac{\partial}{\partial v}\left\{v^2 \nu_m(v)\left[v f^{(0)}(v) + \frac{k_B T_0}{m}\frac{\partial f^{(0)}}{\partial v}\right]\right\}. \tag{5.25}$$

Next, we calculate $\left(\frac{\partial \mathbf{f}^{(1)}}{\partial t}\right)_{\text{col}\,l}$ by substituting $\phi(\mathbf{v}) = \mathbf{v}\psi(v)$ in the general expression (Equation 5.23), where now $\psi(v)$ is an arbitrary function of speed, and taking a cold gas as before:

$$\int d^3v\, \mathbf{v}\psi(v)\left(\frac{\partial f}{\partial t}\right)_{\text{col}}$$

$$= n_0 \int d^3v\, f(\mathbf{r},\mathbf{v},t) \int d^2\Omega_{\mathbf{v}'}\, v\sigma(v,\chi)\left[\mathbf{v}'\psi(v') - \mathbf{v}\psi(v)\right]$$

$$\approx n_0 \int d^3v\, f(\mathbf{r},\mathbf{v},t)\, \psi(v) \int d^2\Omega_{\mathbf{v}'}\, v\sigma(v,\chi)\left[\mathbf{v}' - \mathbf{v}\right],$$

where we have neglected energy exchange in the second step, effectively assuming a neutral particle of infinite mass, that is, the calculation is done to zero order in m/m_0. (N.B. We can not do this in the calculation of $\left(\frac{\partial f^{(0)}}{\partial t}\right)_{\text{col}}$, since this is clearly of first order in m/m_0.) It is left as an exercise to show that

$$n_0 \int d^2\Omega_{\mathbf{v}'}\, v\sigma(v,\chi)\left[\mathbf{v}' - \mathbf{v}\right]$$

$$= -n_0 \mathbf{v} \int d^2\Omega_{\mathbf{v}'}\, v\sigma(v,\chi)(1 - \cos\chi) = -\mathbf{v}\nu_m(v)$$

and thus

$$\int d^3v\ \mathbf{v}\psi(v)\left(\frac{\partial f}{\partial t}\right)_{col} \equiv \int d^3v\ \mathbf{v}\psi(v)\,\hat{\mathbf{v}}\cdot\left(\frac{\partial \mathbf{f}^{(1)}}{\partial t}\right)_{col}$$
$$= -\int d^3v\, f(\mathbf{v})\mathbf{v}\psi(v)\nu_m(v)$$
$$\equiv -\int d^3v\ \mathbf{v}\psi(v)\nu_m(v)\hat{\mathbf{v}}\cdot\mathbf{f}^{(1)}(v).$$

Since $\psi(v)$ is arbitrary this expression can be satisfied if and only if

$$\left(\frac{\partial \mathbf{f}^{(1)}}{\partial t}\right)_{col} = -\nu_m(v)\mathbf{f}^{(1)}(v) \tag{5.26}$$

which holds to zero order in m/m_0 even allowing for thermal motion of the neutrals.

5.3.3.3 Two-term decomposition of the Boltzmann equation (elastic collisions)

We now turn to the left hand side of the Boltzmann equation,

$$Df \equiv \frac{\partial f}{\partial t} + \mathbf{v}\cdot\nabla f + \mathbf{a}\cdot\frac{\partial f}{\partial \mathbf{v}}, \tag{5.27}$$

where $\mathbf{a} = q\,[\mathbf{E} + \mathbf{v}\times\mathbf{B}]\,/m$ now explicitly includes electric and magnetic fields. This can be broken into its scalar and vector parts by direct substitution of Equation 5.15 into 5.27. After some lengthy algebra, we find

$$Df = \left\{\frac{\partial f^{(0)}}{\partial t} + \frac{v}{3}\nabla\cdot\mathbf{f}^{(1)} + \frac{q\mathbf{E}}{3mv^2}\cdot\frac{\partial}{\partial v}[v^2\mathbf{f}^{(1)}]\right\}$$
$$+\hat{\mathbf{v}}\cdot\left\{\frac{\partial \mathbf{f}^{(1)}}{\partial t} + v\nabla f^{(0)} + \frac{q\mathbf{E}}{m}\frac{\partial f^{(0)}}{\partial v} + \boldsymbol{\Omega}_L\times\mathbf{f}^{(1)}\right\} + \ldots,$$

where

$$\boldsymbol{\Omega}_L = \frac{q\mathbf{B}}{m}$$

is the cyclotron frequency, and the dots indicate that there is also a contribution from a tensor of second rank. Equating the scalar term with

Equation 5.25 and the vector term with Equation 5.26, leads to the coupled pair of differential equations:

$$\frac{\partial f^{(0)}}{\partial t} + \frac{v}{3}\nabla \cdot \mathbf{f}^{(1)} + \frac{q\mathbf{E}}{3mv^2} \cdot \frac{\partial}{\partial v}[v^2\mathbf{f}^{(1)}]$$

$$= \frac{m}{m_0}\frac{1}{v^2}\frac{\partial}{\partial v}\left\{v^2\nu_m(v)\left[vf^{(0)}(v) + \frac{k_B T_0}{m}\frac{\partial f^{(0)}}{\partial v}\right]\right\} \tag{5.28}$$

$$\frac{\partial \mathbf{f}^{(1)}}{\partial t} + v\nabla f^{(0)} + \frac{q\mathbf{E}}{m}\frac{\partial f^{(0)}}{\partial v} + \mathbf{\Omega}_L \times \mathbf{f}^{(1)}$$

$$= -\nu_m(v)\mathbf{f}^{(1)}(v). \tag{5.29}$$

These "two-term" kinetic equations provide a reasonable basis for studying electrons, positrons, and muons undergoing elastic collisions in gases. Inelastic collisions (see Section 5.4), however, require not only modification of the collision term but also extension of the spherical expansion (Equation 5.15) beyond just two spherical harmonics.

In general, representation of f through a "multi-term" expansion in spherical harmonics is necessary for all types of charged particles, both light and heavy, and the analysis becomes correspondingly more complex.

5.4 Inelastic Collisions

5.4.1 Wang Chang–Uhlenbeck–de Boer collision term

Suppose that the charged particles are still assumed to be structureless, but now we allow for the possibility that the gas constituents have internal structure, for example, electronic, rotational, and vibrational states, characterized by quantum numbers (or, in reality, sets of quantum numbers) j, with energy levels ε_j of degeneracy ω_j. From equilibrium statistical mechanics, we know that the distribution function for the neutrals is Maxwell–Boltzmann [18]

$$f_{0j}(\mathbf{v}_0) = n_{0j}\, w(\alpha_0, v_0) = n_{0j}\left(\frac{m_0}{2\pi k_B T_0}\right)^{3/2} \exp\left[-\frac{mv_0^2}{2k_B T_0}\right], \tag{5.30}$$

where

$$n_{0j} = \frac{\omega_j}{Z}\exp\left(-\frac{\epsilon_j}{k_B T_0}\right) \tag{5.31}$$

is the relative populations in the state j and $Z = \sum_j \omega_j \exp\left(-\frac{\epsilon_j}{k_B T_0}\right)$ is the partition function. The general picture for an *inelastic collision* between a charged particle and a neutral, which undergoes a transition from state $j \Longrightarrow j'$, is shown in Figure 5.4, and examples of several types of excitations are shown in Figure 5.5.

The CM velocity is still conserved, that is, $\mathbf{G}'=\mathbf{G}$, but the relative speed is not, since total energy conservation requires

$$\frac{1}{2}\mu g^2 + \epsilon_j = \frac{1}{2}\mu g'^2 + \epsilon_{j'}. \tag{5.32}$$

In Figure 5.4, we show $j' > j$, but $j' < j$ (a "superelastic" collision) is also possible, in which the relative speed is actually increased following a collision. Moreover, as explained by Ross et al. [65] microscopic reversibility is represented by

$$\omega_{j'} g'^2 \sigma(j', j; \mathbf{g}', \mathbf{g}) = \omega_j g^{\,2} \sigma(j, j'; \mathbf{g}, \mathbf{g}'), \tag{5.33}$$

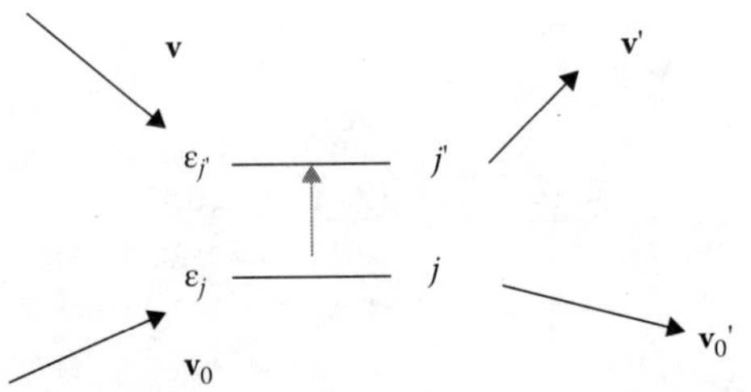

Figure 5.4 An inelastic collision between a structureless point particle of initial velocity $\mathbf{v}$ and a neutral gas atom or molecule with initial velocity $\mathbf{v}_0$, initially in an internal state j and energy level ϵ_j. The final velocities are $\mathbf{v}'$ and $\mathbf{v}_0'$, and the final state of the neutral is j'.

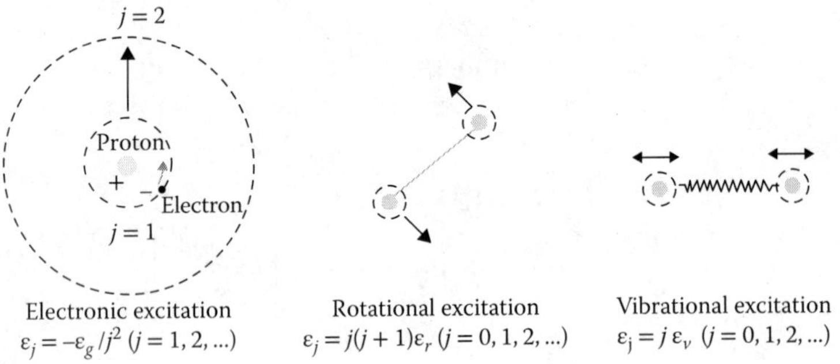

Figure 5.5 Examples of possible excitation of atomic and molecular states in collisions in which j is a generic quantum number and (from left to right) ϵ_g is the atomic ground state energy (e.g., 13.6 eV for a hydrogen atom), $\epsilon_r = h^2/(8\pi^2 I)$ and $\epsilon_v = h\nu$ are constants, characterizing energies of rotations of the diatomic molecule of moment of inertia I and vibrations of molecules with natural frequency ν, respectively. In addition (not shown), simultaneous excitation of rotational and vibrational states ("ro-vibration") is possible.

where $\sigma(j,j';\mathbf{g},\mathbf{g}')$ is the differential cross section for the process shown in Figure 5.5, and $\sigma(j',j;\mathbf{g}',\mathbf{g})$ is the corresponding quantity for the inverse process. This reduces to Equation 3.8 in the case of an elastic collision $j'=j$.

To get the total change in some property $\phi(\mathbf{v})$ of the charged particles, one has to sum over all possible transitions $j\longrightarrow j'$, that is, Equation 5.4 generalizes to

$$\int d^3v\,\phi(\mathbf{v})\left(\frac{\partial f}{\partial t}\right)_{\text{col}}=\sum_{j,j'}\int d^3v\int d^3v_0\,g\,f(\mathbf{r},\mathbf{v},t)f_{0j}(v_0)$$
$$\times\int d^2\Omega_{\mathbf{g}'}\,\sigma(j,j';\mathbf{g},\mathbf{g}')\left[\phi(\mathbf{v}')-\phi(\mathbf{v})\right]. \tag{5.34}$$

Note that as a generalization of Equation (3.11), we may define

$$d^8\vartheta(j,j')=g\sigma(j,j';\mathbf{g},\mathbf{g}')d^2\Omega_{\mathbf{g}'}d^3g\,d^3G$$

which has the property

$$d^8\vartheta'(j',j)=d^8\vartheta(j,j'). \tag{5.35}$$

The proof, which follows from Equations 5.32 and 5.33, is left as an exercise. Using similar manipulations to those previously, one arrives at a generalization of the Boltzmann equation (Equation 5.1),

$$\frac{\partial f}{\partial t}+\mathbf{v}\cdot\nabla f+\mathbf{a}\cdot\frac{\partial f}{\partial \mathbf{v}}=\left(\frac{\partial f}{\partial t}\right)_{\text{col}}=\sum_{j,j'}\int d^3v_0\int d^2\Omega_{\mathbf{g}'}\,g\sigma(j,j';\mathbf{g},\mathbf{g}')$$
$$\times\left[f(\mathbf{r},\mathbf{v}',t)f_{0j'}(v_0')-f(\mathbf{r},\mathbf{v},t)f_{0j}(v_0)\right]. \tag{5.36}$$

This is the *semi-classical* Wang Chang–Uhlenbeck–de Boer (WUB) kinetic equation [2], which is used extensively in gas transport analysis. Note that as with the classical Boltzmann equation (Equation 5.1), the inelastic cross sections in WUB are assumed to depend only upon the polar angle χ, that is

$$\sigma(j,j';\mathbf{g},\mathbf{g}')=\sigma(j,j';g,g',\chi). \tag{5.37}$$

5.4.2 Semi-classical and quantum collision operators

5.4.2.1 Questions concerning the WUB kinetic equation

Equation 5.36 introduces an element of quantum mechanics insofar as the energy levels of the molecules are quantized, but several points require further discussion:

- WUB makes no physical distinction between processes of a vector or scalar nature: rotational, vibrational, or electronic excitation are treated in the same way, differing only in the magnitudes of the respective energy exchanges
- WUB does not explicitly mention the possibility of degeneracy in energy levels, except through the factor ω_j in Equation 5.31. Thus, for example, while rotational energy levels $\epsilon_j = h^2/(8\pi^2 I)j(j+1)$ are $2j+1$-fold degenerate ($\omega_j = 2j+1$), but the projection quantum numbers $m_j = -j, , ..., -1, 0, 1, ...j$ do not appear.
- Electron scattering from rotating diatomic molecules is governed by non-central forces, so should the cross section not depend on both the polar and azimuthal angles, χ and ϕ, respectively?

These questions prompt us to ask if the quantum kinetic equation of Waldmann [4,66] and Snider [67,68] (WS) would be more appropriate to describe particles in molecular gases.

5.4.2.2 Reconciliation with the WS kinetic equation

Although the explicit form of the WS kinetic equation is not needed for the present discussion, we observe that, like WUB, it treats collisions as local, is linear in f, and conserves particle number. The chief differences between it and WUB are as follows:

- In the WS equation, the Wigner distribution density matrix generally appears in place of the scalar distribution function f; and
- Collisions are accounted for through quantum scattering amplitudes $a^{j,j'}_{m_j,m_{j'}}(\hat{\mathbf{g}},\hat{\mathbf{g}}')$ which, for non-central forces depend upon $\hat{\mathbf{g}}$ and $\hat{\mathbf{g}}'$ separately, not their scalar product, and are thus functions of both the polar and azimuthal angles, χ and ϕ

On the other hand, there are several points of convergence:

- For inelastic collisions involving scattering from a molecule with energy levels separated by a gap $\Delta\varepsilon$, quantum interference effects are negligible if

$$\tau\Delta\varepsilon \gg h$$

where τ is the average time between collisions. While this condition is satisfied for vibrational levels, which are widely spaced, it is clearly not the case for the $2j+1$ degenerate rotational sub-levels $m_j = -j, ..., j$, which only the WS kinetic equation can, strictly speaking, treat correctly:

- However, the Wigner matrix for point particles is a scalar, and the Wigner matrix of the neutral molecules is diagonal in, and independent of the m-indices, and it can be shown that WS $\rightarrow$ WUB [4,66], with the differential cross section

$$\sigma(j,j';\mathbf{g},\mathbf{g}') = \sigma(j,j';g,g',\chi) \equiv \frac{1}{2j+1} \sum_{m_j=-j}^{j} \sum_{m'_j=-j'}^{j'} |a_{m_j,m_{j'}}^{j,j'}(\hat{\mathbf{g}},\hat{\mathbf{g}}')|^2. \tag{5.38}$$

By virtue of the summation of squares of scattering amplitudes over projection quantum numbers, this is a function of only one angle χ, defined by $\cos\chi = \hat{\mathbf{g}} \cdot \hat{\mathbf{g}}'$.

The semi-classical WUB kinetic equation (Equation 5.36) is thus consistent with the quantum kinetic WS equation, as long as the differential cross section (Equation 5.37) is interpreted as in Equation 5.38. Importantly, from the point of view of consistency, this is also the differential cross section determined by *ab initio* quantum mechanical calculations and measured in beam experiments.

5.4.3 Inelastic collision term for light particles

On the basis of the above discussion, the kinetic Equation 5.36, therefore, provides a general platform for calculating transport properties of charged particles of all types and masses in any gas, atomic, or molecular, and accounts for both elastic ($j' = j$) and inelastic ($j' \neq j$) collisions. In addition, for light particles $m/m_0 \ll 1$, the general collision operator $\left(\frac{\partial f}{\partial t}\right)_{\text{col}}$ and its tensor components may, if desired, be approximated through an expansion in powers of m/m_0.

1. To *zero order* in m/m_0, the spherical component of the collision operator derives entirely from inelastic processes, and as shown in Appendix B can be expressed in terms of a sum over inelastic and superelastic processes,

$$\left(\frac{\partial f^{(0)}}{\partial t}\right)_{\text{inel}} = \left(\frac{2}{m\epsilon}\right)^{\frac{1}{2}} \sum_{j<j'} n_{0j}\Big[(\epsilon+\epsilon_{jj'})f^{(0)}(\epsilon+\epsilon_{jj'}) \times\sigma^{(0)}(jj';\epsilon+\epsilon_{jj'}) - \epsilon f^{(0)}(\epsilon)\,\sigma^{(0)}(jj';\epsilon)\Big] \tag{5.39}$$
$$+ n_{0j'}\Big[(\epsilon-\epsilon_{jj'})f^{(0)}(\epsilon-\epsilon_{jj'})\sigma^{(0)}(j'j;\epsilon-\epsilon_{jj'}) - \epsilon f^{(0)}(\epsilon)\,\sigma^{(0)}(j'j;\epsilon)\Big]$$

in which $\epsilon = \frac{1}{2}\mu g^2 \approx \frac{1}{2}mv^2$ is used as the independent variable rather than speed, and $\epsilon_{jj'} \equiv \epsilon_{j'} - \epsilon_j$. A similar expression was obtained by Frost and Phelps [3] starting from first principles. The expression (Equation 5.33) for time-reversal symmetry can be written in terms of partial cross sections as

$$\omega_{j'}\left(\epsilon-\epsilon_{jj'}\right)\sigma^{(0)}(j'j;\epsilon-\epsilon_{jj'}) = \omega_j\epsilon\sigma^{(0)}(jj';\epsilon)$$

and allows the right hand side of Equation 5.39 to be expressed entirely in terms of excitation cross sections.

2. To *first order* in m/m_0, the main contribution to $\left(\frac{\partial f^{(0)}}{\partial t}\right)_{\text{col}}$ comes from elastic collisions $j' = j$, as given by the differential operator (Equation 5.25), and the effects of inelastic processes are often neglected.

3. Likewise, although both elastic and inelastic collisions contribute to the the vector and higher order tensor components $\left(\frac{\partial \mathbf{f}^{(l)}}{\partial t}\right)_{\text{col}}$ $(l = 1, 2, 3, \ldots)$ to zero order in m/m_0, the elastic contribution again usually dominates, and the expression (Equation 5.26) applies to a good approximation.

Further details can be found in Appendix B.

5.5 Non-Conservative, Reactive Collisions

5.5.1 Classification of reactive collisions

Suppose that a collision between a charged particle and a neutral molecule results in a "reaction,"

$$\text{particle} + \text{neutral} \rightarrow \text{products} \tag{5.40}$$

in which the "products" may include new particles and different molecular species, and the number of particles may not be conserved. For the purposes of this book, we may classify such reactive collisions into three types:

- *Conservative collisions*, in which the original charged particle emerges in the "products," along with new particles. For the sake of being definite, we take for example, positron impact ionization of a molecule or atom A,

$$e^+ + A \rightarrow A^+ + e^+ + e^-. \tag{5.41}$$

The threshold energy ϵ_I which is equal to the ionization energy of A is lost, and the remaining kinetic energy is shared between the scattered positron and the electron and excited state molecule products. If the electron comes off with zero energy, then from the perspective of the positron, this may be classified as just another type of inelastic collision. Since positron number is conserved, the inelastic collision term of the previous section applies. In general, however, the ejected electron and the positron share the energy, and a new form of the collision operator is required to handle the partitioning of the energy between the scattered and ejected particles. This is discussed below. The reader is referred to Ref. [69] for details of the positron impact ionization operator.

- *Non-conservative collisions, in which the original charged particle is lost,* and does not appear in the products. Examples include electron attachment in an electronegative gas,*

  $$e^- + A \rightarrow A^-,$$

  positron annihilation with a bound atomic electron to form two back-to-back gamma rays,

  $$e^+ + e^- \rightarrow \gamma + \gamma,$$

  or a muon displacing a bound atomic electron to form a muonic atom,

  $$\mu^- + H \rightarrow p\mu + e^-.$$

  A general kinetic collision term is obtained below representing all such cases. Note that the "products" of the collision may (as in the muon-catalyzed fusion cycle of Chapter 19) or may not play any further role in the physical process under consideration.

- *Non-conservative collisions, in which the particle number is increased.* An example is electron impact ionization of an atom or molecule,

  $$e^- + A \rightarrow A^+ + e^- + e^- \tag{5.42}$$

  in which a positive ion and one or more additional electrons are created, all of which contribute further to electron transport properties. In developing the corresponding kinetic collision term, the question arises as to how the energy is partitioned between the two emerging electrons. Sometimes, ionization is treated as just another inelastic process, with the ionization energy ϵ_I of the gas atom or molecule playing the role of an inelastic threshold energy. However, this is a dubious approximation.

5.5.2 Notation

The generic notation for the rate of change of f due to reactive collisions of all types we will define to be $\left(\frac{\partial f}{\partial t}\right)_R$. However, we need to distinguish carefully between particle loss and gain processes, and to that end, we adopt the notation $R = *$ and $R = I$, respectively.

* A third body, or "spectator" molecule M must be present to conserve momentum and energy.

5.5.3 Particle loss collision term

We first consider dilute particles undergoing particle loss reactions. Since the probability of collisions between the products of Equation 5.40 regenerating the charged particles in a reverse of reaction is negligibly small, the reactive collision term therefore is simply a scattering "out" term, of a form similar to the last member of Equation 5.36 with an appropriate reactive cross section σ_*:

$$\begin{aligned}\left(\frac{\partial f}{\partial t}\right)_* &= -\sum_{j,j'}\int d^3v_0 \int d^2\Omega_{\mathbf{g}'} g\sigma_*(j,j';g,g') f(\mathbf{r},\mathbf{v},t) f_{0j}(\mathbf{v}_0) \\ &= -\left\{\sum_{j,j'}\int d^3v_0\, g\sigma_*(j,j';g) f_{0j}(\mathbf{v}_0)\right\} f(\mathbf{r},\mathbf{v},t) \\ &= -\nu_*(v) f(\mathbf{r},\mathbf{v},t), \end{aligned} \tag{5.43}$$

where

$$\sigma_*(j,j';g) \equiv \int d^2\Omega_{\mathbf{g}'}\, \sigma_*(j,j';g,g')$$

is the total or integral reactive cross section, and

$$\nu_*(v) \equiv \sum_{j,j'}\int d^3v_0\, g\sigma_*(j,j';g) f_{0j}(\mathbf{v}_0)$$

is the reactive collision frequency.

The reactive collision term does not vanish when integrated over all velocities,

$$\int d^3v \left(\frac{\partial f}{\partial t}\right)_* = -\int d^3v\, \nu_*(v) f(\mathbf{r},\mathbf{v},t) \neq 0$$

reflecting the fact that there is a net particle loss from phase space, in contrast to particle-conserving collisions (see Figure 2.5).

Thus, the kinetic equation including reactive losses is

$$\begin{aligned}\frac{\partial f}{\partial t} + \mathbf{v}\cdot\nabla f + \mathbf{a}\cdot\frac{\partial f}{\partial \mathbf{v}} = \left(\frac{\partial f}{\partial t}\right)_{\mathrm{col}} &= \sum_{j,j'}\int d^3v_0 \int d^2\Omega_{\mathbf{g}'}\, g\sigma(j,j';g,g') \\ &\times \left[f(\mathbf{r},\mathbf{v}',t) f_{0j'}(v_0') - f(\mathbf{r},\mathbf{v},t) f_{0j}(\mathbf{v}_0)\right] - \nu_*(v) f(\mathbf{r},\mathbf{v},t).\end{aligned}$$

In simplified treatments of reactive effects, the reactive collision frequency and cross section are assumed to be related by

$$\nu_*(v) = n_0 v \sigma_*(v) \tag{5.44}$$

but this is strictly speaking valid for a cold ($T_0 = 0$ K) monatomic gas only.

5.5.4 Electron impact ionization

Unlike the loss processes described above, the collision term for Equation 5.42 must be obtained from first principles. For simplicity, it is assumed that there is only one ionization channel corresponding to an ionization energy ϵ_I. If it is assumed that the gas is cold and that the recoil of the ion M^+ can be neglected, energy conservation can be expressed as

$$\epsilon = \epsilon_I + \epsilon' + \epsilon'' \tag{5.45}$$

where $\epsilon = \frac{1}{2}mv^2$ is the initial electron energy and $\epsilon' = \frac{1}{2}mv'^2$ and $\epsilon'' = \frac{1}{2}mv''^2$ are the energies of the two emerging electrons. To zero order in m/m_0 and for equal partitioning of energy, the scalar part of the ionization kinetic collision term is found to be (see Appendix B)

$$\left(\frac{\partial f^{(0)}}{\partial t}\right)_I = n_0\sqrt{\frac{2}{m\epsilon}}\left[(2\epsilon+\epsilon_I)\sigma_I(2\epsilon+\epsilon_I)f^{(0)}(2\epsilon+\epsilon_I) - \epsilon\sigma_I(\epsilon)f^{(0)}(\epsilon)\right] \tag{5.46}$$

where $\sigma_I(\epsilon)$ is the cross section for the process.

5.6 Two-Term Kinetic Equations for a Lorentz Gas

While both inelastic and non-conservative collisions may contribute significantly to the scalar part $\left(\frac{\partial f^{(0)}}{\partial t}\right)_{\text{col}}$ of the collision term, their effect on the vector part $\left(\frac{\partial \mathbf{f}^{(1)}}{\partial t}\right)_{\text{col}}$ can often be neglected, since the corresponding cross sections are generally much smaller than their elastic counterparts. On that basis, we are now in a position to write down a set of kinetic equations for light particles in the two-term approximation:

$$\begin{aligned}\frac{\partial f^{(0)}}{\partial t} + \frac{v}{3}\nabla\cdot\mathbf{f}^{(1)} + \frac{q\mathbf{E}}{3mv^2}\cdot\frac{\partial}{\partial v}[v^2\mathbf{f}^{(1)}] &= \left(\frac{\partial f^{(0)}}{\partial t}\right)_{\text{col}}\\ &\approx \frac{m}{m_0}\frac{1}{v^2}\frac{\partial}{\partial v}\left\{v^2\nu_m(v)\left[vf^{(0)}(v) + \frac{k_BT_0}{m}\frac{\partial f^{(0)}}{\partial v}\right]\right\} + \left(\frac{\partial f^{(0)}}{\partial t}\right)_{\text{inel}}\\ &\quad + \left(\frac{\partial f^{(0)}}{\partial t}\right)_I - \nu_*(v)f^{(0)}\end{aligned} \tag{5.47}$$

$$\frac{\partial \mathbf{f}^{(1)}}{\partial t} + v\,\nabla f^{(0)} + \frac{q\mathbf{E}}{m}\frac{\partial f^{(0)}}{\partial v} + \mathbf{\Omega}_L\times\mathbf{f}^{(1)} = \left(\frac{\partial \mathbf{f}^{(1)}}{\partial t}\right)_{\text{col}} - \nu_m(v)\mathbf{f}^{(1)}(v) \tag{5.48}$$

where $\left(\frac{\partial f^{(0)}}{\partial t}\right)_{\text{inel}}$ $\nu_*(v)$ and $\left(\frac{\partial f^{(0)}}{\partial t}\right)_I$ are defined by Equations 5.39, 5.44, and 5.46, respectively. While these equations form the basis for many investigations of electron and positron transport properties in gases, we must sound a note of caution. The accuracy of the two-term representation (Equation 5.15) of the distribution function, and therefore of the above equations, depends upon the near isotropy of the velocity distribution function. While elastic collisions between a light particle and a neutral atom or molecule promote such quasi-isotropy, since their effect is mainly to randomize velocities, with small fractional energy exchange $\sim m/m_0$, this is not generally the case in an inelastic collision, where the particle exchanges a fixed amount of energy ϵ_I in exciting a neutral molecule to another internal state, and may therefore mitigate against isotropy in velocity space.

For that reason, *multi-term* representations of the distribution function, which include higher order spherical components in the expansion (Equation 5.15), have become established in the literature and are generally required for accurate transport property calculations for light particles, as discussed in Chapter 12. However, the collision terms in these more general equations remain essentially the same as above—it is *the left hand side* of the equations which must be modified.

5.7 Concluding Remarks

In this chapter, we have formulated the kinetic theory of dilute particles in a gas, allowing for elastic, inelastic, and non-conservative collisions. The classical Boltzmann collision integral was adapted to model idealized charge exchange between an ion and its parent gas, and approximated in differential form for both very light and very heavy particles. Inelastic collisions were treated through the semi-classical kinetic equation of WUB, and collision terms were derived for both reactive loss and gain processes.

A special formulation for light particles, using a two-term spherical harmonic representation of the distribution function is generally valid for scattering dominated by elastic processes, but it has been pointed out that a "multi-term" representation may be required when inelastic processes are important. We also note that for heavier particles, a multi-term representation is *always* required, even when collisions are predominantly elastic.

CHAPTER 6

Charged Particles in Condensed Matter

6.1 Charge Carriers in Crystalline Semiconductors

The one-to-one correspondence between ions and electrons interacting with atoms in a gaseous medium, and charge carriers (holes and electrons) scattered by phonons in a crystalline semiconductor has long been known [17]. Charge carriers interacting with the acoustic branch of lattice vibrations, for which energy transfer in a collision is small, can be analyzed in a similar fashion to light particles* undergoing elastic collisions in a gas and, in particular, the distribution function in momentum space can be reasonably approximated by the first two terms of a spherical harmonic expansion.

When scattering from optical phonons is significant, energy exchange in collisions may be large. The situation is analogous to inelastic scattering of electrons in a molecular gas or heavy ions in a gas, where an accurate kinetic theory analysis requires a multi-term representation of the distribution function.

The overlap between transport theory in gaseous media and semiconductors is explored further in the context of fluid analysis in Chapter 10. For the present, we concentrate on amorphous and soft-condensed matter for which more substantial modifications to the transport theory are required.

6.2 Amorphous Materials

6.2.1 Trapping and the relaxation function

In amorphous semiconductors, the general picture is one in which charge carriers can move in the conduction band and be scattered into localised states, where they remain trapped for finite times τ, before de-trapping back into the conduction band (see Figure 6.1).† In effect, collisions take

* While gaseous ions are of the order of 10^4 more massive than electrons, holes in a semiconductor are much less so, about 25 times greater in CdS, for example. This means that, to a reasonable approximation, the analysis for light particles in a gas may be applied to both holes and electrons in a semiconductor.

† There are various different models for electron conduction in amorphous materials including thermal activated/assisted hopping and tunneling, the latter of which will not be considered in this work. We suggest the reader refer to Refs. [70,71] for further details.

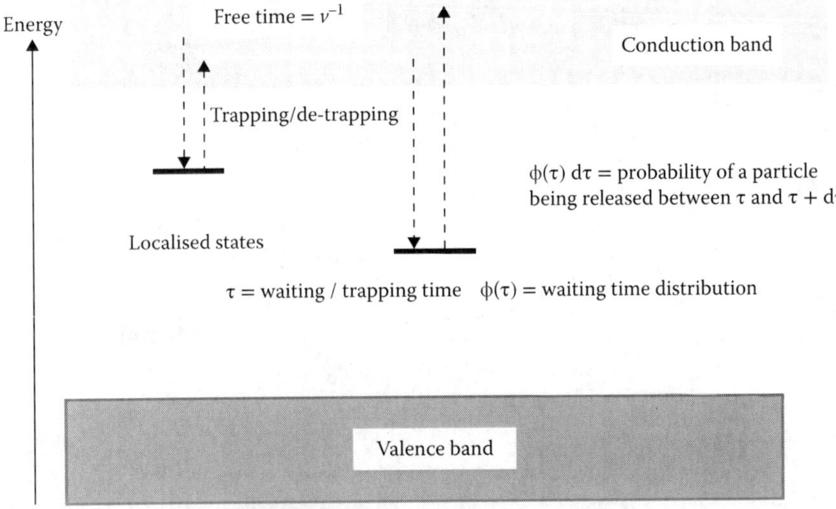

Figure 6.1 Schematic representation of the scattering of charge carriers from the "conduction" band to localised states, where they are trapped for finite times τ before being detrapped back to the conduction band.

finite times and are not instantaneous as they are in gases, and the charge carrier kinetic equation must be modified accordingly. In particular, the kinetic collision operator must incorporate the *relaxation function* or *waiting time distribution* $\phi(\tau)$, which is defined such that $\phi(\tau)\,d\tau$ is the probability of a particle being trapped for a time between τ and $\tau + d\tau$. Since all particles are eventually de-trapped, $\phi(\tau)$ must be normalised,

$$\int_0^\infty \phi(\tau)\,d\tau = 1. \tag{6.1}$$

but situations can arise where the first moment may not exist, that is, $\int_0^\infty \tau\phi(\tau)\,d\tau \to \infty$. Such materials have unusual transport properties [17,37,70], as discussed in Chapter 10.

6.2.2 The kinetic equation for amorphous materials

Let $f(\mathbf{r}, \mathbf{v}, t)$ be the phase space distribution function of free charge carriers, and suppose that at time t they are scattered out of a region of phase space at a rate $\nu f(\mathbf{r}, \mathbf{v}, t)$, where ν is an assumed constant scattering frequency. They then become trapped in localised states, represented as another part of phase space lying vertically below the region in question (see Figure 6.2). The loss term in $\left(\frac{\partial f}{\partial t}\right)_{col}$ (represented by the downward arrow in the diagram) at time t is thus simply $-\nu f(\mathbf{r}, \mathbf{v}, t)$. The gain term in $\left(\frac{\partial f}{\partial t}\right)_{col}$ (the upwardly directed arrow in the diagram) at time t

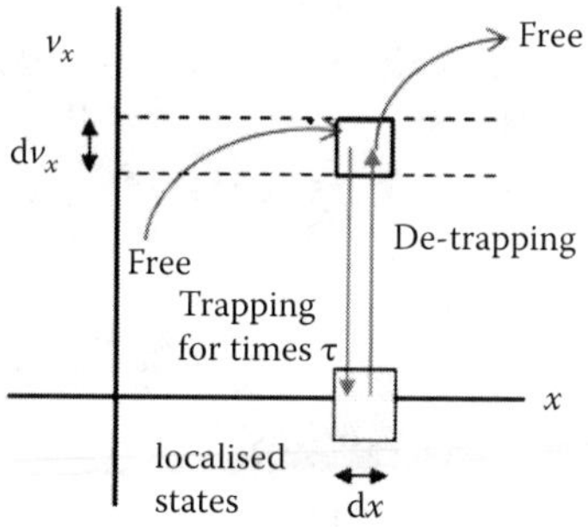

Figure 6.2 The phase space picture corresponding to the energy level diagram of Figure 6.1 showing trapping and de-trapping from localised states. In contrast to instantaneous collisions in a gaseous medium, the interactions in this medium effectively last a finite time. However, they are still local, and transitions take place vertically, as they do for gases.

arises from de-trapping of particles which had become trapped at earlier times $t-\tau$. The *total* number of such particles released per unit time from the localised state is $\int_0^t d\tau\phi(\tau)\int d^3v\ \ \nu f(\mathbf{r},\mathbf{v},t-\tau)=\nu\int_0^t d\tau\ \phi(\tau)\ \ n(\mathbf{r},t-\tau)\equiv \nu\phi{*}n$, where a star denotes a convolution. If the particles are de-trapped with a Maxwellian distribution of velocities at the medium temperature T_0, then the rate at which particles enter the region of phase space is $w(\alpha_0,v)\ \nu\phi{*}n$, where $w(\alpha_0,v)=\left(\frac{\alpha_0^2}{2\pi}\right)^{3/2}\exp(-\frac{1}{2}\alpha_0^2v^2)$.

The collision term is given by the net rate of gain of particles in the phase space region, that is,

$$\left(\frac{\partial f}{\partial t}\right)_{\text{col}}=\nu\left[w(\alpha_0,v)\ \phi{*}n-f(\mathbf{r},\mathbf{v},t)\right]. \qquad (6.2)$$

Note that "collisions" are not instantaneous, and hence unlike gases,

$$\int d^3v\ \left(\frac{\partial f}{\partial t}\right)_{\text{col}}\neq 0$$

that is, the number of particles is not conserved. Physically, this corresponds to a net loss of particles to trapping. Nevertheless, collisions are *local*, as represented by the vertical transitions in Figure 6.2.

Finally, note that for instantaneous scattering $\phi(\tau)=\delta(\tau)$, $\phi{*}n=n(\mathbf{r},t)$, and Equation 6.2 reduces to

$$\left(\frac{\partial f}{\partial t}\right)_{\text{col}}=\nu\left[w(\alpha_0,v)\ n(\mathbf{r},t)-f(\mathbf{r},\mathbf{v},t)\right],$$

which is of the same mathematical form as the relaxation time or Bhatnagar, Gross, and Krook (BGK) model introduced in Chapter 5 is in connection with gases. This model is also well known in the crystalline semiconductor literature [72].

6.3 Coherent Scattering in Soft-Condensed Matter

6.3.1 A model of coherent scattering

The study of the properties of soft-condensed matter *per se* has advanced considerably in recent times [73], but the kinetic theory of charged particles in soft matter has for long been based on the approximate formulation of Cohen and Lekner [24,74]. In this chapter, we consider the problem more rigorously and derive a suitably modified form of Boltzmann's equation accounting for coherent scattering from the medium.

We will assume that the coherent scattering is associated with *elastic collisions* only. Inelastic or reactive collisions scattering is assumed to be incoherent, and the corresponding contributions to the collision operator $\left(\frac{\partial f}{\partial t}\right)_{\text{col}}$ are therefore the same as for a gaseous medium.

We start with a physical picture of the processes under consideration. Suppose a particle is moving in a structured medium—that is, there is a correlation between the positions of the molecules, and some resulting short-range order. If the de Broglie wavelength of the particle is comparable to or larger than the average intermolecular distance, there will be interference effects as the particle undergoes simultaneous scattering from an effectively ordered array of molecules (see Figure 6.3). The *static structure function* (related to the pair correlation function) characterises the order within the medium, and the scattering is said to be *coherent*. At higher energies, when the de Broglie wavelength of the particle becomes significantly smaller than the average intermolecular spacing, interference effects are negligible and the particle scatters incoherently, from one molecule at a time, as is the case at all energies for dilute gases. The situation is analogous to the regimes of physical and geometrical optics, respectively, as we explore this analogy further below.

Scattering is influenced not only by the structure but also by the properties of the individual molecules comprising the medium.* There is an interplay between interference effects produced by an array of scatterers and single scattering effects. This is analogous to the diffraction of light from a ruled grating, where the overall pattern is a combination of

* The interaction potential of the charged particle with the constituent atoms is modified from the "dilute gas" or binary interaction potential through a variety of different processes including the screening. For an *ab-initio* treatment of the extracting cross sections in the soft-condensed phase, the reader is referred to Ref. [75].

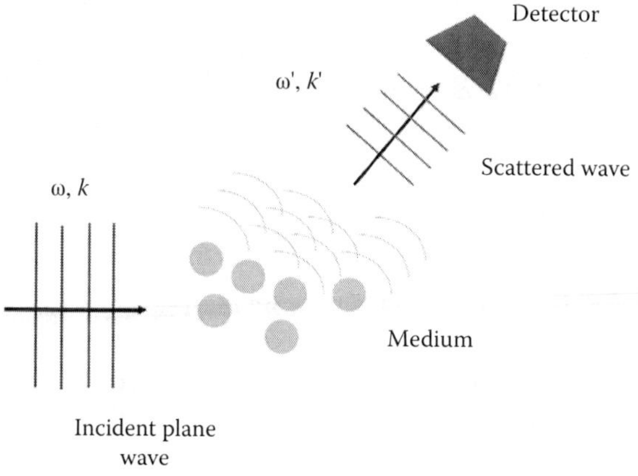

Figure 6.3 Coherent scattering of a particle (represented as a plane wave of de Broglie wavelength $\lambda = 2\pi/k$) from a soft-condensed matter medium with short-range order. If the average particle spacing $d \sim \lambda$, waves scattered from individual molecules interfere to produce a "diffraction" pattern at the detector.

the diffraction pattern from a single slit (analogous to scattering from an individual molecule) and the interference pattern from an array of slits (the effects of structure). Thus, we expect the modified Boltzmann collision term to contain a combination of single-particle scattering cross sections and the structure function of the medium.

To pursue the optical analogy further, if the slits of the grating are ruled at precise intervals, the situation can be likened to scattering from a crystalline solid. At the other extreme, where the slits are randomly distributed, there is no order, interference effects vanish and the overall light distribution is effectively just the diffraction pattern of a single slit, which in the present analogy corresponds to the dilute gas limit. The intermediate scenario, where the distribution of the slits is neither completely ordered nor completely random, corresponds to soft matter, the focus of this chapter.

While the constituent molecules may be correlated in both space and time, we assumed that their internal quantised states are not. Likewise, transitions between these states due to inelastic collisions proceed in an uncorrelated way. We therefore focus on the way the Boltzmann collision operator is to be modified for coherent effects arising from *elastic* scattering only, while treating inelastic collisions as incoherent, in the same way as they are for gases. On the other hand, the dipole moment vectors $\mathbf{p}$ of polar gas molecules, as well as their positions, may be correlated, and the structure function is then tensorial in nature. We briefly touch on this possibility, but for the greater part deal with non-polar media, which corresponds to the limit $p \to 0$.

Note that the structure function, like the cross sections for scattering from individual molecules, is assumed to be provided, either from experimental data (e.g., neutron and x-ray scattering) or theoretical calculations [75].

6.3.2 Scattering theory

Figure 6.4 shows a plane wave representing a particle incident on a volume V, and scattered into spherical waves by molecules of the medium located within a volume d^3r of position vector $\mathbf{r}$, with permanent dipole moments $\mathbf{p}$ directed into solid angle $d^2\hat{p}$. The scattered signal is measured by a detector at D, situated at a large distance R from V, that is, $R \gg r$. The angle between the wave vectors, $\mathbf{k}$ and $\mathbf{k}'$ of the incoming and outgoing beams, respectively is θ. All coordinates are measured with respect to an arbitrary origin O.

The total scattered beam observed at D consists of the contributions from many such scattered spherical waves from throughout V, which interfere to effectively produce a diffraction pattern of the medium. Like other analyses, we consider only single scattering events, and neglect multiple scattering.

The incoming plane wave of amplitude A_0 is represented by

$$\psi_I = A_0 e^{i(\mathbf{k}\cdot\mathbf{r}-\omega t)}, \tag{6.3}$$

while the outgoing wave is assumed to be a spherical wave centred on $\mathbf{r}$, with amplitude proportional to

- A factor $\tilde{f}(\mathbf{k}, \mathbf{k}', \hat{\mathbf{p}})$ accounting for scattering from a *single* molecule,
- The amplitude ψ_I of the incoming wave,

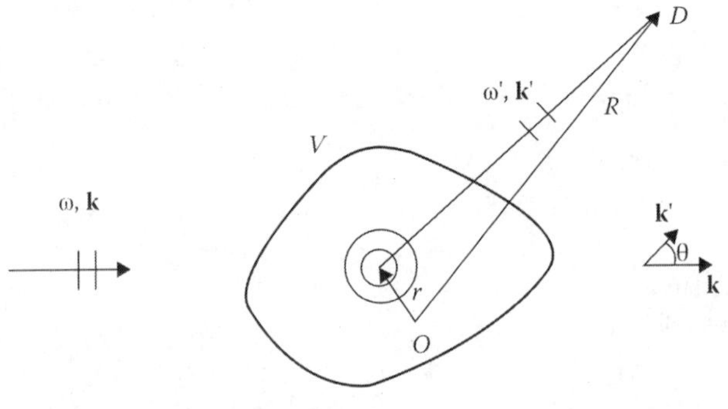

Figure 6.4 Schematic representation of the scattering of a plane wave from a region of a condensed matter medium into a spherical wave which is received at a detector D at a large distance R. The angle between incident and scattered waves is denoted by θ.

- The local fluctuation $\eta(\mathbf{r}, t, \mathbf{p})$ in the density of scatterers at position $\mathbf{r}$ and with dipole moments $\mathbf{p}$, and
- The volume d^3r from which the scattering takes place, and the solid angle $d^2\hat{p}$ defining the direction of polarisation.

In this model, the scattering is assumed to take place from all regions of V instantaneously, that is, we neglect the transit time of the wave across V and subsequently on to the detector. In Section 6.3.3, we shall comment on both time and length scales in the context of transport theory.

According to the model, the contribution to the scattered wave received at D due to scattering from molecules within d^3r and dipole moments directed within $d^2\hat{p}$ is

$$d\psi_S = C\tilde{f}(\mathbf{k}, \mathbf{k}', \hat{\mathbf{p}})\, \eta(\mathbf{r}, \mathbf{p}, t)\, A_0\, e^{i(\mathbf{k}\cdot\mathbf{r}-\omega t)} \frac{e^{ik'|\mathbf{R}-\mathbf{r}|}}{|\mathbf{R}-\mathbf{r}|}\, d^3r\, d^2\hat{p},$$

where C is a constant of proportionality. Thus, the contribution to the scattered amplitude from *all* molecules in the scattering region, with all polarisation vectors, is

$$\psi_S(t) = CA_0 \int d^2\hat{p}\, \tilde{f}(\mathbf{k}, \mathbf{k}', \hat{\mathbf{p}}) \int_V d^3r\, \eta(\mathbf{r}, \mathbf{p}, t)\, e^{i(\mathbf{k}\cdot\mathbf{r}-\omega t)} \frac{e^{ik'|\mathbf{R}-\mathbf{r}|}}{|\mathbf{R}-\mathbf{r}|}\, d^3r.$$

Since $R \gg r$, then

$$k'\,|\mathbf{R} - r| \approx kR - \mathbf{k}' \cdot \mathbf{r}$$

and hence

$$\psi_S(t) \approx A'(t)\frac{e^{ik'R}}{R}, \tag{6.4}$$

where the amplitude of the spherical wave is given by

$$A'(t) = CA_0 \int d^2\hat{p}\, \tilde{f}(\mathbf{k}, \mathbf{k}', \hat{\mathbf{p}}) \int_V d^3r\, \eta(\mathbf{r}, \mathbf{p}, t)\, e^{i[(\mathbf{k}-\mathbf{k}')\cdot\mathbf{r}-\omega t)]}. \tag{6.5}$$

These amplitudes, like the properties of the scatterers in V itself, are randomly fluctuating quantities and appropriate averaging procedures $\langle \ldots \rangle$ are required. Note that the detector measures the average intensity $\langle |A'(t)|^2 \rangle$.

6.3.3 Structure function

6.3.3.1 Correlation and structure functions

A measure of the statistics of the fluctuating properties of the scattered field is provided by the autocorrelation function

$$B(\tau) = \left\langle A'(t)\, A'(t-\tau)^* \right\rangle. \tag{6.6}$$

The spectral power density is just the Fourier transform of Equation 6.6, that is,

$$\Phi(\omega') = \frac{1}{2\pi} \int_{-\infty}^{\infty} d\tau\, e^{i\omega'\tau}\, B(\tau) \tag{6.7}$$

for which the inverse relation is

$$B(\tau) = \int_{-\infty}^{\infty} d\omega'\, e^{-i\omega'\tau}\, \Phi(\omega'). \tag{6.8}$$

Substitution of Equation 6.5 into 6.6 gives

$$\begin{aligned} B(\tau) = |C|^2 |A_0|^2 \int d^2\hat{p} \int d^2\hat{p}_1 \tilde{f}(\mathbf{k}, \mathbf{k}', \hat{\mathbf{p}}) \tilde{f}^*(\mathbf{k}, \mathbf{k}', \hat{\mathbf{p}}_1) \\ \times \int_V d^3r \int_V d^3r_1 \left\langle \eta(\mathbf{r}, \mathbf{p}, t)\, \eta^*(\mathbf{r}_1, \mathbf{p}_1, t-\tau) \right\rangle e^{i[\Delta\mathbf{k}\cdot(\mathbf{r}-\mathbf{r}_1) - \omega\tau]}, \end{aligned} \tag{6.9}$$

where

$$\Delta\mathbf{k} = \mathbf{k} - \mathbf{k}'$$

is the change in wave number, and $\hat{\mathbf{p}}$ and $\hat{\mathbf{p}}_1$ denote the directions of dipoles at the different positions. (N.B. The magnitudes of the dipoles are the same everywhere, $|\mathbf{p}_1| = |\mathbf{p}_1| = p$.) If the density fluctuations are assumed to be both stationary and homogeneous then the space-time autocorrelation function

$$B(\mathbf{r} - \mathbf{r}_1, \mathbf{p}, \mathbf{p}_1, \tau) = \left\langle \eta(\mathbf{r}, \mathbf{p}, t)\, \eta^*(\mathbf{r}_1, \mathbf{p}_1, t-\tau) \right\rangle$$

depends only upon the relative displacement $\rho \equiv \mathbf{r} - \mathbf{r}_1$ and time difference τ. The spatial integrations can be reduced by transforming to the new variables

$$\begin{aligned} \rho &= \mathbf{r} - \mathbf{r}_1 \\ \Gamma &= \frac{1}{2} (\mathbf{r} + \mathbf{r}_1), \end{aligned}$$

so that Equation 6.9 becomes

$$B(\tau) = |C|^2 |A_0|^2 \int d^2\hat{p} \int d^2\hat{p}_1 \tilde{f}(\mathbf{k}, \mathbf{k}', \hat{\mathbf{p}}) \tilde{f}^*(\mathbf{k}, \mathbf{k}', \hat{\mathbf{p}}_1) \times \int_V d^3\Gamma \int_V d^3\rho\, B(\rho, \mathbf{p}, \mathbf{p}_1, \tau) e^{i[\Delta\mathbf{k}\cdot\rho - \omega\tau]}.$$

The Jacobian of this transformation is unity, and since the integrand depends on ρ only, the integral over Γ can be done immediately, to give

$$B(\tau) = V\ |C|^2 |A_0|^2 \int d^2\hat{p} \int d^2\hat{p}_1 \tilde{f}(\mathbf{k}, \mathbf{k}', \hat{\mathbf{p}})\ \tilde{f}^*(\mathbf{k}, \mathbf{k}', \widehat{\mathbf{p}_1}) \times \int_V d^3\rho\, B(\rho, \mathbf{p}, \mathbf{p}_1, \tau)\, e^{i[\Delta\mathbf{k}\cdot\rho - \omega\tau]}. \tag{6.10}$$

The spectral power density (Equation 6.7) is then

$$\Phi(\omega') = (2\pi)^3 V\ |A_0|^2\ |C|^2 \int d^2\hat{p} \int d^2\hat{p}_1 \tilde{f}(\mathbf{k}, \mathbf{k}', \hat{\mathbf{p}}) \times \tilde{f}^*(\mathbf{k}, \mathbf{k}', \widehat{\mathbf{p}_1})\ \tilde{S}(\Delta\mathbf{k}, \Delta\omega \,|\, \mathbf{p}, \mathbf{p}_1), \tag{6.11}$$

where

$$\Delta\omega = \omega - \omega' \tag{6.12}$$

is the shift in angular frequency, and

$$\tilde{S}(\mathbf{K}, \Omega \,|\, \mathbf{p}, \mathbf{p}_1) \equiv \frac{1}{(2\pi)^4} \int_V d^3\rho \int_{-\infty}^{\infty} d\tau\, B(\rho, \mathbf{p}, \mathbf{p}_1, \tau)\, e^{i(\mathbf{K}\cdot\rho - \Omega\tau)} \tag{6.13}$$

is a generalised structure function. Certain general properties of these functions may be inferred solely on the basis of symmetry considerations:

1. Any physical property of the system, in particular $B(\rho, \mathbf{p}, \mathbf{p}_1, \tau)$, must be invariant under a reflection of coordinates, $\rho \to -\rho$, $\mathbf{p} \to -\mathbf{p}$, and a similar property must also hold in K-space, that is

$$\tilde{S}(-\mathbf{K}, \Omega \,|\, -\mathbf{p}, -\mathbf{p}_1) = \tilde{S}(\mathbf{K}, \Omega \,|\, \mathbf{p}, \mathbf{p}_1). \tag{6.14}$$

2. Since there is no preferred direction of orientation of the molecules, the average of $B(\rho, \mathbf{p}, \mathbf{p}_1, \tau)$ over all directions $\hat{\rho}$ must be independent of whether the molecules are polarised or not, that is, $\frac{1}{4\pi}\int d\hat{\rho}\, B(\rho, \mathbf{p}, \mathbf{p}_1, \tau) = B(\rho, \tau)$. A similar property holds for the average of the Fourier transformed quantity in **K**-space,

$$\frac{1}{4\pi} \int d\hat{\mathbf{K}}\, \tilde{S}(\mathbf{K}, \Omega \,|\, \mathbf{p}, \mathbf{p}_1) = S(K, \Omega). \tag{6.15}$$

3. The only way a scalar quantity like $\widetilde{S}$ can be constructed from the available vectors $\mathbf{K}, \mathbf{p}$ and $\mathbf{p}_1$, consistent with the above requirements, is through a combination of functions of K, $\mathbf{p}\cdot\mathbf{p}_1$ and $(\mathbf{p}\cdot\mathbf{K})(\mathbf{p}_1\cdot\mathbf{K})$, as follows:

$$\widetilde{S}(\mathbf{K},\Omega\,|\,\mathbf{p},\mathbf{p}_1) = S(K,\Omega) + S_P(K,\Omega)\left[(\mathbf{p}\cdot\mathbf{K})(\mathbf{p}_1\cdot\mathbf{K}) - \frac{1}{3}K^2\mathbf{p}\cdot\mathbf{p}_1\right]. \tag{6.16}$$

The second rank tensor in square brackets vanishes when integrated over $\widehat{\mathbf{K}}$, as required by Equation 6.15. We refer to $S(K,\Omega)$ as the scalar part of the generalised structure function, while $S_P(K,\Omega)$ arises explicitly from the dipole properties of the constituent molecules.

4. For a non-polar system, $p=0$, the second term on the right hand side vanishes, and

$$\widetilde{S}(\mathbf{K},\Omega\,|\,\mathbf{p},\mathbf{p}_1) = S(K,\Omega) \qquad \text{(non-polar gas).} \tag{6.17}$$

After substituting Equation 6.16 into 6.11, there follows

$$\Phi(\omega') = (2\pi)^3 V\,|A_0|^2\,|C|^2 \times\left\{|f|^2\, S(|\Delta\mathbf{k}|,\Delta\omega) + p^2\left[\left|\mathbf{f}_p\cdot\Delta\mathbf{k}\right|^2 - \frac{1}{3}\left|\mathbf{f}_p\right|^2|\Delta\mathbf{k}|^2\right] S_P(|\Delta\mathbf{k}|,\Delta\omega)\right\}, \tag{6.18}$$

where

$$f(\mathbf{k},\mathbf{k}') = \int d^2\hat{p}\,\widetilde{f}(\mathbf{k},\mathbf{k}',\hat{\mathbf{p}}) \tag{6.19}$$

$$\mathbf{f}_p(\mathbf{k},\mathbf{k}') = \int d^2\hat{p}\,\hat{\mathbf{p}}\,\widetilde{f}(\mathbf{k},\mathbf{k}',\hat{\mathbf{p}}). \tag{6.20}$$

Using further general arguments, it can be shown that

$$\left[\left|\mathbf{f}_p\cdot\Delta\mathbf{k}\right|^2 - \frac{1}{3}\left|\mathbf{f}_p\right|^2|\Delta\mathbf{k}|^2\right] = |\Delta\mathbf{k}|^2\left(1-\left|\widehat{\mathbf{k}}\cdot\widehat{\mathbf{k}'}\right|^2\right)\mathcal{F}\left(k,k\prime,\widehat{\mathbf{k}}\cdot\widehat{\mathbf{k}'}\right), \tag{6.21}$$

where the dependence of $\mathcal{F}$ of $k,k\prime$ and $\widehat{\mathbf{k}}\cdot\widehat{\mathbf{k}'} = \cos\theta$ is prescribed by the details of the interaction of the wave with an individual molecule. On the other hand, the leading term on the right hand side, $1-\cos^2\theta$, is the same for all types of interaction, and vanishes for both backward and forward scattering, $\theta=\pi$ and 0, respectively.

In summary, the structure function of a medium comprised of molecules with permanent dipole moments consists of two distinct contributions, as indicated by terms involving S and S_P in Equation 6.16 and

6.18, and *both* contribute to the scattering of waves from the medium. However, the contribution to scattering from polarisation effects is always zero for forward and backward scattering and is otherwise $O(p^2)$.

6.3.4 Non-polar molecules

6.3.4.1 Dynamic and static structure functions

For *non-polar molecules or atoms*, the structure function is given by Equation 6.17, and Equation 6.18 may then be expressed as

$$\Phi(\omega') = (2\pi)^3 V\,|A_0|^2\,|C|^2\,|f(\mathbf{k},\mathbf{k}')|^2\,S(|\Delta\mathbf{k}|,\Delta\omega). \tag{6.22}$$

It is emphasised that while $S(K,\Omega)$ is an intrinsic property of the medium, with a *functional* dependence upon K and Ω determined completely by the constituent molecules and their interactions, the *arguments* K and Ω themselves are specified by the context of the application, either by the nature of the experimental arrangement set up to probe the properties of the medium, or in theoretical applications, for example, incorporation into the Boltzmann kinetic equation.

We call $S(K,\Omega)$ the *dynamic structure function*, to distinguish it from the integral quantity,

$$S(K) \equiv \int d\Omega\; S(K,\Omega), \tag{6.23}$$

which is referred to as the *static structure function*, and which plays an important role in the following discussion. Further integral properties are detailed below.

6.3.4.2 Integral properties of $S(K,\Omega)$

For light particles of mass m scattered from a medium of temperature T_0 whose molecules have mass $m_0 \gg m$, the following integral identities hold for *independent* K and Ω [76]:

$$\begin{aligned}
&\int_{-\infty}^{\infty} d\Omega\; S(K,\Omega) = S(K)\\
&\int_{-\infty}^{\infty} d\Omega\; \Omega S(K,\Omega) = \frac{K^2}{2m_0}\\
&\int_{-\infty}^{\infty} d\Omega\; \Omega^2 S(K,\Omega) \approx 2k_B T_0 \frac{K^2}{2m_0},
\end{aligned} \tag{6.24}$$

where in the right hand side of the last relation a term of the order m/m_0 relative to the first has been neglected [76], in anticipation of application to light particles below ($m/m_0 \ll 1$).

When formulating the Boltzmann scattering term below, we make the substitutions $\mathbf{K} \rightarrow \Delta\mathbf{k} = \mathbf{k}' - \mathbf{k}$ and $\Omega \rightarrow \Delta\omega = \left(k^2 - k'^2\right)/2m$ for the arguments of the structure function. Consequently, the above identities cannot be applied directly, because $\Delta\omega = \left(k^2 - k'^2\right)/2m$ and $\Delta\mathbf{k}$ are not independent.

6.3.5 Cross sections

6.3.5.1 Single scattering differential cross section

The differential scattering cross section in the *centre-of-mass (CM) reference frame* was introduced in Chapter 3. Here we identify

$$|f(\mathbf{k}, \mathbf{k}')|^2 = n_0 \left(\frac{d\sigma}{d\Omega_{\mathbf{k}'}}\right)^{(\text{lab})} \tag{6.25}$$

as the differential cross section (DCS) in the *laboratory frame* for scattering from a *single molecule*, where $d\Omega_{\mathbf{k}'}$ is the element of solid angle centred on the outgoing wave vector $\mathbf{k}'$. The number density, n_0 of scattering particles has been made explicit at this point to ensure both correct dimensions in the ensuing formulas and also for consistency with the normalisation of the structure factor to unit particle number.

6.3.5.2 Bulk DCS

On the other hand, the *bulk DCS* for *all* such molecules in the region, per unit volume, is defined by the ratio of scattered to incident intensities, divided by V, that is,

$$\frac{d\sigma}{d\Omega_{\mathbf{k}'}} = \frac{\langle |A'(t)|^2 \rangle}{V\,|A_0|^2}. \tag{6.26}$$

The right hand side can be evaluated by using Equations 6.6, 6.8, and 6.22, to obtain

$$\begin{aligned}
\langle |A'(t)|^2 \rangle = B_S(0) &= \int_{-\infty}^{\infty} d\omega'\, \Phi(\omega') \\
&= (2\pi)^3\, V\, |A_0|^2\, |C|^2\, |f(\mathbf{k}, \mathbf{k}')|^2 \int_{-\infty}^{\infty} d(\Delta\omega)\, S(|\Delta\mathbf{k}|, \Delta\omega) \\
&= (2\pi)^3\, V\, |A_0|^2\, |C|^2\, |f(\mathbf{k}, \mathbf{k}')|^2\, S(|\Delta\mathbf{k}|).
\end{aligned}$$

The constant C, which to this point is completely arbitrary, is chosen to satisfy

$$(2\pi)^3\ |C|^2 = 1 \tag{6.27}$$

in order to produce consistency with established results in the dilute gas (no structure) limit. Substituting in Equation 6.26 and using the definition (Equation 6.25) then gives the relationship between the single and bulk DCS,

$$\frac{d\sigma}{d\Omega_{\mathbf{k}'}} = n_0 \left(\frac{d\sigma}{d\Omega_{\mathbf{k}'}}\right)^{(\mathrm{lab})} \int_{-\infty}^{\infty} d\omega'\ S(|\Delta\mathbf{k}|, \Delta\omega) \tag{6.28}$$

$$\equiv n_0 \left(\frac{d\sigma}{d\Omega_{\mathbf{k}'}}\right)^{(\mathrm{lab})} S(|\Delta\mathbf{k}|). \tag{6.29}$$

6.3.5.3 Double DCS

We also define the double differential scattering cross section per unit scattered frequency interval $d\omega'$, per unit volume, by

$$\frac{d\sigma}{d\Omega_{\mathbf{k}'}} \equiv \int_{-\infty}^{\infty} d\omega' \frac{d^2\sigma}{d\Omega_{\mathbf{k}'} d\omega'} \tag{6.30}$$

and hence by Equation 6.28, it follows that

$$\frac{d^2\sigma}{d\Omega_{\mathbf{k}'}\ d\omega'} = n_0 \left(\frac{d\sigma}{d\Omega_{\mathbf{k}'}}\right)^{(\mathrm{lab})} S(\Delta\mathbf{k}, \Delta\omega). \tag{6.31}$$

6.3.5.4 Lab versus CM frame

It is to be emphasised that in the above discussion *all* expressions relate to the *laboratory reference frame*. Since the scattering is from many particles simultaneously, it is not feasible to give a formulation in terms of CM quantities. This is in contrast to the case for binary scattering from a *single molecule*, for which the laboratory DCS as defined by Equation 6.25 can be transformed to the CM DCS $\left(\frac{d\sigma}{d\Omega_{\mathbf{k}'}}\right)^{(\mathrm{CM})}$.

6.4 Kinetic Equation for Charged Particles in Soft-Condensed Matter

6.4.1 The general expression for collisional rate of change

We now consider a low density swarm of charged particles moving in a condensed matter medium in thermodynamic equilibrium. The density of the particles is so low that interactions between them are negligible, and furthermore to a first approximation, the overall state of equilibrium of the medium is effectively undisturbed. The particle phase space distribution function $f(\mathbf{r}, \mathbf{v}, t)$ is defined in the usual way, such that $f(\mathbf{r}, \mathbf{v}, t)\ \mathrm{d}^3v\ \mathrm{d}^3r$ is the number of particles with velocities in the range $\mathbf{v} - \mathbf{v} + \mathrm{d}\mathbf{v}$ and positions $\mathbf{r} - \mathbf{r} + \mathrm{d}\mathbf{r}$. The number density of particles with velocities in the range $\mathbf{v} - \mathbf{v} + \mathrm{d}\mathbf{v}$ is thus $f(\mathbf{r}, \mathbf{v}, t)\ \mathrm{d}^3v$ and the particle flux is therefore $vf(\mathbf{r}, \mathbf{v}, t)\ \mathrm{d}^3v$. In the first place, we shall derive a general expression the rate of change of some arbitrary property $\Phi(\mathbf{v})$ of particle velocity due to interaction with the medium, which we then use in the calculation of transport properties, and ultimately also use to derive the kinetic equation for $f(\mathbf{r}, \mathbf{v}, t)$.

The discussion in the previous section was framed in terms of wave properties $\mathbf{k}$, ω, but it is straightforward to obtain the corresponding expressions in terms of particulate properties, that is, momentum and energy, respectively, using

$$\mathbf{p} = \hbar\mathbf{k} = m\mathbf{v}$$
$$\epsilon = \hbar\omega = \hbar k^2/2m = \frac{1}{2}mv^2,$$

with post-collision velocities, momenta and energies, $\mathbf{v}'$, $\mathbf{p}'$, and ϵ' being similarly related to $\mathbf{k}'$ and ω'. Often as not, we ignore the presence of $\hbar$ and speak of $\mathbf{k}$ or ω as the momentum and energy, respectively. Thus, we mix the wave and particle notation, but the meaning should always be clear from these relations.

Consider now an arbitrary property of the particles which changes from $\Phi(\mathbf{v})$ to $\Phi(\mathbf{v}')$ in a collision. By starting with the definition of $\frac{\mathrm{d}^2\sigma}{\mathrm{d}\Omega_{\mathbf{k}'}\mathrm{d}\omega'}$ and the fact that $v\,f(\mathbf{v})\ \mathrm{d}^3v$ represents the the initial flux of particles in the velocity range $\mathbf{v} - \mathbf{v} + \mathrm{d}\mathbf{v}$, we may deduce that

$$\left[\Phi(\mathbf{v}') - \Phi(\mathbf{v})\right]\ \frac{\mathrm{d}^2\sigma}{\mathrm{d}\Omega_{\mathbf{k}'}\mathrm{d}\omega'}\ v\,f(\mathbf{v})\ \mathrm{d}^3v\ \mathrm{d}\omega'\ \mathrm{d}\Omega_{\mathbf{k}'}$$

represents the rate at which property $\Phi(\mathbf{v})$ is transferred to the medium due to scattering of particles with initial velocities between $\mathbf{v}$ and $\mathbf{v} + \mathrm{d}\mathbf{v}$, into final energies within the range $\omega' - \omega' + \mathrm{d}\omega'$ and with final directions defined by the solid angle $\mathrm{d}\Omega_{\mathbf{k}'}$. Thus, the rate at which the property is transferred to the medium due to collisions over the entire velocity range, per unit volume, is

$$\left(\frac{\partial\Phi}{\partial t}\right)_{\mathrm{col}} = n_0 \int \mathrm{d}^3v\ vf(\mathbf{v}) \int_0^\infty \mathrm{d}\omega' \int \mathrm{d}\Omega_{\mathbf{k}'}[\Phi(\mathbf{v}') - \Phi(\mathbf{v})] \left(\frac{\mathrm{d}\sigma}{\mathrm{d}\Omega_{\mathbf{k}'}}\right)^{(\mathrm{lab})} S(\Delta\mathbf{k}, \Delta\omega), \tag{6.32}$$

where n_0 is the number of scatterers per unit volume, and we have used Equation 6.31 to express the right hand side in terms of the structure factor. This is the basic expression from which we shall work to set up the kinetic equation and hence obtain all quantities of physical interest.

6.4.2 Kinetic and moment equations

As we have observed in Chapter 2, the kinetic equation for dilute particles of charge q, mass m in either a gaseous or condensed matter medium can be written *formally* in the same way , namely,

$$\left(\partial_t + \mathbf{v}\cdot\nabla + \mathbf{a}\cdot\partial_{\mathbf{v}}\right) f = \left(\frac{\partial f}{\partial t}\right)_{\text{col}}. \tag{6.33}$$

Even though $\left(\frac{\partial f}{\partial t}\right)_{\text{col}}$ is substantially different for the different media, the same "trick" employed in Chapter 3 can be used to find the collision term. In the process, we also obtain the balance equation for velocity moments,

$$\langle\Phi(\mathbf{v})\rangle \equiv \frac{1}{n}\int \mathrm{d}^3v\, f(\mathbf{r},\mathbf{v},t)\Phi(\mathbf{v}), \tag{6.34}$$

where

$$n(\mathbf{r},t) = \int \mathrm{d}^3v\, f(\mathbf{r},\mathbf{v},t).$$

Thus, we multiply the kinetic equation (Equation 6.33) by an arbitrary function $\Phi(\mathbf{v})$ of particle velocity and integrate over all $\mathbf{v}$ to obtain the moment equation

$$\partial_t(n\langle\Phi(\mathbf{v})\rangle) + \nabla\cdot n\langle v\Phi(\mathbf{v})\rangle - n\langle\mathbf{a}\cdot\boldsymbol{\partial}_{\mathbf{v}}\Phi(\mathbf{v})\rangle = \left(\frac{\partial\Phi}{\partial t}\right)_{\text{col}},$$

where

$$\left(\frac{\partial\Phi}{\partial t}\right)_{\text{col}} \equiv \int \mathrm{d}^3v\;\Phi(\mathbf{v})\left(\frac{\partial f}{\partial t}\right)_{\text{col}}. \tag{6.35}$$

Equations 6.32 and 6.35 are then equated, and the collision operator $\left(\frac{\partial f}{\partial t}\right)_{\text{col}}$ can be obtained by following a similar mathematical procedure to that for gases, as described in Chapters 3 and 5. To that end, it is useful to first of all consider the above results in the gas limit, if only to establish consistency with previous results.

6.4.3 Dilute gas limit

For an unstructured, dilute monatomic gas of temperature T_0, atomic mass m_0, whose velocities $\mathbf{v}_0$ are distributed according to a Maxwellian,

$$f_0(\mathbf{v}_0) = n_0 w(v_0) \equiv n_0 \left(\frac{m_0}{2\pi k_B T_0} \right)^{\frac{3}{2}} \exp\left(-\frac{m_0 v_0^2}{2k_B T_0} \right), \tag{6.36}$$

the structure function can be decomposed as follows:

$$S(\mathbf{K}, \Omega) = \int d^3 v_0 \, w(v_0) S_{\mathbf{v}_0}(\mathbf{K}, \Omega). \tag{6.37}$$

After substitution in Equation 6.32 and some rearranging, there follows

$$\left(\frac{\partial \Phi}{\partial t} \right)_{\text{col}} = \int d^3 v \int d^3 v_0 \, v f(\mathbf{v}) f_0(\mathbf{v}_0) \tag{6.38}$$

$$\times \int_{-\infty}^{\infty} d(\Delta\omega) \int d\Omega_{\mathbf{k}'} [\Phi(\mathbf{v}) - \Phi(\mathbf{v}')] \left(\frac{d\sigma}{d\Omega_{\mathbf{k}'}} \right)^{(\text{lab})} S_{\mathbf{v}_0}(\Delta\mathbf{k}, \Delta\omega). \tag{6.39}$$

For binary scattering from a structureless medium in which the atoms recoil with velocity $\mathbf{v}_0'$,

$$S_{\mathbf{v}_0}(\Delta\mathbf{k}, \Delta\omega) = S_{\mathbf{v}_0}(\mathbf{K}) \delta \left(\Delta\omega - \Delta\mathbf{k} \cdot \overline{\mathbf{v}}_0 \right), \tag{6.40}$$

where $S_{\mathbf{v}_0}(\mathbf{K})$ is the static structure function, equal to unity for a dilute gas, and $\overline{\mathbf{v}}_0 = \frac{1}{2}(\mathbf{v}_0 + \mathbf{v}_0')$ is the average atomic velocity before and after a collision. Since $\Delta\omega = \frac{1}{2} m_0 (v_0^2 - v_0'^2)$ and $\Delta\mathbf{k} = m(\mathbf{v} - \mathbf{v}') \equiv m_0(\mathbf{v}_0 - \mathbf{v}_0')$, the delta function in Equation 6.40 implies

$$\Delta\omega = \frac{1}{2} m (v^2 - v'^2) = \frac{1}{2} m_0 (v_0^2 - v_0'^2), \tag{6.41}$$

which simply represents energy conservation in a binary collision. Furthermore, we observe that cross sections in the lab and CM reference frames are related by Ref. [76]

$$v \left(\frac{d\sigma}{d\Omega_{\mathbf{k}'}} \right)^{(\text{lab})} d\Omega_{\mathbf{k}'} = g \left(\frac{d\sigma}{d\Omega_{\mathbf{g}'}} \right)^{(\text{CM})} d\Omega_{\mathbf{g}'} \equiv g\sigma(\mathbf{g}, \mathbf{g}') d\Omega_{\mathbf{g}'}, \tag{6.42}$$

where $\mathbf{g} = \mathbf{v} - \mathbf{v}_0$ is the relative velocity before collision, $d\Omega_{\mathbf{g}'}$ is an element of solid angle corresponding to post-collision relative velocities $\mathbf{g}'$ and the DCS has simply been written as $\sigma(\mathbf{g}, \mathbf{g}')$ to conform with the notation of Chapter 2.

Thus, after doing the integration over $\Delta\omega$ and setting $S_{\mathbf{v}_0}(\mathbf{K})=1$, we arrive at the following expression for the collisional rate of change,

$$\left(\frac{\partial\Phi}{\partial t}\right)_{\mathrm{col}} = \int \mathrm{d}^3 v \int \mathrm{d}^3 v_0\, g f(\mathbf{v}) f_0(\mathbf{v}_0) \int \mathrm{d}\Omega_{\mathbf{g}'} [\Phi(\mathbf{v}') - \Phi(\mathbf{v})] \sigma(\mathbf{g}, \mathbf{g}') \tag{6.43}$$

in which Equation 6.41 is implicit. By comparing this with Equation 6.35 and using time-reversal symmetry (Equation 3.8), we obtain

$$\left(\frac{\partial f}{\partial t}\right)_{\mathrm{col}} = \int \mathrm{d}^3 v \int \mathrm{d}^3 v_0 \int \mathrm{d}\Omega_{\mathbf{g}'} [f(\mathbf{v}') f_0(\mathbf{v}_0') - f(\mathbf{v}) f_0(\mathbf{v}_0)] g \sigma(\mathbf{g}, \mathbf{g}'),$$

which is the Boltzmann kinetic operator for particle-gas atom collisions derived in Chapter 3. This justifies the choice of the arbitrary constant as shown in Equation 6.27.

The procedure is generalised below to obtain the expression for $\left(\frac{\partial f}{\partial t}\right)_{\mathrm{col}}$ for a scattering of light particles from a condensed matter medium.

6.4.4 Light particles

For a medium consisting of condensed matter, it makes no sense to speak about either binary collisions or a transformation to CM coordinates. A particle in effect interacts with the medium as a whole, over a domain corresponding to the range of the pair correlation function. One assumption that does carry over from gaseous kinetic theory is that the length scale characterising this region of interaction is very small compared with any relevant macroscopic length. Likewise, the interaction time is assumed very small compared with any relevant macroscopic time interval. Effectively, the "collision" takes place at a point in space and an instant in time, and the collision term $\left(\frac{\partial f}{\partial t}\right)_{\mathrm{col}}$ and the expression (Equation 6.32) therefore involve only the local, instantaneous value of the particle distribution function. Otherwise, however, there are no significant differences between dilute gases and condensed matter.

6.4.4.1 Rate of change of a scalar

For light particles undergoing binary collisions with a dilute cold *gas* (stationary gas atoms, $T_0=0$), the fractional energy transfer $\Delta\omega/\omega$ is of first order in $m/m_0 \ll 1$, assuming no internal states of the atom are excited. The rate of change of any *scalar* property $\Phi=\Phi(v)$ is then found by keeping terms to first order in $\Delta\omega$ in Equation 6.43. When the gas temperature T_0 is non-zero, the situation is a little more complicated, and one must keep terms up to *second* order in $\Delta\omega$ to express the right side of Equation 6.43 accurate to first order in m/m_0. On the other hand, when $\Phi(\mathbf{v})$ is of higher tensorial rank, that is, a vector or a tensor, it is necessary to keep only to zero order in the mass ratio.

For a condensed matter medium, while the concept of a binary collision is meaningless, energy exchange is still limited by the mass of the constituent atoms, and we still calculate the right hand side of Equation 6.32 to first order m/m_0 when $\Phi(\mathbf{v}) = \Phi(v)$ is a scalar. In what follows, as a matter of mathematical convenience, we choose a particular form,

$$\Phi(v) = e^{-\alpha m v^2/2}, \tag{6.44}$$

where α is an arbitrary parameter. Substitution in Equation 6.32 then gives

$$\left(\frac{\partial \Phi}{\partial t}\right)_{\text{col}} = n_0 \int d^3v\, e^{-\alpha m v^2/2} v f(\mathbf{v}) \int_{-\infty}^{\infty} d(\Delta\omega) \times \int d\Omega_{\mathbf{k}'} [e^{\alpha \Delta\omega} - 1] \left(\frac{d\sigma}{d\Omega_{\mathbf{k}'}}\right)^{(\text{lab})} S(\Delta\mathbf{k}, \Delta\omega) \tag{6.45}$$

$$= n_0 \alpha \int d^3v\, e^{-\alpha m v^2/2} v f(\mathbf{v}) \int d\Omega_{\mathbf{k}'} \int_{-\infty}^{\infty} d(\Delta\omega) \left[\Delta\omega + \frac{1}{2}\alpha(\Delta\omega)^2 + \dots\right] \left(\frac{d\sigma}{d\Omega_{\mathbf{k}'}}\right)^{(\text{lab})} S(\Delta\mathbf{k}, \Delta\omega). \tag{6.46}$$

Just as for dilute gases, we must retain terms to *second order* in $\Delta\omega$ in order for the right hand side to be accurate to first order in m/m_0.

At first sight, one is tempted to perform the $\Delta\omega$-integration immediately by using Equation 6.24, but this would not be valid, as $\Delta\mathbf{k}$ and $\Delta\omega$ are not independent. Thus

$$\Delta\mathbf{k} = \mathbf{k} - \mathbf{k}' = \Delta\mathbf{k}_0 + (k - k')\,\hat{\mathbf{k}}'$$

where

$$\Delta\mathbf{k}_0 = \mathbf{k} - k\,\hat{\mathbf{k}}'$$

corresponds to the case where the wave number is simply rotated from initial to final directions without any change in magnitude, through the same scattering angle θ as in the actual case (see Figure 6.5).

Since to first order in $k - k'$, the energy loss is

$$\Delta\omega = (k^2 - k'^2)/2m \approx k(k - k')/m,$$

then to first order in $\Delta\omega$ the following expansion holds:

$$S(\Delta\mathbf{k}, \Delta\omega) = S(\Delta\mathbf{k}_0, \Delta\omega) + \frac{m}{k}\Delta\omega\hat{\mathbf{k}}' \cdot \left[\frac{\partial}{\partial \mathbf{K}} S(\mathbf{K}, \Delta\omega)\right]_{\mathbf{K}=\Delta\mathbf{k}_0}. \tag{6.47}$$

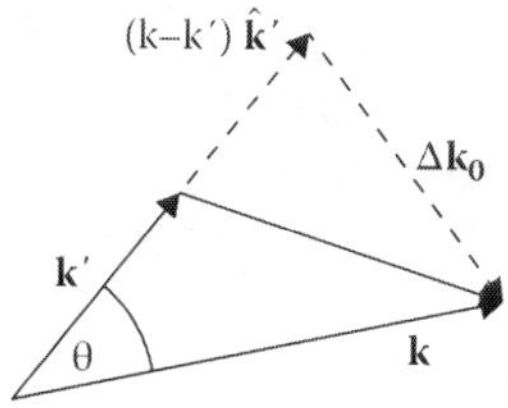

Figure 6.5 Illustration of the definition of $\Delta\mathbf{k}_0$.

When evaluating subsequent expressions arising from this expansion, we use the following identities:

$$|\Delta\mathbf{k}_0| = 2k\sin\frac{\theta}{2}$$
$$\hat{\mathbf{k}}'\cdot\Delta\mathbf{k}_0 = -k(1-\cos\theta).$$

We need also to consider the dependence of $\left(\frac{d\sigma}{d\Omega_{\mathbf{k}'}}\right)^{(\text{lab})}$ upon $\Delta\omega$, which can be deduced from Equation 6.25, plus the fact that for a medium of non-polar atoms interacting through central forces the scattering amplitude depends on $\mathbf{k}$ and $\mathbf{k}'$ through the modulus of their vector difference [76], that is,

$$f(\mathbf{k},\mathbf{k}') = f(|\Delta\mathbf{k}|).$$

Since

$$\begin{aligned}|\Delta\mathbf{k}|^2 &\approx |\Delta\mathbf{k}_0|^2 + 2\frac{m}{k}\Delta\omega\hat{\mathbf{k}}'\cdot\Delta\mathbf{k}_0\\ &= 4m(1-\cos\theta)\left[\omega-\Delta\omega/2\right],\end{aligned} \tag{6.48}$$

to first order in $\Delta\omega$, we may, therefore, write

$$\left(\frac{d\sigma}{d\Omega_{\mathbf{k}'}}\right)^{(\text{lab})} = \sigma(\omega-\Delta\omega/2,\theta) = \sigma(\omega,\theta) - \frac{\Delta\omega}{2}\frac{\partial}{\partial\omega}\sigma(\omega,\theta) \tag{6.49}$$

to the same approximation, where $\sigma(\omega,\theta)$ is a function dependent only upon the initial energy and the scattering angle. In what follows, this is referred to as simply "the differential cross section."

The integration over $\Delta\omega$ in Equation 6.46 can then be done with the help of Equation 6.24:

$$\int_{-\infty}^{\infty} d(\Delta\omega)\left[\Delta\omega + \frac{1}{2}\alpha(\Delta\omega)^2 + \ldots\right]\left(\frac{d\sigma}{d\Omega_{\mathbf{k}'}}\right)^{(\text{lab})} S(\Delta\mathbf{k},\Delta\omega)$$

$$= \int_{-\infty}^{\infty} d(\Delta\omega) \left[\Delta\omega + \frac{1}{2}\alpha(\Delta\omega)^2 + \ldots\right] \left[\sigma(\omega,\theta) - \frac{\Delta\omega}{2}\frac{\partial}{\partial\omega}\sigma(\omega,\theta)\right]$$

$$\times \left\{ S(\Delta\mathbf{k}_0, \Delta\omega) + \frac{m}{k}\hat{\mathbf{k}}' \cdot \left[\frac{\partial}{\partial\mathbf{K}}(\Delta\omega\, S(\mathbf{K}, \Delta\omega))\right]_{\mathbf{K}=\Delta\mathbf{k}_0} \right\}$$

$$= 2\frac{m}{m_0}(1-\cos\theta)\left\{\omega\left[\sigma(\omega,\theta) + k_B T_0\left(\alpha\sigma(\omega,\theta) - \sigma'(\omega,\theta)\right)\right] - k_B T_0 \sigma(\omega,\theta)\right\}, \tag{6.50}$$

where $\sigma' = \frac{\partial}{\partial\omega}\sigma(\omega,\theta)$. Substitution in Equation 6.46, followed by integration over scattering angles, then gives

$$\left(\frac{\partial\Phi}{\partial t}\right)_{\text{col}} = \alpha 2\frac{m}{m_0} n_0 \int d^3v\, e^{-\alpha m v^2/2} v f(\mathbf{v}) \left\{\omega\left[\sigma_m + k_B T_0\left(\alpha\sigma_m - \sigma'_m\right)\right] - k_B T_0 \sigma_m\right\},$$

where

$$\sigma_m(\omega) = \int d\Omega_{\mathbf{k}'}\,(1-\cos\theta)\,\sigma(\omega,\theta) \tag{6.51}$$

is the momentum transfer cross section, and $\sigma'_m = \frac{d}{d\omega}\sigma_m(\omega)$. Note that the solid angle in this integral is given by

$$d\Omega_{\mathbf{k}'} = 2\pi \sin\theta\, d\theta \tag{6.52}$$

and hence in this and other integrals which follow

$$\int d\Omega_{\mathbf{k}'}\ldots = 2\pi \int_0^{\pi} d\theta \sin\theta \ldots. \tag{6.53}$$

Finally, since the term in the integrand multiplying $f(\mathbf{v})$ is a scalar, we may integrate over the directions $\hat{\mathbf{v}}$ of velocity, to obtain

$$\left(\frac{\partial\Phi}{\partial t}\right)_{\text{col}} = \alpha\frac{2m}{m_0} n_0 \int_0^{\infty} dv\, e^{-\alpha m v^2/2}\, v^3 f^{(0)}(v) \left\{\omega\left[\sigma_m + k_B T_0\left(\alpha\sigma_m - \sigma'_m\right)\right] - k_B T_0\, \sigma_m\right\}, \tag{6.54}$$

where

$$f^{(0)}(v) = \int d\hat{\mathbf{v}}\, f(\mathbf{v})$$

is the scalar part of the distribution function.

6.4.4.2 Scalar collision operator

Equation 6.54 is to be equated with the rate of change of a scalar quantity,

$$\left(\frac{\partial \Phi}{\partial t}\right)_{\text{col}} \equiv \int dv\, v^2 e^{-\alpha m v^2/2} \left(\frac{\partial f^{(0)}}{\partial t}\right)_{\text{col}}$$

obtained by integrating the general definition (Equation 6.35) over all directions $\hat{\mathbf{v}}$. After observing that

$$\alpha e^{-\alpha m v^2/2} \equiv -\frac{1}{mv}\frac{\mathrm{d}}{dv} e^{-\alpha m v^2/2}$$

and integrating Equation 6.54 by parts twice, the coefficients of $e^{-\alpha m v^2/2}$ in the respective integrands may be compared to yield

$$\left(\frac{\partial f^{(0)}}{\partial t}\right)_{\text{col}} = \frac{m}{m_0 v^2}\frac{\partial}{\partial v}\left[v^3 \nu_m \left(f^{(0)} + \frac{k_B T_0}{mv}\frac{\partial f^{(0)}}{\partial v}\right)\right] \tag{6.55}$$

which has already appeared in Chapter 5 for an unstructured dilute gas medium, and which is well known in the literature [18,21] Note that for convenience, the right hand side has been written in terms of the momentum transfer collision frequency

$$\nu_m(v) = n_0 v \sigma_m(v). \tag{6.56}$$

6.4.4.3 Rate of change of a vector

We now set

$$\Phi(\mathbf{v}) = \mathbf{v}\Psi(v)$$

in Equation 6.32, where $\Psi(v)$ is an arbitrary scalar function of speed, and find

$$\begin{aligned}\left(\frac{\partial \Phi}{\partial t}\right)_{\text{col}} &= n_0 \int \mathrm{d}^3 v\, v f(\mathbf{v}) \int_0^\infty \mathrm{d}\omega' \int \mathrm{d}\Omega_{\mathbf{k}'}\, [\Psi(v')\mathbf{v} - \Psi(v)\mathbf{v}'] \\ &\qquad \times \left(\frac{\mathrm{d}\sigma}{\mathrm{d}\Omega_{\mathbf{k}'}}\right)^{(\text{lab})} S(\Delta\mathbf{k}, \Delta\omega) \\ &= n_0 \int \mathrm{d}^3 v\, v f(\mathbf{v}) \int_{-\infty}^{\infty} \mathrm{d}(\Delta\omega) \int \mathrm{d}\Omega_{\mathbf{k}'}\, \Psi(v)[\mathbf{v}' - \mathbf{v}]\sigma(\omega, \theta) \\ &\qquad \times S(\Delta\mathbf{k}_0, \Delta\omega) \quad + O(\Delta\omega),\end{aligned}$$

where we have expanded $\Delta\omega$ as before, but now the leading terms in the right hand side are of zero order only, and the working is much simpler. Thus, we have been able to use Equation 6.49 to write

$$\left(\frac{d\sigma}{d\Omega_{\mathbf{k}'}}\right)^{(\text{lab})} = \sigma(\omega, \theta) + O(\Delta\omega)$$

and we can now do the integration over $\Delta\omega$ immediately, using the first of equations (Equation 6.24):

$$\int_{-\infty}^{\infty} d(\Delta\omega)\, S(\Delta\mathbf{k}_0, \Delta\omega) = S(\Delta\mathbf{k}_0)$$

we find that

$$\left(\frac{\partial\Phi}{\partial t}\right)_{\text{col}} = n_0 \int d^3v\, vf(\mathbf{v}) \int d\Omega_{\mathbf{k}'}\, \Psi(v)[\mathbf{v}' - \mathbf{v}]\tilde{\sigma}(v, \theta).$$

The quantity

$$\tilde{\sigma}(v, \theta) = \sigma(\omega, \theta) S(\Delta\mathbf{k}_0) \tag{6.57}$$

is an *effective (structure modified) DCS*, combining the effects of structure with single scattering properties. It is implicitly assumed that the medium is isotropic, in which case, the functional dependence of the static structure factor simplifies:

$$S(\Delta\mathbf{k}_0) = S(|\Delta\mathbf{k}_0|) \equiv S\left(2k \sin\frac{\theta}{2}\right).$$

As a matter of notational convenience, we have chosen to represent the functional dependence of $\tilde{\sigma}$ in terms of v and θ, but alternate representation in terms of either (ω, θ) or (k, θ) is clearly also possible, since $\omega = \frac{1}{2}mv^2 = \frac{k^2}{2m}$.

The integral over scattering angles $\Omega_{\mathbf{k}'}$ then proceeds in a way similar to Chapter 3. Thus

$$\int d\Omega_{\mathbf{k}'}[\mathbf{v}' - \mathbf{v}]\tilde{\sigma}(v, \theta) = -\mathbf{v}\tilde{\sigma}_m(v),$$

where

$$\tilde{\sigma}_m(v) = \int d\Omega_{\mathbf{k}'}\, (1 - \cos\theta)\tilde{\sigma}(v, \theta) \tag{6.58}$$

is the *effective (structure modified) momentum transfer cross section.* This reduces to the usual momentum transfer cross section Equation 6.51 in the limit of a structureless medium, that is, when $S(\Delta\mathbf{k}_0) \to 1$. If we also define the vector part of the distribution function by

$$\mathbf{f}^{(1)}(v) = \int d\hat{\mathbf{v}}\, \hat{\mathbf{v}} f(\mathbf{v}),$$

then

$$\left(\frac{\partial \Phi}{\partial t}\right)_{\text{col}} = -\int_0^\infty dv\, v^2\, \mathbf{f}^{(1)}(v) v \Psi(v) \widetilde{\nu}_m(v), \tag{6.59}$$

where

$$\widetilde{\nu}_m(v) \equiv n_0 v \widetilde{\sigma}_m(v). \tag{6.60}$$

defines an *effective (structure modified) momentum transfer collision frequency*.

6.4.4.4 Vector and higher rank tensor collision operators

Integration of Equation 6.35 over all directions, with $\Phi = \mathbf{v}\Psi(v)$ gives

$$\left(\frac{\partial \Phi}{\partial t}\right)_{\text{col}} = \int_0^\infty dv\, v^2\, v \Psi(v) \left(\frac{\partial \mathbf{f}^{(1)}}{\partial t}\right) \tag{6.61}$$

and hence it follows by comparison with Equation 6.59 that

$$\left(\frac{\partial \mathbf{f}^{(1)}}{\partial t}\right) = -\widetilde{\nu}_m(v) \mathbf{f}^{(1)}. \tag{6.62}$$

The operators for higher order tensors can also be calculated to zero order in $\Delta\omega$, and the discussion proceeds along similar lines, but the working will not be given here. In general, for tensors of rank $l \geq 1$

$$\left(\frac{\partial f^{(l)}}{\partial t}\right)_{\text{col}} = -\widetilde{\nu}_l(v) f^{(l)}(v), \tag{6.63}$$

where

$$\widetilde{\nu}_l(v) = n_0 v\, \widetilde{\sigma}_l(v) \equiv n_0 v \int d\Omega_{\mathbf{k}'}\, [1 - P_l(\cos\theta)] \widetilde{\sigma}(v, \theta) \qquad (l = 1, 2, 3, \ldots), \tag{6.64}$$

and $P_l(\cos\theta)$ is a Legendre polynomial with the integral is to be carried out as in Equation 6.53.

6.5 Concluding Remarks

In this chapter, we first observed the one-to-one correspondence between the kinetic theory of charge carriers in crystalline semiconductors and in gases, and then derived expressions for $\left(\frac{\partial \mathbf{f}}{\partial t}\right)_{\text{col}}$ in the following cases:

- Electrons in an amorphous disordered medium, allowing for trapping to and de-trapping from localised states, as characterised by the relaxation function $\phi(\tau)$.
- Charged particles scattered elastically and coherently in a soft-condensed matter medium, as characterised by the dynamic structure function $S(\mathbf{K}, \Omega)$. For light particles, the scalar component $\left(\frac{\partial f^{(0)}}{\partial t}\right)_{\text{col}}$ is found to be the same as for a gaseous medium (Chapter 5), but in the vector component $\left(\frac{\partial \mathbf{f}^{(1)}}{\partial t}\right)_{\text{col}}$, the effective structure-modified momentum-transfer collision frequency $\widetilde{\nu}_m$, defined by Equation 6.60, replaces the usual momentum-transfer collision frequency ν_m. Similar modifications are found for higher order tensor components of the collision term $\left(\frac{\partial f^{(l)}}{\partial t}\right)_{\text{col}}$ $(l = 2, 3, \ldots)$. Inelastic collisions are considered to be incoherent, and hence their contribution to $\left(\frac{\partial \mathbf{f}}{\partial t}\right)_{\text{col}}$ is the same as for a gaseous medium (see Chapter 5).

We now have kinetic equations describing charge carriers in gases, disordered media and soft-condensed matter, which may be solved with input of cross sections for scattering from individual constituent molecules, the relaxation function, and the structure function of the medium, respectively. There are two levels of approximation:

- Through solution of approximate fluid equations, formed by taking velocity moments of the kinetic equations (Chapters 7–10). Fortunately, much of the hard work has been done already, since the collision operators of the kinetic equations were themselves derived from their velocity moments in the first place! This procedure yields the velocity moments of physical interest directly, without the need to first calculate the charge carrier distribution function f; and
- Through accurate analytical and numerical solution of the kinetic equations for f, from which the required velocity moments can be obtained.

These offer alternative, complementary ways of converting information at the microscopic level to measurable macroscopic transport properties.

Part II

Fluid Modelling in Configuration Space

CHAPTER 7

Fluid Modelling: Foundations and First Applications

7.1 Moment Equations for Gases

In the first place, we formulate moment equations for charged particles undergoing elastic collisions in a gaseous medium, based on the Boltzmann kinetic equation. The results are subsequently carried over to charge carriers in condensed matter, with appropriate substitutions for key quantities.

7.1.1 General moment equation

The Boltzmann equation for a dilute swarm of charged particles undergoing elastic collisions with the neutral gas is (see Chapters 3 and 5)

$$\frac{\partial f}{\partial t} + \mathbf{v}\cdot\nabla f + \mathbf{a}\cdot\frac{\partial f}{\partial \mathbf{v}} = \left(\frac{\partial f}{\partial t}\right)_{\text{col}}$$
$$= \int \mathrm{d}^3 v_0 \int \mathrm{d}^2\Omega_{\mathbf{g}'}\, g\sigma(g,\chi)\left[f(\mathbf{r},\mathbf{v}',t)f_0(\mathbf{v}'_0) - f(\mathbf{r},\mathbf{v},t)f_0(\mathbf{v}_0)\right], \tag{7.1}$$

where

$$\mathbf{a} = \frac{q}{m}(\mathbf{E} + \mathbf{v}\times\mathbf{B}).$$

This is integrated with some function of velocity, $\phi(\mathbf{v})$, over all velocities to form a moment equation. The right hand side is already known from the derivation of the Boltzmann collision term in Chapter 3. Thus

$$\int \mathrm{d}^3 v\, \phi(\mathbf{v})\left(\frac{\partial f}{\partial t}\right)_{\text{col}}$$
$$= \int \mathrm{d}^3 v \int \mathrm{d}^3 v_0\, g f(\mathbf{r},\mathbf{v},t) f_0(\mathbf{v}_0) \int \mathrm{d}^2\Omega_{\mathbf{g}'}\, \sigma(g,\chi)\left[\phi(\mathbf{v}') - \phi(\mathbf{v})\right], \tag{7.2}$$

while the left hand side is

$$\frac{\partial[n\langle\phi(\mathbf{v})\rangle]}{\partial t} + \nabla\cdot\left[n\langle\mathbf{v}\phi(\mathbf{v})\rangle\right] - n\left\langle \mathbf{a}\cdot\frac{\partial\phi(\mathbf{v})}{\partial\mathbf{v}}\right\rangle, \tag{7.3}$$

where $\langle\rangle$ denotes the velocity average, which is defined in Chapter 2. Thus, the general moment equation for particles undergoing elastic collisions in a gas can be written as

$$\frac{\partial[n\langle\phi(\mathbf{v})\rangle]}{\partial t} + \nabla\cdot\left[n\langle v\phi(v)\rangle\right] - n\left\langle \mathbf{a}\cdot\frac{\partial\phi(\mathbf{v})}{\partial\mathbf{v}}\right\rangle$$

$$= \int d^3v \int d^3v_0\, g f(\mathbf{r},\mathbf{v},t) f_0(\mathbf{v}_0) \int d^2\Omega_{\mathbf{g}'}\, \sigma(g,\chi)\left[\phi(\mathbf{v}') - \phi(\mathbf{v})\right]. \quad (7.4)$$

7.1.2 Equation of continuity

Setting $\phi = 1$ in Equation 7.4 gives

$$\frac{\partial n}{\partial t} + \nabla\cdot n\langle\mathbf{v}\rangle = 0, \quad (7.5)$$

where the right hand side is zero because $\phi(\mathbf{v}') - \phi(\mathbf{v}) \equiv 0$ in this case, corresponding to charge particle number conservation in collisions. For non-conservative collisions, such as electron attachment or ionization, the right hand side will contain a source term.

7.1.3 Momentum balance equation

Next, we set $\phi(\mathbf{v}) = m\mathbf{v}$ in Equation 7.4, and obtain

$$\frac{\partial[nm\langle\mathbf{v}\rangle]}{\partial t} + \nabla\cdot\left[nm\langle\mathbf{vv}\rangle\right] - nm\langle\mathbf{a}\rangle = -n\mu\langle\langle\nu_m(g)\mathbf{g}\rangle_0\rangle, \quad (7.6)$$

where $\mu = \frac{mm_0}{m+m_0}$ is the reduced mass, and the right hand side is an average over both ion and neutral velocity distributions, that is, an average $\langle\rangle$ over $\mathbf{v}$ of an average $\langle\rangle_0$ over $\mathbf{v}_0$. The steps in deriving the expression for the right hand side are enumerated below in some detail and can be regarded as the "recipe" to be followed in the general case. We proceed from Equation 7.2 with $\phi(\mathbf{v}) = m\mathbf{v}$:

$$\int d^3v\, m\mathbf{v}\left(\frac{\partial f}{\partial t}\right)_{\mathrm{col}}$$

$$= -m\int d^3v \int d^3v_0\, g f(\mathbf{r},\mathbf{v},t) f_0(\mathbf{v}_0) \int d^2\Omega_{\mathbf{g}'}\sigma(g,\chi)\left[\mathbf{v} - \mathbf{v}'\right]. \quad (7.7)$$

1. Transform to centre of mass (CM) and relative velocities

$$\mathbf{v} = \mathbf{G} + M_0\mathbf{g}$$
$$\mathbf{v}_0 = \mathbf{G} - M\mathbf{g}$$

and substitute in the integral over scattering angles

$$\mathbf{I}=\int d^2\Omega_{\mathbf{g}'}\ \sigma(g,\chi)\ \left[\mathbf{v}-\mathbf{v}'\right]=M_0\int d^2\Omega_{\mathbf{g}'}\sigma(g,\chi)[\mathbf{g}-\mathbf{g}'], \tag{7.8}$$

where we recall that $\mathbf{G}'=\mathbf{G}$, $g'=g$.

2. Evaluate the integral using the coordinate frame shown in Figure 7.1, for which

$$\mathbf{g}=g\ (0,0,1)$$
$$\mathbf{g}'=g\ (\sin\chi\cos\psi, \sin\chi\sin\psi, \cos\chi).$$

Thus

$$I_x=-gM_0\int d^2\Omega_{\mathbf{g}'}\ \sigma(g,\chi)\sin\chi\cos\psi$$
$$=-gM_0\int_0^{2\pi} d\psi\cos\psi\int_0^{\pi} d\chi\sin^2\chi\ \sigma(g,\chi)=0$$
$$I_y=-gM_0\int d^2\Omega_{\mathbf{g}'}\ \sigma(g,\chi)\sin\chi\sin\psi$$
$$=-gM_0\int_0^{2\pi} d\psi\sin\psi\int_0^{\pi} d\chi\sin^2\chi\ \sigma(g,\chi)=0$$
$$I_z=gM_0\int d^2\Omega_{\mathbf{g}'}\ \sigma(g,\chi)[1-\cos\chi]$$
$$=gM_0\int_0^{2\pi} d\psi\int_0^{\pi} d\chi\sin\chi\ \sigma(g,\chi)[1-\cos\chi]=gM_0\ \sigma_m(g),$$

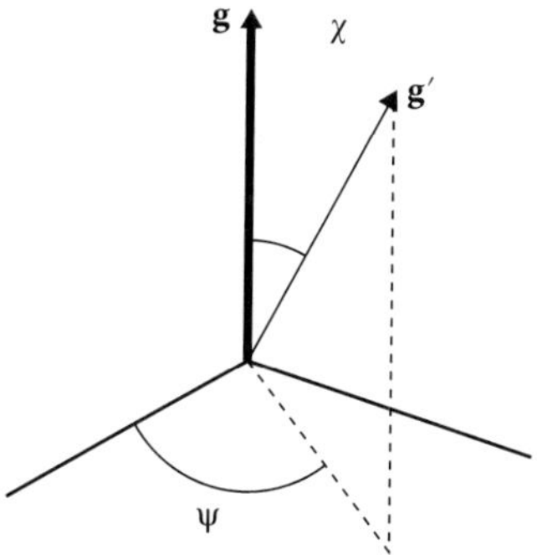

Figure 7.1 The coordinate system used for the calculation of integrals over scattering angles. For elastic scattering, the relative velocity vector is simply rotated through an angle χ, with unchanged magnitude.

where

$$\sigma_m(g) = [\sigma^{(0)}(g) - \sigma^{(1)}(g)] = 2\pi \int_0^{\pi} d\chi \sin\chi\, \sigma(g,\chi)[1-\cos\chi]$$

is the cross section for momentum transfer defined previously. These results may be combined into a single vector equation

$$\mathbf{I} = M_0\sigma_m(g)g(0,0,1) \equiv M_0\sigma_m(g)\mathbf{g}.$$

Substitution in Equation 7.7 then gives

$$\int d^3v\, m\mathbf{v}\left(\frac{\partial f}{\partial t}\right)_{\text{col}} = -mM_0 \int d^3v \int d^3v_0\, gf(\mathbf{r},\mathbf{v},t)f_0(\mathbf{v}_0)\sigma_m(g)\mathbf{g},$$

which can be expressed in terms of an average over the neutral distribution,

$$\int d^3v_0\, gf_0(\mathbf{v}_0)\sigma_m(g)\mathbf{g} \equiv n_0\langle g\sigma_m(g)\mathbf{g}\rangle_0 = \langle \nu_m(g)\mathbf{g}\rangle_0$$

and finally the ion distribution as well,

$$\int d^3v\, m\mathbf{v}\left(\frac{\partial f}{\partial t}\right)_{\text{col}} = -mM_0 \int d^3v\, f(\mathbf{r},\mathbf{v},t)\langle \nu_m(g)\mathbf{g}\rangle_0 = -n\mu\langle\langle \nu_m(g)\mathbf{g}\rangle_0\rangle,$$

where we have also made the substitution $mM_0 = \mu$. This is the term which appears in the right hand side of Equation 7.6.

7.1.4 Energy balance equation

Setting $\phi(\mathbf{v}) = \frac{1}{2}mv^2$ in Equation 7.4 gives

$$\frac{\partial\left[n\left\langle\frac{1}{2}mv^2\right\rangle\right]}{\partial t} + \nabla\cdot\left[n\left\langle\frac{1}{2}mv^2\mathbf{v}\right\rangle\right] - nm\langle\mathbf{a}\cdot\mathbf{v}\rangle$$
$$= n\frac{2\mu}{m+m_0}\langle\langle \nu_m(g)\left[\frac{1}{2}mv^2 - \frac{1}{2}m_0v_0^2 - \frac{1}{2}(m-m_0)\mathbf{v}\cdot\mathbf{v}_0\right]\rangle_0\rangle, \qquad (7.9)$$

which follows from

$$\int d^3v\,\frac{1}{2}mv^2\left(\frac{\partial f}{\partial t}\right)_{\text{col}}$$
$$= -\frac{1}{2}m\int d^3v\int d^3v_0\,g\,f(\mathbf{r},\mathbf{v},t)\,f_0(\mathbf{v}_0)\int d^2\Omega_{\mathbf{g}'}\,\sigma(g,\chi)\left[v^2 - v'^2\right]$$

and the fact that

$$\frac{1}{2}m[v^2 - v'^2] = \mu\mathbf{G}\cdot(\mathbf{g}-\mathbf{g}')$$

together with an integral over scattering angles similar to Equation 7.8.

7.1.5 External force terms

Since in all cases, the external force is specified by $\mathbf{a} = \frac{q}{m}(\mathbf{E}+\mathbf{v}\times\mathbf{B})$, then in the left hand side of the momentum balance equation (Equation 7.6), the velocity averaged force is

$$m\langle\mathbf{a}\rangle = q(\mathbf{E}+\langle\mathbf{v}\rangle\times\mathbf{B}), \tag{7.10}$$

while in the left hand side of the energy balance equation (Equation 7.9)

$$m\langle\mathbf{a}\cdot\mathbf{v}\rangle = q\mathbf{E}\cdot\langle\mathbf{v}\rangle \tag{7.11}$$

reflecting the fact that the magnetic field does no work.

7.1.6 Notation and terminology

7.1.6.1 g versus ϵ−dependence

Normally, cross sections are tabulated as functions of energy $\epsilon = \frac{1}{2}\mu g^2$, rather than of just g itself, and either representation is valid. Similarly, collision frequencies can be regarded as functions of ϵ, rather than g, if desired. We can therefore make the transition in notation

$$\sigma_m(g) \to \sigma_m(\epsilon)$$
$$\nu_m(g) \to \nu_m(\epsilon) \equiv n_0\left(\frac{2\epsilon}{\mu}\right)^{\frac{1}{2}}\sigma_m(\epsilon) \tag{7.12}$$

interchangeably, whenever it suits us.

7.1.6.2 Fluid, balance, and moment equations

Although the moment equations physically represent a conservation or balance of average physical properties, obtained by taking velocity moments of the Boltzmann equation, we may also refer to them as *fluid equations*. A set of such equations constructed for the system under investigation, and usually approximated in some way for practical purposes, is popularly referred to as a *fluid model*.

7.1.7 The problem of closure

The terms on the right hand side of Equations 7.6 and 7.9 involve averages of products of collision frequency with momentum and energy, which could, in principle, be found by taking additional moments of the Boltzmann equation. However, these too would involve further unknown averages, which would require taking additional moment equations, and so on. The higher order moments in the left hand side of the moment equations also present similar difficulties. This is the well-known problem of *closure* in fluid modelling. There are two ways of closing the equations:

1. Averages can be evaluated by *assuming* some expression for the particle distribution function $f(\mathbf{r}, \mathbf{v}, t)$. If this is successively refined in a series of systematic, controlled approximations [33] then the procedure is similar to the moment method of solution of the Boltzmann equation (see Chapter 12) and good accuracies can be achieved. Otherwise, if it is simply an uncontrolled, *ad hoc*, one-off guess then there is no way of estimating the accuracy. A critique of plasma modelling can be found in Ref. [43].

2. In either case, the fundamental issue with assuming the distribution function is that it is equivalent to assuming *all* moments of f, a severe and unnecessary approximation in our opinion. For that reason, we favour a scheme which approximates only a few low order moments, and does so in a systematic way. The method, which builds upon the Maxwell model discussion in Chapter 4, is commonly referred to as "momentum-transfer theory" in the literature [33].

7.2 Constant Collision Frequency Model

7.2.1 The fundamental equations

Suppose that the interaction is through an inverse fourth power law potential, so that the momentum transfer collision frequency is constant, independent of g (or ϵ). Then, since the neutral gas is at rest,

$$\langle \mathbf{v}_0 \rangle_0 = 0 \tag{7.13}$$

with temperature T_0

$$\frac{1}{2} m_0 \langle v_0^2 \rangle_0 = \frac{3}{2} k_B T_0. \tag{7.14}$$

Equations 7.6 and 7.9 simplify to

$$\frac{\partial [nm\langle \mathbf{v} \rangle]}{\partial t} + \nabla \cdot \left[nm\langle \mathbf{v}\mathbf{v} \rangle \right] - nq\,(\mathbf{E} + \langle \mathbf{v} \rangle \times \mathbf{B}) = -n\mu\nu_m \langle \mathbf{v} \rangle \tag{7.15}$$

and

$$\frac{\partial\left[n\left\langle\frac{1}{2}mv^2\right\rangle\right]}{\partial t}+\nabla\cdot\left[n\left\langle\frac{1}{2}mv^2\mathbf{v}\right\rangle\right]-nq\mathbf{E}\cdot\langle\mathbf{v}\rangle$$
$$=-n\frac{2\mu}{m+m_0}\nu_m\left[\left\langle\frac{1}{2}mv^2\right\rangle-\frac{3}{2}k_BT_0\right] \tag{7.16}$$

respectively, where Equations 7.10 and 7.11 have been substituted for the external force terms.

7.2.2 Convective time derivative

While the partial derivative $\partial/\partial t$ describes the time rate of change of a property at a fixed point, the above equations can also be conveniently written in terms of the convective or total time derivative,

$$\frac{\mathrm{d}}{\mathrm{d}t}=\partial_t+\mathbf{v}\cdot\nabla, \tag{7.17}$$

which is the rate of change measured in the reference frame moving with the fluid element at velocity **v**. If ψ denotes some average property, then it is straightforward to show, together with the equation of continuity (Equation 7.5) that the following identity holds:

$$n\frac{\mathrm{d}\psi}{\mathrm{d}t}=\frac{\partial(n\psi)}{\partial t}+\nabla\cdot(n\psi\langle\mathbf{v}\rangle). \tag{7.18}$$

7.2.3 Alternate form of the fluid equations

The equation of continuity (Equation 7.5) can be rewritten as

$$\frac{\mathrm{d}n}{\mathrm{d}t}+n\nabla\cdot\mathbf{v}=0. \tag{7.19}$$

Substituting $\psi=m\langle\mathbf{v}\rangle$ and $\langle\frac{1}{2}mv^2\rangle$ successively in Equation 7.18 and using the expressions for $\frac{\partial[nm\langle\mathbf{v}\rangle]}{\partial t}$ and $\frac{\partial\left[n\langle\frac{1}{2}mv^2\rangle\right]}{\partial t}$ given by Equations 7.15 and 7.16, respectively gives, after some algebraic manipulation, an equivalent set of moment equations:

$$nm\frac{\mathrm{d}\langle\mathbf{v}\rangle}{\mathrm{d}t}+\nabla\cdot\mathsf{P}-nq(\mathbf{E}+\langle\mathbf{v}\rangle\times\mathbf{B})=-n\mu\nu_m\langle\mathbf{v}\rangle \tag{7.20}$$

$$n\frac{\mathrm{d}}{\mathrm{d}t}\left(\left\langle\frac{1}{2}mv^2\right\rangle-\frac{1}{2}m\langle v\rangle^2\right)+\nabla\cdot\mathbf{J}_q+\mathsf{P}:\nabla\mathbf{v}$$
$$=-n\frac{2\mu}{m+m_0}\nu_m\left[\left\langle\frac{1}{2}mv^2\right\rangle-\frac{3}{2}k_BT_0-\frac{1}{2}(m+m_0)\langle v\rangle^2\right], \tag{7.21}$$

where

$$\mathsf{P} = nm\langle \mathbf{VV} \rangle \equiv nk_B \mathsf{T} \tag{7.22}$$

defines the *pressure* and *temperature tensors*, P and T, respectively (both of second rank), and

$$\mathbf{J}_q = n\left\langle \frac{1}{2} mV^2 \mathbf{V} \right\rangle \equiv n\mathbf{Q} \tag{7.23}$$

defines the *heat flux vector* $\mathbf{J}_q$ and heat flux vector $\mathbf{Q}$ per particle, respectively, while $\mathbf{V} = \mathbf{v} - \langle \mathbf{v} \rangle$ is the peculiar velocity. A single dot denotes the usual scalar product, while a double dot: represents a double contraction of second rank tensors:

$$\mathsf{P} : \nabla \mathbf{v} = \sum_{i=1}^{3} \sum_{j=1}^{3} P_{ij} \frac{\partial v_j}{\partial x_i}. \tag{7.24}$$

Even for this simple collision model, the moment equations are not closed, since P and $\mathbf{J}_q$ are unknown. For now, however, we continue with the discussion of approximation of the collision terms.

7.3 Momentum Transfer Approximation

The constant collision frequency model provides a template for dealing with the more general moment Equations 7.6 and 7.9. Suppose now that the effective energy-dependence of $\nu_m(\epsilon)$ is in some sense weak, so that its Taylor series representation

$$\nu_m(\epsilon) = \nu_m(\bar{\epsilon}) + (\epsilon - \bar{\epsilon})\nu_m'(\bar{\epsilon}) + \ldots \tag{7.25}$$

in the neighborhood of some reference energy, $\bar{\epsilon}$, at which the dominant contribution to the averages occurs, converges rapidly. If this is substituted in Equations 7.6 and 7.9 then the *leading terms* on the right hand side are of exactly the same *mathematical form* as their constant collision frequency counterparts Equations 7.15 and 7.16, respectively, with

$$\nu_m \to \nu_m(\bar{\epsilon}) \tag{7.26}$$

and similarly for the alternate forms Equations 7.20 and 7.20. If further it is *assumed* that the reference energy is approximately equal to the mean energy in the CM, that is,

$$\begin{aligned}
\bar{\epsilon} \approx \langle\langle\epsilon\rangle_0\rangle &= \frac{1}{2}\mu\langle\langle g^2\rangle_0\rangle \\
&= \frac{1}{2}\mu\left\{\langle v^2\rangle - 2\langle \mathbf{v}\rangle\cdot\langle \mathbf{v}_0\rangle_0 + \langle v_0^2\rangle_0\right\} \\
&= \frac{m_0\frac{1}{2}m\langle v^2\rangle + m\frac{1}{2}m_0\langle v_0^2\rangle_0}{m+m_0} \\
&= \frac{m_0\frac{1}{2}m\langle v^2\rangle + m\frac{3}{2}k_B T_0}{m+m_0}
\end{aligned} \tag{7.27}$$

then the approximation is known as "momentum transfer theory," and it is clear that it might be expected to work best when the collision frequency is a slowly varying function of energy. In fact it works surprisingly well even if this is not the case. The terms arising from the derivatives in Equation 7.25 are, *generally speaking*, more in the nature of *correction* terms, useful for determining the accuracy of the approximation, and thereby providing an internal consistency check. However, they are not generally of primary importance for representing collision transfer terms for conservative collisions.*

From this point on, we shall use Equations 7.15 and 7.16 with Equations 7.26 and 7.27 implied, and focus mainly (but not exclusively) on cases where forcing term derives entirely from an external electric field **E** only. The emphasis is on derivation of *empirical relationships* between physically interesting quantities rather than on calculation transport properties separately.

7.4 Stationary, Spatially Uniform Case

7.4.1 Drift velocity and Wannier relation

The simplest case to deal with is a steady, spatially uniform state, with constant field, and where all quantities are distinguished by superscripts "(0)." (This will constitute the "unperturbed" state when we begin to deal with spatially varying problems.) The average velocity is traditionally called the "*drift velocity*," $\mathbf{v}_d$, that is,

$$\langle \mathbf{v}\rangle^{(0)} \equiv \mathbf{v}_d,$$

* When dealing with non-conservative collisions (ionization, attachment, ion–atom reactions, and so on) the derivatives in fact become the leading terms, and are therefore of central importance in such cases (see Chapter 9).

while the average energy in the CM (Equation 7.27) is

$$\bar{\epsilon} \equiv \varepsilon = \frac{m_0 \frac{1}{2} m \langle v^2 \rangle^{(0)} + m \frac{3}{2} k_B T_0}{m + m_0}. \tag{7.28}$$

The momentum and energy balance equations greatly simplify, since all time and space derivatives vanish, and then

$$q(\mathbf{E} + \mathbf{v_d} \times \mathbf{B}) = \mu \nu_m(\varepsilon) \mathbf{v}_d \tag{7.29}$$

and

$$q\mathbf{E} \cdot \mathbf{v}_d = \frac{2\mu}{m + m_0} \nu_m(\varepsilon) \left\{ \frac{1}{2} m \langle v^2 \rangle^{(0)} - \frac{3}{2} k_B T_0 \right\} \tag{7.30}$$

respectively. If the scalar product of Equation 7.29 is taken with $\mathbf{v}_d$, and the result is subtracted from Equation 7.30, there follows,

$$\frac{1}{2} m \langle v^2 \rangle^{(0)} = \frac{3}{2} k_B T_0 + \frac{1}{2}(m + m_0) v_d^2 \tag{7.31}$$

an equation first obtained by Wannier in 1953 [25] for electric fields only. An equivalent form for the mean energy in the CM can be obtained by substitution of this last equation into Equation 7.28:

$$\varepsilon = \frac{3}{2} k_B T_0 + \frac{1}{2} m_0 v_d^2. \tag{7.32}$$

Equations 7.29 and 7.32 constitute a closed set of two non-linear equations in the two unknowns $\mathbf{v}_d$ and ε.

The Wannier energy relation is used extensively in an *empirical* way, that is, *measured* values of v_d are substituted into the right hand to produce values of mean energy, which are then used subsequently to estimate other quantities, for example, ion–molecule reaction rate coefficients. Since the Wannier relation is strictly true only for non-reactive, elastic ion–molecule collisions governed by a polarization potential, this procedure looks questionable. However, it turns out that the Wannier relation is surprisingly accurate, typically to within 10% or so, outside this narrow regime.

7.5 Transport in an Electric Field

7.5.1 Mobility coefficient

In the absence of a magnetic field, the *mobility coefficient* K for the charged particles in the gas in question is *defined* by the relation

$$\mathbf{v}_d = K\mathbf{E}. \tag{7.33}$$

It can either be measured in experiment or calculated theoretically using this definition. At the level of momentum transfer theory, it follows from Equation 7.29 with $B=0$ that

$$K=\frac{e}{\mu\nu_m(\varepsilon)}. \tag{7.34}$$

Since ε is a function of v_d and therefore of K itself, this latter expression is really an implicit equation for the mobility coefficient.

A *transport coefficient* is a material coefficient and is generally defined as the constant of proportionality in the relation linking the flow of some property with the "force" which causes it. The mobility coefficient is an example. Equation 7.33 can be rewritten in terms of the electric current density (current per unit area) $\mathbf{J}=ne\mathbf{v}_d$, to give Ohm's law,

$$\mathbf{J}=\sigma_{\text{cond}}\mathbf{E}, \tag{7.35}$$

where $\sigma_{\text{cond}}=neK$ is the electrical conductivity coefficient.

7.5.2 Solution of the moment equations

First, we note that since $\nu_m(\varepsilon)=n_0\sqrt{\frac{2\varepsilon}{\mu}}\sigma_m(\varepsilon)$, Equation 7.34 indicates that $K\sim 1/n_0$, and hence the *reduced mobility*,

$$\mathcal{K}\equiv\frac{n_0}{n_s}K, \tag{7.36}$$

where n_s is Loschmidt's number (the number density of an ideal gas under standard conditions of temperature and pressure), is independent of gas number density.

To find the mean energy ε and mobility K, Equations 7.32 and 7.34 must be solved simultaneously for a given momentum transfer cross section $\sigma_m(\varepsilon)$, and specified gas density, temperature, and electric field. The calculation is simplified, however, if ε is chosen as the independent variable, and the field is considered as a dependent variable. The procedure is shown schematically in Table 7.1. One first selects a value of ε and then calculates drift velocity from $v_d=\sqrt{\frac{2}{m_0}\left(\varepsilon-\frac{3k_BT_0}{2}\right)}$, an equivalent form of the Wannier relation (Equation 7.32), then mobility from Equation 7.34 and finally the electric field from Equation 7.33. Dividing through by number density gives the *reduced electric field* $E/n_0=v_d/n_s\mathcal{K}$. Another value of ε is then selected, and the procedure repeated until the desired range of field (actually E/n_0) is covered.

This strategy was first introduced by Viehland and Mason [27,28] for solution of the Boltzmann equation. The ion distribution function is represented through an expansion about a Maxwellian at a specified "effective"

Table 7.1 Calculating mean energy and mobility variation with E/n_0, where dots indicate that numerical data is to be inserted

ε	$v_d = \sqrt{\frac{2}{m_0}(\varepsilon - \frac{3k_B T_0}{2})}$	$K = \frac{e}{\mu\, \nu_m(\varepsilon)}$	$E/n_0 = v_d/(n_0 K) = v_d/n_s \mathcal{K}$
.	.	.	.

Note: Dots represent numerical values of quantities shown in headings.

temperature T_{eff}, related to the mean energy in the CM by $\varepsilon = \frac{3}{2}k_B T_{eff}$, and E/n_0 found subsequently.

7.5.3 Scaling

It is clear from the above discussion that ε and v_d depend upon E and n_0 solely through their ratio E/n_0. Since the reduced mobility (Equation 7.36) depends upon ε, it too is a function of the reduced field E/n_0, which is normally expressed in units of *townsend*, 1 Td $= 10^{-21}$V m^2. Although this scaling has been demonstrated in the context of approximate momentum transfer theory, it also follows directly from the Boltzmann equation (Equation 7.1).

In the discussion so far, gas temperature T_0 has been implicitly assumed to be constant but, like E/n_0, it too can be varied in experiment. However, in the momentum-transfer approximation, mobility is given by Equation 7.34, and thus depends upon T_0 and E/n_0 in the combination $\varepsilon = \frac{3}{2}k_B T_0 + \frac{1}{2}m_0 v_d^2(E/n_0)$. Thus, ion mobility is in fact the same for two different sets of experimental parameters $(T_0, E/n_0)_1$ and $(T_0, E/n_0)_2$ corresponding to the same value of ε. Solutions of the Boltzmann equation indicate that this scaling law is surprisingly accurate in many cases [33].

7.5.4 Sample calculations

7.5.4.1 Inverse fourth power, polarization potential/constant collision frequency—exact representation of drift velocity

In this case, the equations are exact and Equation 7.29 can be solved immediately for drift velocity without any reference to the energy balance equation:

$$\mathbf{v}_d = \frac{e\mathbf{E}}{\mu\nu_m},$$

that is, the mobility coefficient $K = \frac{e}{\mu\nu_m}$ is a constant, and the drift velocity is simply proportional to the the electric field strength. Wannier's energy relation (Equation 7.32) then shows that the mean energy ε is a quadratic in E. These relationships are *exact* for this particular model.

7.5.4.2 Weak fields, arbitrary interaction

Suppose that now charge particles interact with the neutrals according to some realistic potential, but the field is weak and/or the gas temperature is high, so that

$$\frac{3}{2}k_B T_0 \gg \frac{1}{2}m_0 v_d^2 \tag{7.37}$$

and by virtue of Equation 7.32,

$$\varepsilon \approx \frac{3}{2}k_B T_0.$$

Hence, the mobility coefficient $K = \frac{e}{\mu\nu_m(\frac{3}{2}k_B T_0)}$ is constant and v_d is simply proportional to E/n_0. On a plot of v_d vs E/n_0, the low field part of the curve is therefore always linear, for *any* type of interaction.

7.5.4.3 Strong fields, power law interaction

Assuming a very strong field (or equivalently, a cold gas) such that the inequality of Equation 7.37 is reversed, that is,

$$\frac{3}{2}k_B T_0 \ll \frac{1}{2}m_0 v_d^2 \tag{7.38}$$

then by Equation 7.32, $\varepsilon \approx \frac{1}{2}m_0 v_d^2$. Suppose that the momentum transfer cross section follows some power law, say

$$\sigma_m(\epsilon) \sim \epsilon^p$$

so that the collision frequency (Equation 7.12),

$$\nu_m(\epsilon) \equiv n_0 \left(\frac{2\epsilon}{\mu}\right)^{\frac{1}{2}} \sigma_m(\epsilon) \sim \epsilon^{p+\frac{1}{2}}.$$

From Equation 7.29, it then follows that $v_d \sim (E/n_0)^{\frac{1}{2p+2}}$. Thus, for example, for $p=0$ (constant cross section), $v_d \sim (E/n_0)^{\frac{1}{2}}$, while for $p=-1/6$, corresponding to an inverse twelfth power law interaction potential, $v_d \sim (E/n_0)^{\frac{3}{5}}$. The functional dependence of reduced mobility $\mathcal{K}$ upon reduced field is then $(E/n_0)^{-\frac{1}{2}}$ and $(E/n_0)^{-\frac{2}{5}}$, respectively.

Although the values of the numerical multipliers can be in error by 10% or so, the exponents themselves are *exactly* reproduced by the present approximate theory, and they themselves contain useful information about the nature of the interaction.

7.5.4.4 Realistic interaction potentials

Suppose we now consider ions which interact with neutral gas atoms through the Mason–Schamp potential (see Section 4.6.3), for which the functional form of $\nu_m(\epsilon)$ is shown schematically in Figure 4.5. It can be seen from Equation 7.34, that the mobility coefficient is effectively just the inverse of this, and that (once the field dependence of ε has been determined) the shape of the mobility versus field curve must be qualitatively similar to Figure 7.2.

There are three regions of field to be considered:

1. Low fields: the mobility coefficient is approximately constant, corresponding to the dominance of the long-range polarization potential, as discussed above;
2. High fields: the close range, inverse twelfth power law dominates, leading to the corresponding field dependence as shown; and
3. Intermediate fields: there is a "bump" in the mobility curve, corresponding to the enhanced transparency of the gas at intermediate energies, as attractive and repulsive forces cancel out to some extent.

This is a clear example of the way in which microscopic collision properties are reflected in macroscopic transport phenomena.

7.5.4.5 Extracting cross sections from electron swarm transport data

The fluid equations (Equations 7.34 and 7.32) may be used to estimate the electron momentum transfer cross section σ_m from measurement of electron drift velocities v_d as a function of E/n_0 in an atomic gas. Although not as accurate as an analysis using Boltzmann's equation, the information

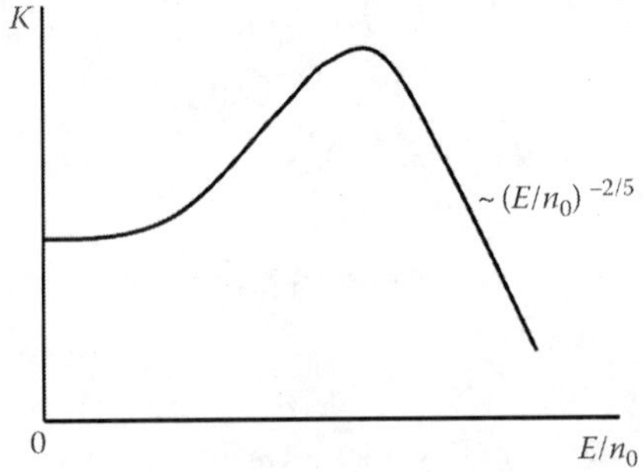

Figure 7.2 Schematic representation of the field dependence of reduced mobility for a 12-6-4 potential.

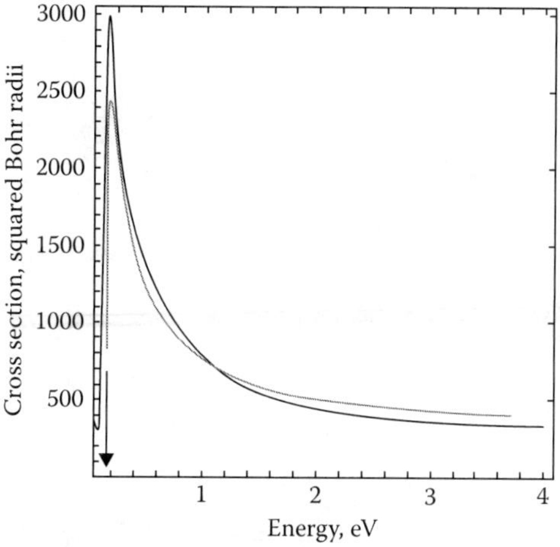

Figure 7.3 Momentum transfer cross sections for electrons in sodium vapour at $T = 803°K$, as calculated (1) theoretically from the Fano profile formula (upper line) and (2) by inverting swarm data using Equations 7.32 and 7.34 (lower line) (P. Nicoletopoulos, http:/arxiv.org/abs/physics/0307081, 2003). The arrow indicates the thermal energy $3k_BT_0/2 = 0.1\text{eV}$.

provided is nevertheless useful from a semi-quantitative perspective. An example for sodium vapour is shown in Figure 7.3.

7.5.5 Higher order moments

Spatially uniform moment equations can be formed with $\phi = m\mathbf{vv}$ and $\frac{1}{2}mv^2\mathbf{v}$, and solved for the temperature tensor

$$k_B\mathsf{T} = m\langle\mathbf{VV}\rangle^{(0)}, \tag{7.39}$$

with $\mathbf{V} = \mathbf{v} - \mathbf{v}_d$, and the heat flux per particle respectively.

$$\mathbf{Q} = \frac{1}{2}m\langle V^2\mathbf{V}\rangle^{(0)}. \tag{7.40}$$

In a coordinate system where the field defines the z-axis, that is, where $\mathbf{E} = (0, 0, E)$, the heat flux vector and temperature tensor have the form

$$\mathbf{Q} = (0, 0, Q)$$

and

$$k_B\mathbf{T} = \begin{bmatrix} k_BT_\perp & 0 & 0 \\ 0 & k_BT_\perp & 0 \\ 0 & 0 & k_BT_\parallel \end{bmatrix}, \tag{7.41}$$

respectively, with

$$Q = B\, m_0 v_d^3 \tag{7.42}$$

$$k_B T_\perp = k_B T_0 + A_\perp m_0 v_d^2 \tag{7.43}$$

$$k_B T_\parallel = k_B T_0 + A_\parallel m_0 v_d^2. \tag{7.44}$$

The constants are obtained after lengthy algebra:

$$\begin{aligned}
A_\perp &= \frac{(1+\frac{m}{m_0})\frac{\nu_v}{3\nu_m}}{\frac{2m}{m_0}+\frac{\nu_v}{\nu_m}} \\
A_\parallel &= \frac{\frac{2m}{m_0}(1-\frac{\nu_v}{3\nu_m})+\frac{\nu_v}{3\nu_m}}{\frac{2m}{m_0}+\frac{\nu_v}{\nu_m}} \\
B &= \frac{(1+\frac{m}{m_0})^2\left[A_\parallel+\frac{1}{2}+\frac{3m}{2m_0}\right]}{1+\frac{3m^2}{m_0^2}+\frac{4m\ \nu_v}{3m_0\ \nu_m}} - \left(A_\parallel+\frac{1}{2}+\frac{m}{2m_0}\right).
\end{aligned} \tag{7.45}$$

Here, ν_m and ν_v are the the collision frequencies for momentum transfer and "viscosity" (the integral over the differential cross section with $1 - P_2(\cos\chi)$), respectively, both of which are evaluated at the average energy ε.

Note that since $A_\parallel + 2A_\perp = 1$, taking the trace of the temperature tensor gives

$$\begin{aligned}
\text{Trace } k_B \mathsf{T} &= m\langle V^2\rangle^{(0)} \\
&= m[\langle v^2\rangle^{(0)} - v_d^2] \\
&= 2k_B T_\perp + k_B T_\parallel \\
&= 3k_B T_0 + m_0 v_d^2,
\end{aligned}$$

which is consistent with the expression (Equation 7.31), as it should be.

7.5.6 Simplifications for very light particles

Taking $m \ll m_0$ in Equation 7.45 shows that

$$\begin{aligned}
A_\perp \approx A_\parallel &\approx 1/3 \\
B &\approx 0
\end{aligned} \tag{7.46}$$

from which it follows that

$$\begin{aligned}
Q &\approx 0 \\
k_B T_\perp \approx k_B T_\parallel &\approx k_B T_0 + \frac{1}{3} m_0 v_d^2.
\end{aligned} \tag{7.47}$$

The scalar nature of the temperature tensor is also implicit in the two-term representation of the velocity distribution for electrons and other light particles, as discussed in Chapter 5.

7.5.7 A short note on tensor representation

It is important to understand the structure of the tensors which are now beginning to appear. In the spatial uniform situation, there is only one preferred direction in the system, $\hat{\mathbf{E}}$, and hence any *vector,* that is, tensor of rank 1, *must* be of the form $\alpha\hat{\mathbf{E}}$, where α is a scalar which may depend upon the magnitude of the field. Both $\mathbf{v}_d$ and $\mathbf{Q}$ have this property. If $\hat{\mathbf{E}}$ is directed along the z-axis of a system of coordinates, then so are these vectors. We can construct tensors of *rank 2* from both the second order dyadic $\hat{\mathbf{E}}\hat{\mathbf{E}}$ and the unit tensor I, for example, the temperature tensor $k_B\mathsf{T} = \alpha\mathrm{I} + \beta\hat{\mathbf{E}}\hat{\mathbf{E}}$, where again α and β may depend upon the magnitude of the field. If $\hat{\mathbf{E}}$ is directed along the z-axis of a system of coordinates, then *any* second rank tensor must have the structure shown in Equation 7.41.

This process can be extended to third and higher order tensors if desired, but with obviously increasing complexity. Likewise, when another vector, such as magnetic field $\mathbf{B}$, enters into to the system, the process becomes more complicated, but it is nevertheless clear why drift velocity has a contribution from the vector $\hat{\mathbf{E}}\times\hat{\mathbf{B}}$, for example.

7.6 Spatial Variations, Hydrodynamic Regime, and Diffusion Coefficients

7.6.1 Linearized moment equations, generalized Einstein relations

We now consider systems subject to a uniform applied electric field, but no magnetic field. In the weak-gradient hydrodynamic regime, the space-time dependence of any quantity can be represented though a density gradient expansion. Rather than adopting a formal approach, we simply suppose that the spatially uniform situation described above is slightly perturbed by *weak* gradients ∇n in number density, and write

$$\begin{aligned}\langle \mathbf{v}\rangle &= \mathbf{v}_d + \delta\mathbf{v}\\ \bar{\epsilon} &= \varepsilon + \delta\varepsilon\end{aligned} \tag{7.48}$$

and so on, where quantities $\delta\mathbf{v}$ and $\delta\varepsilon$ denote small deviations of order ∇n from the spatially uniform state. The equation of continuity (Equation 7.5) then becomes without further approximation

$$\frac{\partial n}{\partial t} + \mathbf{v}_d\cdot\nabla n + \nabla\cdot(n\delta\mathbf{v}) = 0, \tag{7.49}$$

but it remains to find an expression for $\delta\mathbf{v}$ before this assumes a useful form. Note that time and space variations of small quantities, or products of small quantities will be regarded as of second order, and therefore neglected. Thus, while there are perturbations $\delta\mathbf{Q}$ and $\delta\mathsf{T}$, to $\mathbf{Q}$ and T, respectively, these occur under a gradient operation, and hence to first order in ∇n, they do not contribute, that is,

$$\nabla \cdot \mathsf{P} = \nabla \cdot nk(\mathsf{T}+\delta\mathsf{T}) \approx k_B\mathsf{T}\cdot\nabla n \tag{7.50}$$
$$\nabla \cdot \mathbf{J}_q = \nabla \cdot n(\mathbf{Q}+\delta\mathbf{Q}) \approx \mathbf{Q}\cdot\nabla n.$$

Similarly, the time derivative of density is, to first order,

$$\frac{\partial n}{\partial t} \approx -\mathbf{v}_d \cdot \nabla n.$$

After carrying out these approximations in the *left hand side* of Equations 7.15 and 7.16, the following set of linearised balance equations result:

- *Momentum balance*

$$-k_B\mathsf{T}\cdot\nabla n + nm\mathbf{a} = n\mu\nu_m(\bar{\epsilon})\langle\mathbf{v}\rangle \tag{7.51}$$

- *Energy balance*

$$-\frac{1}{n\nu_e(\varepsilon)}\mathbf{Q}\cdot\nabla n = \bar{\epsilon} - \frac{3}{2}k_B T_0 - \frac{1}{2}m_0\langle\mathbf{v}\rangle^2, \tag{7.52}$$

where

$$\nu_e(\epsilon) = \frac{2m}{m+m_0}\nu_m(\epsilon)$$

may be considered as the collision frequency for energy transfer.

We now approximate the respective *right hand sides* by substituting Equation 7.48, and again linearize in small quantities. To zero order, we regain the unperturbed, spatially uniform state balance Equations 7.29 and 7.32, while to first order:

$$-k_B\mathsf{T}\cdot\nabla n = n\mu\{\nu_m(\varepsilon)\delta\mathbf{v} + \mathbf{v}_d\nu_m'(\varepsilon)\delta\varepsilon\} \tag{7.53}$$

$$-\frac{1}{n\nu_e(\varepsilon)}\mathbf{Q}\cdot\nabla n = \delta\varepsilon - m_0\mathbf{v}_d\cdot\delta\mathbf{v}. \tag{7.54}$$

Elimination of $\delta\varepsilon$ between these equations gives

$$n\delta\mathbf{v} = -\mathsf{D}\cdot\nabla n \tag{7.55}$$

where for $\mathbf{E}$ directed along the z-axis, the *diffusion tensor* D is of a structure similar to Equation 7.41, that is,

$$\mathbf{D} = \begin{bmatrix} D_\perp & 0 & 0 \\ 0 & D_\perp & 0 \\ 0 & 0 & D_\parallel \end{bmatrix} \tag{7.56}$$

with components

$$D_\perp = \frac{k_B T_\perp}{\mu \nu_m(\varepsilon)} \qquad \text{(transverse diffusion coefficient)} \tag{7.57}$$

$$D_\parallel = \frac{k_B T_\parallel - \frac{1}{2} m_0 v_d \, Q \, \frac{\nu'_m(\varepsilon)}{\nu_m(\varepsilon)}}{\mu \nu_m(\varepsilon) \left[1 + m_0 v_d^2 \frac{\nu'_m(\varepsilon)}{\nu_m(\varepsilon)}\right]} \qquad \text{(longitudinal diffusion coefficient).} \tag{7.58}$$

These relations can be written in a more useful form by recalling the definition of the mobility coefficient (Equation 7.34), from which the following identity can be proved:

$$\frac{\frac{\mathrm{d}\ln K}{\mathrm{d}\ln E}}{1 + \frac{\mathrm{d}\ln K}{\mathrm{d}\ln E}} = -m_0 v_d^2 \frac{\nu'_m(\varepsilon)}{\nu_m(\varepsilon)}.$$

Thus, we have the *generalized Einstein relations (GERs)*,

$$\frac{D_\perp}{K} = \frac{k_B T_\perp}{e} \tag{7.59}$$

$$\frac{D_\parallel}{K} = \frac{k_B T_\parallel}{e} \left[1 + (1 + \Delta) \frac{\mathrm{d}\ln K}{\mathrm{d}\ln E}\right] \tag{7.60}$$

where

$$\Delta = \frac{Q}{2 k_B T_\parallel v_d}. \tag{7.61}$$

The algebraic details are left as an exercise for the student.

Note that a GER of the form (Equation 7.60) with $\Delta = 0$ was first conjectured in 1953 by Wannier [25].

7.6.2 Example for light particles

The GER can best be viewed as empirical relations in which measurement or calculation of the mobility coefficient over a range of fields enables the diffusion coefficients to be estimated. By far the simplest cases to deal with involve either electrons, positrons, or muons, for which $m \ll m_0$, $T_{\|} \approx T_{\perp}$, and by virtue of Equation 7.47, both Q and Δ are small. The ratio of the two GER is then

$$\frac{D_{\|}}{D_{\perp}} = \left[1 + (1+\Delta)\frac{\mathrm{d}\ln K}{\mathrm{d}\ln E}\right] \approx 1 + \frac{\mathrm{d}\ln K}{\mathrm{d}\ln E} = \frac{\mathrm{d}\ln v_d}{\mathrm{d}\ln E}. \tag{7.62}$$

This means that the ratio of the two diffusion coefficients can be calculated directly from the slope of the v_d versus E curve on a log-log plot, be that of experimental or theoretical origin. To give just one example: for the constant cross section model, we have already shown that for strong fields $v_d \sim E^{\frac{1}{2}}$, and by Equation 7.62,

$$\frac{D_{\|}}{D_{\perp}} = 0.5\,.$$

This compares well with more accurate solution of the Boltzmann equation, which yields a value of 0.491 for the ratio. The expression (Equation 7.62) works well for other elastic cross section models and for electrons in noble gases, where collisions are predominantly elastic. For molecular gases where inelastic processes may be important, the parameter Δ plays an important role, and can no longer be neglected, as discussed in Chapter 8.

7.6.3 Anisotropy in configuration and velocity spaces

For many years, until the experiment of Wagner et al. and associated kinetic theory [77], it was axiomatic in the gaseous electronics literature that near-isotropy in velocity space for electrons, that is, $T_{\|} \approx T_{\perp}$ implied the same for configuration space, and in particular, $D_{\|} \approx D_{\perp}$. It is clear, however, from Equations 7.59 and 7.60 that anisotropic diffusion $\frac{D_{\|}}{D_{\perp}} \neq 1$ arises from *both* anisotropy in velocity space $\frac{T_{\|}}{T_{\perp}} \neq 1$ *and* the differential mobility $\frac{\mathrm{d}\ln K}{\mathrm{d}\ln E}$. The latter is responsible for anisotropy in configuration space, even when near-isotropy exists in velocity space, as is the case for light particles, as can be seen from Equations 7.43 and 7.44 with $m/m_0 \ll 1$.

7.6.4 Fick's law and the diffusion equation

If Equation 7.55 is substituted in Equation 7.49, there results *Fick's law of diffusion*,

$$\Gamma = n\langle \mathbf{v}\rangle = n\mathbf{v}_d - \mathsf{D}\cdot\nabla n, \tag{7.63}$$

which together with the equation of continuity (Equation 7.5) leads to an equation for the number density alone, called the *diffusion equation*:

$$\frac{\partial n}{\partial t} + \mathbf{v}_d\cdot\nabla n - \mathsf{D} : \nabla\nabla n = 0, \tag{7.64}$$

where : represents a double contraction over tensor indices. For the case where **E** defines the z-axis of a system of coordinates, it can be written in Cartesian coordinates as

$$\frac{\partial n}{\partial t} + v_d\frac{\partial n}{\partial z} - D_\perp\left(\frac{\partial^2}{\partial x^2} + \frac{\partial^2}{\partial y^2}\right)n - D_\parallel\frac{\partial^2 n}{\partial z^2} = 0, \tag{7.65}$$

but the choice of the most suitable coordinate system obviously hinges on what symmetries are inherent in the problem under investigation.

Fick's law is similar in form to many other well-known flux-gradient relations, for example, Fourier's law of heat conduction, Newton's law of viscosity, and so on, in which transport coefficients appear as constants of proportionality, or more precisely, as constant coefficients in a gradient expansion, valid in the hydrodynamic regime.

In this regime, the mean energy may also be represented by a density gradient expansion. Using Equations 7.48, 7.54, and 7.55, it follows that

$$\bar{\varepsilon}(\mathbf{r},t) = \varepsilon + \gamma\cdot\frac{1}{n(\mathbf{r},t)}\nabla n(\mathbf{r},t), \tag{7.66}$$

where

$$\gamma = -m_0\mathbf{v}_d\cdot\mathsf{D} - \frac{1}{\nu_e(\varepsilon)}\mathbf{Q} \tag{7.67}$$

is called the "gradient energy parameter" [78].

The Boltzmann equation itself can also be solved by making a similar density gradient expansion of the velocity distribution function. Note that a first order density gradient expansion is sufficient to analyze experiments which involve particle conserving collisions between charged particles and neutral gas atoms and molecules, but that for non-conservative collisions, for example, ionization, attachment, ion-atom reactions, it is necessary to go to *second* order in ∇n, as explained in Chapter 9.

7.6.5 Local field approximation

The above analysis is for the hydrodynamic regime in which a density gradient expansion is applied to represent transport quantities. It is strictly valid for spatially uniform fields $a = qE/m$ only. However, it is common to see the same mathematical expressions applied outside this regime of validity, with the non-uniform field $E(\mathbf{r})$ substituted for E in the expressions for transport coefficients. This is called the "local field" approximation and, while it holds to a reasonable approximation when the field varies sufficiently slowly, a "non-local" analysis is generally required for non-uniform fields. Here, the kinetic or fluid equations are solved without the density gradient expansion, and without any reference to transport coefficients, though an Ansatz is required to achieve closure [43]. For the present, however, the analysis continues with spatially uniform fields.

7.7 Diffusion of Charge Carriers in Semiconductors

The one-to-one correspondence between the kinetic theory of charge carriers in gases and crystalline semiconductor has already been noted, and the same can be said for fluid analysis. In particular, when the charge carriers interact primarily with the acoustic branch of lattice vibrations of a semiconductor, energy transfer in a collision with a phonon is small, and we can use the fluid equations for light particles undergoing elastic collisions in a gas. Thus, Equations 7.32 and 7.34 apply to a semiconductor, with m_0 and $n_0\sigma_m(\varepsilon)$ to be interpreted as k_BT_0/s^2, where s is the speed of sound, and the inverse mean free path

$$l^{-1}(\varepsilon) \equiv \frac{k_BT_0}{4hs\varepsilon^2}\int_0^{h/\pi(2m\varepsilon)^{\frac{1}{2}}} dq\, q^2 VC_q$$

respectively. Here, V is the volume of the crystal, q is the wave number of a phonon, and C_q is the matrix element corresponding to the scattering process.

Just as for a gaseous medium, the Einstein relations (Equations 7.59 and 7.60) follow, for which $T_{\|} \approx T_{\perp}$ yield the ratio of diffusion coefficients

$$\frac{D_{\|}}{D_{\perp}} \approx 1 + \frac{\mathrm{d}\ln K}{\mathrm{d}\ln E} = \frac{\mathrm{d}\ln v_d}{\mathrm{d}\ln E}$$

in the case of charge carriers scattered from acoustic phonons.

As an example, consider the deformation potential, for which the mean free path l is a constant, independent of energy. This is directly comparable with the constant cross section model for electrons in a gas, and hence it follows immediately that $v_d \sim E^{\frac{1}{2}}$ and $\frac{D_{\|}}{D_{\perp}} \approx 0.5$.

CHAPTER 8

Fluid Models with Inelastic Collisions

8.1 Introduction

This chapter extends the fluid analysis of Chapter 7 to include the effects of inelastic collisions, in which a charged particle gives up part of its kinetic energy to excite a neutral molecule to a higher energy level. Superelastic collisions, in which the neutral makes a transition from a higher to a lower energy state and the charged particle emerges with an increased kinetic energy, are also possible. The total energy, consisting of the translational kinetic energies of the particle and neutral molecule, plus the internal energy of the neutral, must be conserved, and the energy balance equation modified accordingly.

Although the analysis is carried out in the context of a gaseous medium, many of the results apply to soft-condensed matter since, in contrast to elastic processes, scattering associated with inelastic processes may be regarded as incoherent.

8.2 Moment Equations with Inelastic Collisions

8.2.1 The general moment equation

In Chapter 5, the semi-classical inelastic Wang Chang–Uhlenbeck–de Boer (WUB) collision operator was introduced. For point charged particles in a neutral gas of atoms or molecules with internal states characterized by the quantum number j, the rate at which some particle property $\phi(\mathbf{v})$ changes per unit volume and time, due to interaction with the neutral gas, is given by Equation 5.34. The general moment equation of Chapter 7 (Equation 7.4) extended to include inelastic processes is thus

$$\frac{\partial[n\langle\phi(\mathbf{v})\rangle]}{\partial t}+\nabla\cdot[n\langle\mathbf{v}\phi(\mathbf{v})\rangle]-nm\left\langle\mathbf{a}\cdot\frac{\partial\phi(\mathbf{v})}{\partial\mathbf{v}}\right\rangle$$
$$=\sum_{j,j'}\int d^3v\int d^3v_0\, g f(\mathbf{r},\mathbf{v},t) f_{0j}(\mathbf{v}_0)$$
$$\times\int d^2\Omega_{g'}\,\sigma(j,j';g,\chi)\left[\phi(\mathbf{v}')-\phi(\mathbf{v})\right]. \tag{8.1}$$

Here j, j' denote the initial and final neutral states, $\sigma(j,j';g,\chi)$ is the differential cross section, while the distribution function of the neutrals is

Maxwellian in velocity space, and Maxwell–Boltzmann in internal states j, that is

$$f_{0j}(\mathbf{v}_0) = n_{0j}\, w(\alpha_0, v_0) = n_{0j} \left(\frac{m_0}{2\pi k_B T_0}\right)^{3/2} \exp\left[-\frac{m v_0^2}{2k_B T_0}\right], \tag{8.2}$$

where

$$n_{0j} = n_0 \frac{\omega_j}{Z} \exp\left(-\frac{\epsilon_j}{k_B T_0}\right) \tag{8.3}$$

and the energy and degeneracy of state j are denoted by ϵ_j and ω_j, respectively, and $Z = \sum_j \omega_j \exp\left(-\frac{\epsilon_j}{k_B T_0}\right)$ is the partition function. In this formulation, $\sigma(j, j'; g, \chi)$ is really an average over degenerate levels (if any). For rotational states, for example, it depends on angular momentum quantum numbers j, but not the projection quantum numbers m_j.

As for elastic collisions, momentum is conserved in a collision, that is,

$$\mathbf{G}' = \mathbf{G},$$

but now internal states must be accounted for in the conservation of total energy,

$$\frac{1}{2}\mu g^2 + \epsilon_j = \frac{1}{2}\mu g'^2 + \epsilon_{j'} \tag{8.4}$$

and consequently $g' \neq g$ in general. Only for an elastic collision with $j' = j$ is the kinetic energy conserved, that is, $g' = g$.

8.2.2 Equation of continuity

Setting $\phi = 1$ in Equation 8.1 gives zero for the right hand side, since $\phi' - \phi = 0$, reflecting the conservative nature of the inelastic collision operator and hence we obtain the familiar equation of continuity:

$$\frac{\partial n}{\partial t} + \nabla \cdot n\langle \mathbf{v} \rangle = 0.$$

8.2.3 Momentum balance

Setting $\phi = m\mathbf{v}$ in Equation 8.1 gives

$$\begin{aligned}\frac{\partial[nm\langle \mathbf{v} \rangle]}{\partial t} + \nabla \cdot [nm\langle \mathbf{vv} \rangle] - nm\langle \mathbf{a} \rangle = \\ -m \sum_{j,j'} \int d^3v \int d^3v_0\, g\, f(\mathbf{r}, \mathbf{v}, t) f_{0j}(\mathbf{v}_0) \\ \times \int d^2\Omega_{\mathbf{g}'}\, \sigma(j, j'; g, \chi)(\mathbf{v} - \mathbf{v}'),\end{aligned} \tag{8.5}$$

where the integral over the scattering angles

$$\mathbf{I}_{j,j'} = \int d^2\Omega_{\mathbf{g}'}\ \sigma(j,j';g,\chi)\left[\mathbf{v}-\mathbf{v}'\right]$$

can be evaluated in exactly the same way as in Chapter 7, by transforming to relative and centre-of-mass (CM) velocities. The result is similar, but of course the internal states of the neutral must now be taken into account. Thus, we find

$$\begin{aligned}\mathbf{I}_{j,j'} &= M_0 \int d^2\Omega_{\mathbf{g}'}\ \sigma(j,j';g,\chi)[\mathbf{g}-\mathbf{g}'] \\ &= M_0\ \sigma_m(j,j';g)\mathbf{g},\end{aligned} \tag{8.6}$$

where

$$\sigma_m(j,j';g) = 2\pi\int_0^{\pi} d\chi\ \sin\chi\ \sigma(j,j';g,\chi)[1-\frac{g'}{g}\cos\chi] \tag{8.7}$$

is the cross section for momentum transfer associated with the transition $j \to j'$. Upon substituting Equation 8.6 into 8.5, we find an equation of the same *mathematical form* as for elastic collisions,

$$\frac{\partial[nm\langle\mathbf{v}\rangle]}{\partial t} + \nabla\cdot\left[nm\langle\mathbf{vv}\rangle\right] - nm\mathbf{a} = -n\mu\langle\langle\nu_m(g)\mathbf{g}\rangle_0\rangle, \tag{8.8}$$

where, however, the collision frequency for momentum transfer is now given by

$$\nu_m(g) \equiv \sum_{j,j'} n_{0j} g\sigma_m(j,j';g). \tag{8.9}$$

While it is reasonable to assume that the contributions to the right hand side of Equation 8.9 arising from *elastic* processes $(j'=j)$ are slowly varying functions of g, no such approximation can be made for the inelastic terms $(j' \neq j)$, since the inelastic cross sections typically rise sharply from zero at the threshold energy. However, since these inelastic cross sections are usually several orders of magnitude less than the elastic cross sections, the total momentum collision frequency $\nu_m(g)$ can, to a good approximation, be considered to be a slowly varying function of g, or equivalently, of the energy $\epsilon = \frac{1}{2}\mu g^2$ in the CM. Thus, the right hand side of Equation 8.8 can be approximated as in momentum transfer theory:

$$\langle\langle\nu_m(g)\mathbf{g}\rangle_0\rangle \approx \nu_m(\bar{\epsilon})\langle\mathbf{v}\rangle,$$

where $\bar{\epsilon}$ is the average energy in the CM,

$$\begin{aligned}\bar{\epsilon} &= \langle\langle \frac{1}{2}\mu g^2\rangle_0\rangle = \frac{1}{2}\mu\left\{\langle v^2\rangle + \langle v_0^2\rangle_0\right\}\\ &= M_0\frac{1}{2}m\langle v^2\rangle + M\frac{3}{2}k_BT_0\end{aligned} \tag{8.10}$$

and, as before, $\langle \mathbf{v}_0\rangle = 0$. Thus, the approximate momentum balance equation including inelastic collisions at the level of first order momentum transfer theory is

$$\frac{\partial[nm\langle\mathbf{v}\rangle]}{\partial t} + \nabla\cdot\left[nm\langle\mathbf{vv}\rangle\right] - nm\mathbf{a} = -n\mu\nu_m(\bar{\epsilon})\langle\mathbf{v}\rangle \tag{8.11}$$

which is of exactly the same form as the corresponding balance equation for elastic collisions obtained in Chapter 7.

8.2.4 Energy balance equation

In contrast to the first two balance equations, the energy balance equation *does* change significantly due to inelastic collisions. Thus, setting $\phi = \frac{1}{2}mv^2$ in Equation 8.1 gives

$$\begin{aligned}&\frac{\partial\left[n\langle\frac{1}{2}mv^2\rangle\right]}{\partial t} + \nabla\cdot\left[n\langle\frac{1}{2}mv^2\mathbf{v}\rangle\right] - nm\,\langle\mathbf{a}\cdot\mathbf{v}\rangle\\ &\qquad = -\sum_{j,j'}\int d^3v\int d^3v_0\, g f(\mathbf{r},\mathbf{v},t) f_{0j}(\mathbf{v}_0)\,\mathcal{I}_{j,j'},\end{aligned} \tag{8.12}$$

where

$$\mathcal{I}_{j,j'} = \frac{1}{2}m\int d^2\Omega_{\mathbf{g}'}\ \sigma(j,j';g,\chi)\left[v^2 - v'^2\right].$$

Since $\mathbf{v} = \mathbf{G} + M_0\mathbf{g}$, $\mathbf{G}' = \mathbf{G}$ and g' is related to g through Equation 8.4, there follows

$$\begin{aligned}\mathcal{I}_{j,j'} &= \int d^2\Omega_{\mathbf{g}'}\ \sigma(j,j';g,\chi)\{[\mu\mathbf{G}\cdot(\mathbf{g}-\mathbf{g}') + M_0[\epsilon_{j'} - \epsilon_j]\}\\ &= \sigma_m(j,j';g)\mu\mathbf{G}\cdot\mathbf{g} + \sigma_T(j,j';g)M_0(\epsilon_{j'} - \epsilon_j),\end{aligned} \tag{8.13}$$

where $\sigma_m(j,j';g)$ is defined by Equation 8.7 and

$$\sigma_T(j,j';g) = 2\pi\int_0^{\pi} d\chi\ \sin\chi\ \sigma(j,j';g,\chi) \tag{8.14}$$

is the total cross section. Substitution of Equation 8.13 into 8.12 gives

$$\frac{\partial\left[n\langle\frac{1}{2}mv^2\rangle\right]}{\partial t}+\nabla\cdot\left[n\langle\frac{1}{2}mv^2\mathbf{v}\rangle\right]-nm\langle\mathbf{a}\cdot\mathbf{v}\rangle=$$
$$-n\frac{2\mu}{m+m_0}\langle\langle\nu_m(g)\,\delta\epsilon\rangle_0\rangle$$
$$-nM_0\sum_{j,j'}\langle\langle\nu_T(j,j';g)\rangle_0\rangle(\epsilon_{j'}-\epsilon_j), \tag{8.15}$$

where

$$\delta\epsilon\equiv\frac{1}{2}mv^2-\frac{1}{2}m_0v_0^2-\frac{1}{2}(m-m_0)\,\mathbf{v}\cdot\mathbf{v}_0$$

and

$$\nu_T(j,j';g)\equiv n_{0j}g\sigma_T(j,j';g) \tag{8.16}$$

is the total collision frequency for the inelastic process $j\rightarrow j'$.

The *first term* on the right hand side of the exact Equation 8.15 can be satisfactorily approximated using momentum transfer theory and contributes a term of the same *mathematical form* as the right hand of the *elastic* energy balance equation. We can, however, make no such approximation in the *second term*, as inelastic processes are characterized by rapid variations in the cross section near threshold. Some other means of approximation has to be found, and this will be discussed presently.

For now, we cast the second term in an alternate form, by rewriting the summation as

$$\sum_{j,j'}\langle\langle\nu_T(j,j';g)\rangle_0\rangle(\epsilon_{j'}-\epsilon_j)$$
$$=\sum_{j<j'}\langle\langle\nu_T(j,j';g)\rangle_0\rangle(\epsilon_{j'}-\epsilon_j)$$
$$+\sum_{j>j'}\langle\langle\nu_T(j,j';g)\rangle_0\rangle(\epsilon_{j'}-\epsilon_j)$$
$$=\sum_{j<j'}\left[\langle\langle\nu_T(j,j';g)\rangle_0\rangle-\langle\langle\nu_T(j',j;g)\rangle_0\rangle\right](\epsilon_{j'}-\epsilon_j)$$
$$\equiv\sum_i\left[\langle\vec{\nu}_i(\epsilon)\rangle-\langle\overleftarrow{\nu}_i(\epsilon)\rangle\right]\epsilon_i^*. \tag{8.17}$$

Notation has been simplified in several respects:

- The inelastic excitation processes $j\rightarrow j'$ has been characterized by a forward overhead arrow, and represented through a single index i, with the threshold energy denoted by $\epsilon_i^* = \epsilon_{j'} - \epsilon_j$. Thus

$$\vec{\nu}_i(\epsilon) \equiv \nu_i\,(j, j'; g);\ \vec{\sigma}_i(\epsilon) = \sigma_T\,(j, j'; g)$$

and

$$\vec{\nu}_i(\epsilon) = n_{0j}\sqrt{\frac{2\epsilon}{\mu}}\vec{\sigma}_i(\epsilon)$$

- Averaging over both particle and neutrals in the double averaging process $\langle\langle \ldots \rangle_0\rangle$ has been replaced by an equivalent single averaging process over the CM energy $\epsilon = \frac{1}{2}\mu g^2$.
- Superelastic collisions $j' \to j$ are denoted by a reverse overhead arrow, and written in similar abbreviated notation, that is

$$\overleftarrow{\nu}_i(\epsilon) \equiv \nu_T\,(j',\, j; g);\ \overleftarrow{\sigma}_i(\epsilon) = \sigma_T\,(j',\, j; g) \tag{8.18}$$

and

$$\overleftarrow{\nu}_i(\epsilon) = n_{0j'}\sqrt{\frac{2\epsilon}{\mu}}\,\overleftarrow{\sigma}_i(\epsilon). \tag{8.19}$$

With this notation, the energy balance Equation 8.15 becomes

$$\frac{\partial\left[n\langle\frac{1}{2}mv^2\rangle\right]}{\partial t} + \nabla\cdot\left[n\langle\frac{1}{2}mv^2\mathbf{v}\rangle\right] - nm\,\langle\mathbf{a}\cdot\mathbf{v}\rangle$$
$$= -n\frac{2\mu}{m+m_0}\nu_m(\bar{\epsilon})\left[\langle\frac{1}{2}mv^2\rangle - \frac{3}{2}k_BT_0\right]$$
$$- nM_0\sum_i\left[\langle\vec{\nu}_i\rangle - \langle\overleftarrow{\nu}_i\rangle\right]\epsilon_i^*. \tag{8.20}$$

While the last term on the right hand side representing energy loss due to inelastic collisions is *exact*, it is not in form suitable for practical applications, since $\langle\vec{\nu}_i(\epsilon)\rangle$ and $\langle\overleftarrow{\nu}_i(\epsilon)\rangle$ must be expressed in terms of the mean energy $\bar{\epsilon}$, and related to the corresponding average cross sections. Since inelastic cross sections vary rapidly near threshold, it is not appropriate to use momentum transfer theory, and another prescription must be found.

8.2.4.1 Electric fields only

As in Chapter 7, we limit the discussion that follows to the electric fields only case. The effects of a magnetic field are discussed in Chapter 18.

8.3 Representation of the Average Inelastic Collision Frequencies

8.3.1 Definition of averages

The average collision frequencies for the ith excitation and de-excitation processes are defined by

$$\langle \vec{\nu}_i(\epsilon)\rangle = \int_0^\infty \vec{\nu}_i(\epsilon) f(\epsilon)\epsilon^{\frac{1}{2}}\,d\epsilon = n_{0j}\int_{\epsilon_i^*}^\infty \sqrt{\frac{2\epsilon}{\mu}}\vec{\sigma}_i(\epsilon) f(\epsilon)\epsilon^{\frac{1}{2}}\,d\epsilon \tag{8.21}$$

$$\langle \overleftarrow{\nu}_i(\epsilon)\rangle = \int_0^\infty \overleftarrow{\nu}_i(\epsilon) f(\epsilon)\epsilon^{\frac{1}{2}}\,d\epsilon = n_{0j'}\int_0^\infty \sqrt{\frac{2\epsilon}{\mu}}\overleftarrow{\sigma}_i(\epsilon) f(\epsilon)\epsilon^{\frac{1}{2}}\,d\epsilon, \tag{8.22}$$

respectively, where $f(\epsilon)$ is the (unknown) distribution function of CM energies ϵ, for a given value of the mean CM energy $\bar{\epsilon}$. Notice that the lower limits of the integrals are different, reflecting the fact that inelastic cross sections are non-zero only above threshold, while there is no such restriction for superelastic cross sections. Our first task is to relate these two frequencies.

8.3.2 Relationship between inelastic and superelastic collision frequencies

Collisions are described by the laws of mechanics, which are invariant under time-reversal. This leads to the "principle of detailed balance" [65]

$$\omega_{j'}\epsilon'\overleftarrow{\sigma}_i(\epsilon') = \omega_j\epsilon\vec{\sigma}_i(\epsilon) \tag{8.23}$$

where ω_j and $\omega_{j'}$ are the degeneracies of the initial and final states j and j', respectively, and

$$\epsilon' = \epsilon - \epsilon_i^*$$

is the energy in the CM after a collision. This may be written in an equivalent form as

$$\omega_{j'}\epsilon\overleftarrow{\sigma}_i(\epsilon) = \omega_j(\epsilon + \epsilon_i^*)\vec{\sigma}_i(\epsilon + \epsilon_i^*). \tag{8.24}$$

The integrals (Equations 8.21 and 8.22) may be evaluated approximately by assuming some particular form of the distribution function $f(\epsilon)$, for example,

$$f(\epsilon) = A\exp\left(-\frac{\epsilon}{k_B T_{\text{eff}}}\right), \tag{8.25}$$

where A is a normalization constant, and the "effective temperature" T_{eff} is defined by

$$\bar{\epsilon} = \frac{3}{2}k_B T_{\text{eff}}.$$

Substituting Equation 8.25 in 8.22 and using the microscopic reversibility condition (Equation 8.24) gives, after a little algebra,

$$\langle \overleftarrow{\nu}_i \rangle = \langle \overrightarrow{\nu}_i \rangle \exp\left[-\epsilon_i^* \left(\frac{1}{k_B T_0} - \frac{1}{k_B T_{\text{eff}}} \right) \right]. \tag{8.26}$$

It is useful to examine this expression in limiting cases:

1. Near equilibrium (e.g., for very weak fields), $\frac{1}{2}\, m\langle v^2 \rangle \approx \frac{1}{2} m_0 \langle v_0^2 \rangle_0 = 3k_B T_0/2$, and hence by Equation 8.10 $T_{\text{eff}} \approx T_0$. Equation 8.26 then indicates that the forward and reverse rates are the same, $\langle \overleftarrow{\nu}_J \rangle \approx \langle \overrightarrow{\nu}_J \rangle$ for all processes i (principle of detailed balancing), and hence the last term in Equation 8.20 vanishes.
2. On the other hand, for highly non-equilibrium situations (e.g., in strong fields), where $T_{\text{eff}} \gg T_0$, Equation 8.26 indicates that $\langle \overleftarrow{\nu}_J \rangle \approx \langle \overrightarrow{\nu}_J \rangle \exp\left[-\frac{\epsilon_J}{k_B T_0} \right]$. For inelastic processes for which threshold energy is larger than thermal energy, $k_B T_0 \sim 1/40eV$), that is, $\frac{\overline{\epsilon}_J}{k_B T_0} > 1$, it follows that $\langle \overleftarrow{\nu}_i \rangle$ is much smaller than $\langle \overrightarrow{\nu}_i \rangle$. This is the case, for example, in excitation of the vibrational states of a typical diatomic molecule, but not necessarily so for rotational states.
3. For a cold gas, $T_0 = 0$ all neutrals are in the ground state, $\langle \overleftarrow{\nu}_i \rangle \to 0$, and superelastic collisions can be neglected altogether.

8.3.3 The smoothing function

Momentum transfer theory at the basic level effectively approximates an average of a function of energy by the same function of the average energy. If this prescription had been applied to the inelastic collision frequency $\langle \overrightarrow{\nu}_i(\epsilon) \rangle$ defined by Equation 8.19, it would have given $\langle \overrightarrow{\nu}_i(\epsilon) \rangle \approx \overrightarrow{\nu}_i(\overline{\epsilon}) = n_{0j} \sqrt{\frac{2\overline{\epsilon}}{\mu}} \overrightarrow{\sigma}_i(\overline{\epsilon})$. Since

$$\begin{aligned} \overrightarrow{\sigma}_i(\overline{\epsilon}) &= 0 \quad (\overline{\epsilon} \le \epsilon_i^*) \\ &> 0 \quad (\overline{\epsilon} > \epsilon_i^*) \end{aligned}$$

this would imply that inelastic processes are completely absent for mean energies below ϵ_i^*, but are suddenly "switched on" in every collision for mean energies above ϵ_i^*. This is unphysical, since particles have a wide range of energies, and even for $\overline{\epsilon} < \epsilon_i^*$ there are always some collisions with $\epsilon > \epsilon_i^*$, enabling neutrals to be excited inelastically. On physical grounds, we can therefore see why $\langle \overrightarrow{\nu}_i(\epsilon) \rangle$ should vary continuously with $\overline{\epsilon}$, not abruptly according to lowest order momentum transfer.

On the other hand, it can be seen from Equation 8.24 that the *superelastic* cross section $\overleftarrow{\sigma}_i(\epsilon)$ has no abrupt behaviour and varies continuously for $\epsilon \geq 0$. Thus, it is reasonable to represent the averages of Equations 8.19 and 8.24 as

$$\omega_{j'}\bar{\epsilon}\overleftarrow{\sigma}_i(\bar{\epsilon}) = \omega_j(\bar{\epsilon}+\epsilon_i^*)\vec{\sigma}_i(\bar{\epsilon}+\epsilon_i^*) \tag{8.27}$$

and

$$\langle \overleftarrow{\nu}_i(\epsilon)\rangle \approx \overleftarrow{\nu}_i(\bar{\epsilon}) = n_{0j}\sqrt{\frac{2\bar{\epsilon}}{\mu}}\overleftarrow{\sigma}_i(\bar{\epsilon}), \tag{8.28}$$

respectively.

If Equation 8.28 is substituted in the left hand side of Equation 8.26, and $\overleftarrow{\sigma}_i(\bar{\epsilon})$ is eliminated using Equation 8.27, there follows (see Exercises)

$$\langle \vec{\nu}_i \rangle = S(\bar{\epsilon})\, n_{0,j}\sqrt{\frac{2\bar{\epsilon}}{\mu}}\vec{\sigma}_i(\bar{\epsilon}+\epsilon_i), \tag{8.29}$$

where

$$S(\bar{\epsilon}) = \left(1+\frac{\epsilon_i^*}{\bar{\epsilon}}\right)\exp\left(-\frac{3\epsilon_i^*}{2\bar{\epsilon}}\right). \tag{8.30}$$

Note that S varies continuously from near zero for mean energies well below threshold to unity well above threshold, and thus introduces the required element of smoothing in the representation of $\langle\vec{\nu}_J\rangle$. The effect of the "smoothing factor" on the average cross section is portrayed schematically in Figure 8.1 for a constant cross section model.

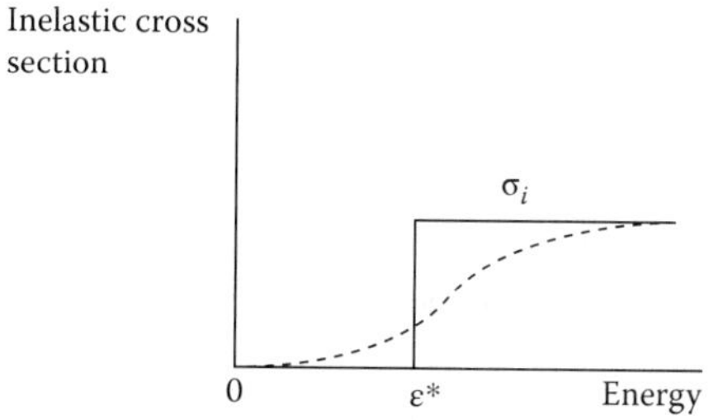

Figure 8.1 Schematic representation of the effect of the smoothing factor defined by Equation 8.30 upon an inelastic cross section $\vec{\sigma}_i \equiv \sigma_i$ constant above a threshold energy $\epsilon_i^* \equiv \varepsilon^*$. The average cross section, in this case, is simply $\sigma_i\ S(\bar{\epsilon}) = \sigma_i\left(1+\frac{\varepsilon^*}{\bar{\epsilon}}\right)\exp\left(-\frac{3\varepsilon^*}{2\bar{\epsilon}}\right)$ and is represented by the dashed line.

8.4 Hydrodynamic Regime

8.4.1 Weak-gradient fluid equations

Inelastic collisions do not alter the mathematical form of either the equation of continuity or the momentum balance equation. Their effect appears *explicitly* only through the addition of an extra term in the energy balance equation (Equation 8.20). Therefore, fluid equations including inelastic collisions may be obtained simply by adding this term to the corresponding elastic fluid equations of Chapter 7. The complete set of fluid equations in the weak-gradient, hydrodynamic regime is thus

- *Equation of continuity*

$$\frac{\partial n}{\partial t} + \nabla \cdot n\langle \mathbf{v} \rangle = 0. \tag{8.31}$$

- *Momentum balance*

$$-k_B \mathrm{T} \cdot \nabla n + nm\mathbf{a} = n\mu\nu_m(\bar{\epsilon})\langle \mathbf{v} \rangle. \tag{8.32}$$

- *Energy balance*

$$-\frac{1}{n\nu_e(\varepsilon)}\mathbf{Q} \cdot \nabla n = \bar{\epsilon} - \frac{3}{2}k_B T_0 - \frac{1}{2}m_0\langle \mathbf{v} \rangle^2 + \Omega(\bar{\epsilon}), \tag{8.33}$$

where

$$\nu_e \equiv \frac{2m}{m + m_0}\nu_m$$

the quantity

$$\Omega(\bar{\epsilon}) \equiv M_0 \sum_i \frac{\langle \vec{\nu}_i \rangle - \langle \overleftarrow{\nu}_i \rangle}{\nu_e(\bar{\epsilon})}\epsilon_i^* \tag{8.34}$$

accounts for the relative importance of inelastic energy losses.

8.4.2 Spatially uniform case

In the spatially uniform case, $\nabla n \to 0$, $\langle \mathbf{v} \rangle \to \mathbf{v}_d$, $\bar{\epsilon} \to \varepsilon$, and the momentum and energy balance equations become

$$\mathbf{v}_d = \frac{\mathbf{a}}{M_0\nu_m(\varepsilon)} \tag{8.35}$$

and

$$\varepsilon = \frac{3}{2}k_B T_0 + \frac{1}{2}m_0 \mathbf{v}_d{}^2 - \Omega(\varepsilon), \tag{8.36}$$

respectively. These are generalisations of the expressions for mobility and Wannier energy relations. These two equations may then may be solved for $\mathbf{v}_d$ and ε for specified momentum transfer and inelastic cross sections. The same numerical procedure may be used as described in Chapter 7, by taking ε as the independent variable.

8.4.2.1 Diffusion coefficients and heat flux "correction" term

Spatial variations are handled in exactly the same way as in Chapter 7, by considering small perturbations to $\mathbf{v}_d$ and ε of order ∇n, and linearizing the equations accordingly. Once again, Fick's law emerges,

$$\Gamma = n\mathbf{v}_d - \mathsf{D}\cdot\nabla n,$$

where the diffusion tensor D has transverse and longitudinal components given by the same generalized Einstein relations (GER) as for the purely elastic collision case, namely

$$\frac{D_\perp}{K} = \frac{k_B T_\perp}{e} \tag{8.37}$$

$$\frac{D_\parallel}{K} = \frac{k_B T_\parallel}{e}\left[1 + (1+\Delta)\frac{\mathrm{d}\ln K}{\mathrm{d}\ln E}\right]. \tag{8.38}$$

The quantity,

$$\Delta = \frac{Q}{2k_B T_\parallel v_d} \tag{8.39}$$

is of the same mathematical form as for the elastic collision case discussed in Chapter 7, but it may now be far more significant than a mere "correction factor." Thus, in some circumstances, Δ plays a crucial role in preserving the physical integrity of the GER, as discussed below.

8.4.3 Light particles, cold gas

Consider now a spatially uniform system of light particles $m \ll m_0$ in a cold gas ($T_0 = 0$), for which $M_0 \to 1$, $\nu_m(\varepsilon) = n_0\sqrt{\frac{2\varepsilon}{m}}\sigma_m(\varepsilon)$, $\nu_e(\varepsilon) = 2\frac{m}{m_0}\nu_m(\varepsilon)$, and ε is the average particle energy. With $\mathbf{a} = e\mathbf{E}/m$, the drift velocity (Equation 8.35) becomes

$$\mathbf{v}_d = \frac{e\mathbf{E}}{m\nu_m(\varepsilon)}. \tag{8.40}$$

If $T_0 = 0$, the excited levels of the neutrals are not populated and, as we have observed, no superelastic processes are possible. For simplicity, suppose there is just one inelastic channel with threshold energy ϵ^*, governed by an inelastic cross section $\sigma_i(\varepsilon)$. In these circumstances, Equation 8.34 simplifies to

$$\Omega(\varepsilon) \equiv \frac{m_0}{2m} \epsilon^* \frac{\sigma_i(\varepsilon)}{\sigma_m(\varepsilon)} S(\varepsilon), \tag{8.41}$$

where the approximation (Equation 8.29) has been applied, and $S\left(\frac{3\varepsilon^*}{2\varepsilon}\right)$ is the smoothing function (Equation 8.30).

Since $k_B T_{\parallel} \approx 2\varepsilon/3$ the heat flux "correction factor" (Equation 8.39) approximates to

$$\Delta = \frac{3Q}{4\varepsilon v_d}.$$

8.5 Negative Differential Conductivity

8.5.1 NDC criterion

For electrons in certain gases, negative differential conductivity (NDC) is observed over a range of fields, where the drift velocity actually *decreases* as the field is increased, that is,

$$\frac{dv_d}{dE} < 0. \tag{8.42}$$

Note that in contrast, the average energy is *always* a monotonically increasing function of field, that is,

$$\frac{d\varepsilon}{dE} > 0. \tag{8.43}$$

If Equation 8.36 is differentiated with respect to E, and conditions (Equations 8.42 and 8.43) are applied, then it follows that NDC corresponds to those mean energies ε for which

$$1 + \Omega'(\varepsilon) < 0, \tag{8.44}$$

where the dash is an energy derivative.

The physical origin of NDC is as follows, assuming for the sake of being definite an elastic collision frequency $\nu_m(\varepsilon)$ which is a monotonically increasing function of ε, and appropriate elastic and inelastic cross sections: With increasing E/n_0 inelastic collisions soak up the energy of the electron swarm, and ε therefore rises only slowly. The collision frequency $\nu_m(\varepsilon)$, being a monotonic function of ε, thus also rises slowly, permitting v_d to continue to increase, according to Equation 8.40. At some point, inelastic collisions decline in *relative* importance, as measured by the *ratio* (Equation 8.41) and no longer consume as much energy from the electron swarm overall. Then, as the inelastics are further reduced in importance, ε *increases* dramatically, as does $\nu_m(\varepsilon)$. Since Equation 8.40 shows that v_d is inversely proportional to ν_m, the drift velocity then *declines*, provided that Equation 8.44 holds. The *signature* of NDC is, thus, a rapidly rising mean energy ε.

From this, it can be seen that it is not the energy dependence of $\sigma_i(\varepsilon)$ alone which is the deciding factor, but rather it is the *ratio* of cross sections, as measured by $\Omega(\varepsilon)$, which determines whether or not NDC takes place. We shall focus on both these quantities in the example which follows.

8.5.2 Model calculation

The cross section models shown in Figure 8.2 both furnish a form of $\Omega(\varepsilon)$ satisfying the NDC criterion (Equation 8.44) over a range of energies (see Figure 8.3). The corresponding drift velocity is shown in Figure 8.4.

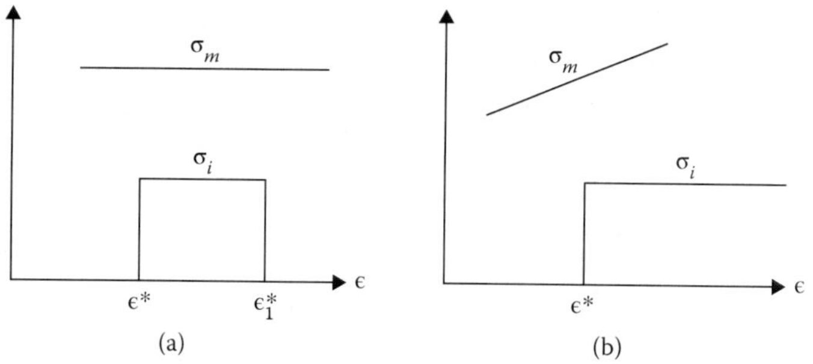

Figure 8.2 Just two types of the many possible combinations of cross sections leading to NDC, (a) the 'hat' model and (b) the 'step function' model. For a comprehensive discussion of these and other types of models, see Ref. [79].

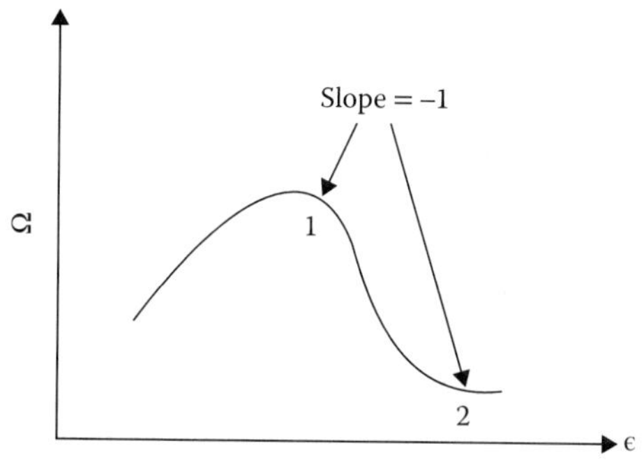

Figure 8.3 A schematic portrayal of the function $\Omega(\varepsilon)$ for cross section combinations which lead to NDC. The NDC region corresponds to mean energies ε lying in the region for which $\Omega'(\varepsilon) < 1$.

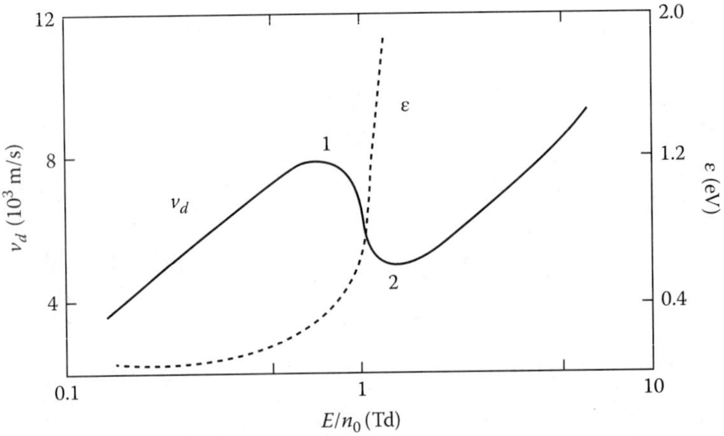

Figure 8.4 Drift velocity and mean energy for the "hat" model (a) of Figure. 8.2 [80].

8.5.3 GERs in the presence of NDC

For light particles $k_B T_\parallel \approx k_B T_\perp \approx \frac{2\varepsilon}{3}$ and the GERs (Equations 7.59 and 7.60) for the transverse and longitudinal diffusion coefficient become

$$\frac{D_\perp}{K} = \frac{2\varepsilon}{3e} \tag{8.45}$$

$$\frac{D_\parallel}{K} = \frac{2\varepsilon}{3e}\left[1 + (1+\Delta)\frac{\mathrm{d}\ln K}{\mathrm{d}\ln E}\right] \tag{8.46}$$

and hence

$$\frac{D_\parallel}{D_\perp} = \left[1 + (1+\Delta)\frac{\mathrm{d}\ln K}{\mathrm{d}\ln E}\right]. \tag{8.47}$$

Since the mobility coefficient K is defined by $v_d = KE$, an equivalent definition for the NDC regime (Equation 8.42) is

$$1 + \frac{\mathrm{d}\ln K}{\mathrm{d}\ln E} < 0.$$

This means that if the heat flux Q and hence Δ were neglected, the right hand side of Equation 8.46 would be negative, implying $D_\parallel < 0$, in violation of the second law of thermodynamics! Thus, it is vital to retain Q in the fluid equations in an NDC regime, since Δ, however small, plays an absolutely crucial role in maintaining the physical integrity of the fluid equations. Thus, although, Equation 8.46 predicts that $D_\parallel$ may indeed become small, it never becomes negative, as long as Δ is retained.

CHAPTER 9

Fluid Modelling with Loss and Creation Processes

9.1 Sources and Sinks of Particles

9.1.1 Non-conservative collisions in gases

We now consider collisions between charged particles and neutral atoms or molecules which do not conserve particle number, that is,

$$\text{Charged particle} + \text{neutral} \rightarrow \text{products}, \tag{9.1}$$

where the products on the right hand side generally include a different type and/or number of particles from the left hand side. Examples were given in Chapter 5. The aim of this chapter is to understand their effects on transport properties using moment equations.

The moment equation for property $\phi(\mathbf{v})$ is modified through the addition of a reactive collision term

$$\left(\frac{\partial \phi}{\partial t}\right)_R \equiv \frac{1}{n}\int \mathrm{d}^3 v\, \phi(\mathbf{v}) \left(\frac{\partial f}{\partial t}\right)_R, \tag{9.2}$$

where $\left(\frac{\partial f}{\partial t}\right)_R$ is the rate of change of the particle distribution function due to collisions of the type (Equation 9.1). The first and most obvious modification is to the equation of continuity (moment equation with $\phi = 1$), which must include a source (sink) term to account for a gain (loss) of particles. Thus

$$\partial_t n + \nabla \cdot \Gamma = S, \tag{9.3}$$

where

$$S = \int \mathrm{d}^3 v \left(\frac{\partial f}{\partial t}\right)_{\text{col}} = \int \mathrm{d}^3 v \left(\frac{\partial f}{\partial t}\right)_R$$

denotes the rate of creation (loss) of particle per unit time and volume. The presence of S modifies the diffusion equation and, as we shall see, results in corrections to the mobility and diffusion coefficients, with important ramifications for interpretation of experiment. Understanding the nature of these *explicit* reactive effects is one of the two main tasks of this chapter.

The other task is to account for the *implicit* effects of reactions on transport properties, arising from the fact that particles lost and/or gained in reactive collisions carry momentum and energy, and hence the momentum and energy balance fluid equations must be modified accordingly through terms like Equation 9.2, with $\phi = m\mathbf{v}$ and $\frac{1}{2}mv^2$, respectively.

A discussion of gaseous discharges, including attachment and ionization phenomena in particular, can be found in Ref. [13].

9.1.2 Non-conservative processes in condensed matter

Although the bulk of the discussion in this chapter is for a gaseous medium, the results carry over to soft-condensed matter, if reactive scattering events like Equation 9.1 can be considered to be incoherent.

The physical picture is quite different for transport processes in amorphous materials. Nevertheless, as remarked in Section 6.2, trapping and de-trapping of charge carriers to and from localized states are mathematically similar to reactions in gases. Thus, the finite duration of a collision or trapping time τ gives the appearance of a loss process, even though particle number is conserved per se. Integrating Equation 9.2 over all velocities gives Equation 9.3, with

$$S = \int d^3v \left(\frac{\partial f}{\partial t}\right)_{\text{col}} = -\nu(1 - \phi(\tau)*)n\,,$$

where $\phi(\tau)$ is the relaxation function, ν is the rate of trapping of free particles, and $*$ denotes a convolution. Straightforward algebraic manipulation then gives,

$$\tilde{\partial}_t n + \nabla \cdot \Gamma = 0\,, \tag{9.4}$$

where

$$\tilde{\partial}_t \equiv \partial_t + \nu(1 - \phi(\tau)*) \tag{9.5}$$

is an operator accounting for loss and creation of particles by trapping and de-trapping. For certain classes of relaxation function, $\tilde{\partial}_t$ takes the form of a fractional time derivative and Equation 9.4 then becomes what is known as the fractional diffusion equation, which forms the basis of so-called "fractional kinetics" [37].

For such materials, there are also "recombination" processes operative where electrons and holes recombine, and the charged particles are lost. These processes can occur in the "conduction" band, as well as in traps, so-called trap-based recombination. The physical mechanism of this recombination can be monomolecular and bimolecular in nature. The reader is referred to Refs. [45,81,82] for further details.

We shall return to this topic in Chapter 14, but for now our attention is confined to reactions in gaseous media.

9.2 Reacting Particle Swarms in Gases

9.2.1 Balance equation including non-conservative collisions

9.2.1.1 Notation

To distinguish between processes, the reactive index of Equation 9.2 is written as $R = *$ for a generic particle loss, and $R = I$ for ionization.

9.2.1.2 Collision term for particle loss

Consider a reactive loss process, like electron attachment in an electronegative gas,

$$e^- + A + M \Longrightarrow A^- + M, \tag{9.6}$$

where M is a "spectator" molecule required in order to conserve momentum and energy. Other examples of reactive loss processes include ion–molecule reactions, positron annihilation and muonic atom formation.

In such cases, the charged particle is lost completely in the collision and, since $n \ll n_0$, subsequent collisions of charged particles with the "products" can be assumed to be negligible. In these circumstances, we therefore need to focus only on the effect of the reactions on the average properties of the original particles.

In a reactive collision like Equation 9.6, only the "scattering out" part of the reactive collision term is relevant in the kinetic equation as explained in Section 5.5. The contribution to the loss term on the right hand side of the corresponding moment equation for some arbitrary function $\phi(\mathbf{v})$ is therefore

$$\begin{aligned}
\left(\frac{\partial \phi}{\partial t}\right)_* &= \sum_{j,j'} \int d^3v \int d^3v_0\, g f(\mathbf{r},\mathbf{v},t) f_{0j}(\mathbf{v}_0) \int d^2\Omega_{\mathbf{g}'}\, \sigma_*(j,j';g,\chi)\phi(\mathbf{v}) \\
&= \int d^3v\, \phi(\mathbf{v}) f(\mathbf{r},\mathbf{v},t) \int d^3v_0\, w(\alpha,\mathbf{v}_0) \sum_{j,j'} n_{0j} g \sigma_*(j,j';g) \\
&= n\langle\langle \phi(\mathbf{v}) \nu_*(\epsilon)\rangle_0\rangle,
\end{aligned} \tag{9.7}$$

where $\sigma_*(j,j';g) = \int d^2\Omega_{\mathbf{g}'}\, \sigma_*(j,j';g,\chi)$ is the integral reactive loss cross section in which the internal state of the neutral changes from $j \to j'$,

$$\nu_*(\epsilon) \equiv \sum_{j,j'} n_{0j} g \sigma_*(j,j';g) \tag{9.8}$$

is the total loss collision frequency for the reaction, and $\epsilon = \frac{1}{2}\mu g^2$ is the energy in the centre-of-mass (CM) frame. In what follows the right hand side of Equation 9.7 is approximated using momentum transfer theory.

9.2.1.3 Collision term for ionization

On the other hand, production of electrons ionization

$$e^- + A \Longrightarrow A^+ + 2e^- \tag{9.9}$$

is somewhat more complicated, because the products, that is, the two electrons emerging from the collision, must be accounted for in subsequent collisions.

While we could take appropriate moments of the ionization collision operator obtained in Section 5.5, it is more straightforward to proceed from first principles [83]. Consider an electron of mass m, velocity $\mathbf{v}$ impacting on a neutral atom or molecule whose ionisation energy is ϵ_I, which results in the release of another electron from the valence shell. Although there are many possible ways in which an ionising event can take place, since the resulting ion can be left in any one of its internal states, we shall, for simplicity, proceed as if there is just one such channel. It is also assumed for the purposes of dealing with ionisation that the gas is cold, that is, the neutrals are stationary, and that any recoil is negligible. Under these conditions, the energy in the CM frame, $\epsilon = \frac{1}{2}mv^2$, arises entirely from the kinetic energy of the electron.

The average rate at which the electrons as a whole lose a property $\phi(\mathbf{v})$ through ionizing collisions is [84]

$$\left(\frac{\partial \phi}{\partial t}\right)_I = n\left\langle \nu_I(\epsilon)\left[\phi(\mathbf{v}) - \phi(\mathbf{v}') - \phi(\mathbf{v}'')\right]\right\rangle, \tag{9.10}$$

where the average is only over $\mathbf{v}$, since the neutrals are assumed stationary. Here $\mathbf{v}'$ and $\mathbf{v}''$ denote velocities of the two electrons emerging after the collision,

$$\nu_I(\epsilon) = n_0\sqrt{\frac{2\epsilon}{m}}\sigma_I(\epsilon) \tag{9.11}$$

is the rate at which ionizing collisions occur, and $\sigma_I(\epsilon)$ is the cross section for ionisation.

9.2.1.4 General moment equation with reactions

The general moment equation, making only the reactive collision terms explicit, is thus

$$\begin{aligned}
&\frac{\partial[n\langle\phi(\mathbf{v})\rangle]}{\partial t} + \nabla\cdot\left[n\langle\mathbf{v}\phi(\mathbf{v})\rangle\right] - n\left\langle \mathbf{a}\cdot\frac{\partial\phi(\mathbf{v})}{\partial\mathbf{v}}\right\rangle \\
&= -\left(\frac{\partial\phi}{\partial t}\right)_* - \left(\frac{\partial\phi}{\partial t}\right)_I \cdots \\
&= -n\langle\langle\phi(\mathbf{v})\nu_*(\epsilon)\rangle_0\rangle - n\left\langle \nu_I(\epsilon)\left[\phi(\mathbf{v}) - \phi(\mathbf{v}') - \phi(\mathbf{v}'')\right]\right\rangle - \ldots
\end{aligned} \tag{9.12}$$

where ... denotes the contributions from elastic and inelastic collision terms, which will be added in due course.

9.2.2 Basic balance equations

- *Equation of continuity*: Setting $\phi = 1$ in Equation 9.12 gives the equation of continuity,

$$\frac{\partial n}{\partial t} + \nabla \cdot n\langle \mathbf{v}\rangle = n\langle \nu_I(\epsilon)\rangle - n\langle \nu_*(\epsilon)\rangle,$$

  where we have written the double average like $\langle\langle \nu_*(\epsilon)\rangle_0\rangle$ simply as an average over CM energy $\langle \nu_*(\epsilon)\rangle$.

- *Momentum balance equation:* Now setting $\phi = m\mathbf{v}$, we observe that since motion of the neutral molecule and recoil of the ion are both assumed negligible, momentum is conserved among electrons, that is, $m\mathbf{v} \approx m\mathbf{v}' + m\mathbf{v}''$ and hence $\phi(\mathbf{v}) - \phi(\mathbf{v}') - \phi(\mathbf{v}'') \approx 0$. Thus, ionisation contributes negligibly, and only the particle loss term contributes to the reactive loss term on the right hand side of Equation 9.12

$$\frac{\partial [nm\langle \mathbf{v}\rangle]}{\partial t} + \nabla \cdot \left[nm\langle \mathbf{v}\mathbf{v}\rangle\right] - n\,\langle \mathbf{a}\rangle = -n\langle\langle m\mathbf{v}\nu_*(\epsilon)\rangle_0\rangle - \ldots .$$

- *Energy balance equation:* On the other hand with $\phi = \frac{1}{2}mv^2$, energy conservation associated with ionisation may be written as $\phi(\mathbf{v}) = \phi(\mathbf{v}') + \phi(\mathbf{v}'') + \epsilon_I$, and therefore, the contribution to the right hand side of Equation 9.12 from ionisation is $-n\langle \nu_I(\epsilon) > \epsilon_I$. The energy balance equation is thus

$$\frac{\partial \left[n\langle \frac{1}{2}mv^2\rangle\right]}{\partial t} + \nabla \cdot \left[n\frac{1}{2}m\langle \mathbf{v}v^2\rangle\right] - n\,\langle \mathbf{a}\cdot\mathbf{v}\rangle$$
$$= -n\left\langle\left\langle \frac{1}{2}mv^2\ \nu_*(\epsilon)\right\rangle_0\right\rangle - n\langle \nu_I(\epsilon)\rangle\epsilon_I + \ldots .$$

9.2.3 Approximation of the reactive terms

9.2.3.1 Average ionization rate

Since the ionization cross section $\sigma_I(\epsilon)$ generally rises sharply from zero at the threshold energy ϵ_I, a smoothing factor should be applied to the averaging process, as in Equation 8.29. Thus, the average ionization rate is

$$\langle \nu_I(\epsilon)\rangle = S(\bar{\epsilon})n_0\sqrt{\frac{2\bar{\epsilon}}{m}}\sigma_I(\bar{\epsilon} + \epsilon_I) \equiv \nu_I(\bar{\epsilon}), \tag{9.13}$$

where $S(\bar{\epsilon}) = \left(1 + \frac{\epsilon_I}{\bar{\epsilon}}\right)\exp\left(-\frac{3\epsilon_I}{2\bar{\epsilon}}\right)$.

9.2.3.2 Average loss rates

If the reactive loss cross sections are smoothly varying, we can apply the approximation of momentum transfer theory and assume a rapidly convergent Taylor expansion [83],

$$\nu_*(\epsilon) = \nu_*(\bar{\epsilon}) + (\epsilon - \bar{\epsilon})\,\nu'_*(\bar{\epsilon}) + \ldots \tag{9.14}$$

about the mean energy $\bar{\epsilon}$ in the CM frame. However, unlike the analysis for particle-conserving collisions, where just the first term (the lowest order of momentum transfer theory) of the expansion for $\nu_m(\epsilon)$ is normally retained, it is *absolutely essential* to retain the second term, involving the derivative $\nu'_*(\bar{\epsilon})$ to produce any effect whatever on transport properties. It is only in the special case where reactions are not energy selective, that is, $\nu'_*(\epsilon) = 0$, that transport properties are unaffected: the overall *number* of swarm particles decays in time at a rate governed by ν_* but *averages*, such as drift velocity and mean energy, remain unaffected by such a special reaction.

Thus, the reactive loss contributions to the equation of continuity, momentum, and energy balance equations are

$$\langle \nu_*(\epsilon) \rangle \approx \nu_*(\bar{\epsilon})$$

$$\langle \langle m\mathbf{v}\,\nu_*(\epsilon) \rangle_0 \rangle \approx \nu_*(\bar{\epsilon}) \langle \langle m\mathbf{v} \rangle_0 \rangle + \nu'_*(\bar{\epsilon}) \langle \langle m\mathbf{v}[\epsilon - \bar{\epsilon}] \rangle_0 \rangle$$

and

$$\left\langle \left\langle \frac{1}{2} m v^2 \nu_*(\epsilon) \right\rangle_0 \right\rangle \approx \nu_*(\bar{\epsilon}) \left\langle \left\langle \frac{1}{2} m v^2 \right\rangle_0 \right\rangle + \nu'_*(\bar{\epsilon}) \left\langle \left\langle \frac{1}{2} m v^2 [\epsilon - \bar{\epsilon}] \right\rangle_0 \right\rangle.$$

The higher order moments, thus introduced, have to be calculated approximately. For example, by assuming a Maxwellian velocity distribution function, it can be shown that

$$\langle \langle m\mathbf{v}[\epsilon - \bar{\epsilon}] \rangle_0 \rangle \approx \xi m \langle \mathbf{v} \rangle \tag{9.15}$$

$$\left\langle \left\langle \frac{1}{2} m v^2 [\epsilon - \bar{\epsilon}] \right\rangle_0 \right\rangle \approx \xi \left[\frac{1}{2} m \langle v^2 \rangle + \frac{1}{2} m \langle \mathbf{v} \rangle^2 \right], \tag{9.16}$$

where

$$\xi \equiv \frac{2}{3} M_0 \left[\frac{1}{2} m \langle v^2 \rangle - \frac{1}{2} m \langle \mathbf{v} \rangle^2 \right]. \tag{9.17}$$

9.2.4 Full set of fluid equations

By adding these reactive terms to the right hand side of the particle-conserving moment equations of Chapters 7 and 8, we get the following set of fluid equations:

- *Equation of continuity*

$$\frac{\partial n}{\partial t} + \nabla \cdot n\langle \mathbf{v} \rangle = n\left(\nu_I - \nu_*\right) \tag{9.18}$$

- *Momentum balance*

$$\frac{\partial [nm\langle \mathbf{v} \rangle]}{\partial t} + \nabla \cdot \left[nm\langle \mathbf{vv} \rangle\right] - n\langle \mathbf{a} \rangle = -n\nu_m \mu \langle \mathbf{v} \rangle - nm\langle \mathbf{v} \rangle \left\{\nu_* + \xi \nu_*'\right\} \tag{9.19}$$

- *Energy balance*

$$\begin{aligned}
&\frac{\partial\left[n\langle \frac{1}{2}mv^2\rangle\right]}{\partial t} + \nabla \cdot \left[n\frac{1}{2}m\langle \mathbf{v}v^2\rangle\right] - n\langle \mathbf{a}\cdot\mathbf{v}\rangle \\
&= -n\frac{2\mu\nu_m}{m+m_0}\left(\frac{1}{2}m\langle v^2\rangle - \frac{3}{2}k_BT_0\right) - nM_0\sum_i\left[\langle \overrightarrow{\nu}_i\rangle - \langle \overleftarrow{\nu}_i\rangle\right]\epsilon_i^* \\
&\quad - n\left\{\nu_*\frac{1}{2}m\langle v^2\rangle + \xi\left[\frac{1}{2}m\langle v^2\rangle + \frac{1}{2}m\langle \mathbf{v}\rangle^2\right]\right\} - n\nu_I\epsilon_I.
\end{aligned} \tag{9.20}$$

The average inelastic and superelastic collision frequencies for process i, $\langle \overrightarrow{\nu}_i\rangle$ and $\langle \overleftarrow{\nu}_i\rangle$ may be represented in terms of the inelastic cross sections $\overrightarrow{\sigma}_i\,(\bar{\epsilon}\,)$ as outlined in Chapter 8.

It is implicit in these equations that ν_m, ν_I and ν_* are all functions of the CM mean energy

$$\bar{\epsilon} = M_0\frac{1}{2}m\langle v^2\rangle + M\frac{3}{2}k_BT_0. \tag{9.21}$$

9.2.5 Closing the moment equations

We now have a set of three general fluid equations for the three physical quantities of interest n, $\langle \mathbf{v}\rangle$ and $\langle \frac{1}{2}mv^2\rangle$, but the left hand side still involves the unknown higher order moments $\langle \mathbf{vv}\rangle$ and $\langle \mathbf{v}v^2\rangle$. We could establish balance equations for the latter, but they, in turn, contain even higher order moments, and so on. That is, the equations are, in general, not closed, as already highlighted in Section 7.1.7.

However, there are special cases where the problem is avoided or reduced to some extent:

1. Since the problem moments $\langle \mathbf{vv} \rangle$ and $\langle \mathbf{v}v^2 \rangle$ are associated with spatial derivatives, they do not appear for spatially uniform systems, and the fluid equations are automatically closed.

2. Similarly, in the weak-gradient hydrodynamic regime, the problem terms can be simplified, since spatial dependence of all moments is projected onto $n(\mathbf{r}, t)$, which is found from solution of the diffusion equation, and hence the fluid equations can also be closed in this case.

3. For light particles $m \ll m_0$, properties in velocity space may be nearly isotropic if elastic processes dominate, and hence it is reasonable to make the approximation $\langle \mathbf{vv} \rangle \approx \frac{1}{3}\langle v^2 \rangle$. An *Ansatz* or postulate is available for particle-conserving cases [43,85] expressing the energy flux $\mathbf{J}_E = n\frac{1}{2}m\langle \mathbf{v}v^2 \rangle$ in terms of n, $\langle \mathbf{v} \rangle$ and $\langle \frac{1}{2}mv^2 \rangle$, thereby closing the equations. However, an Ansatz for light particles including the effects of non-conservative collisions has yet to be developed. The situation is even more complicated for heavier charged particles, for which an Ansatz for each of the moments $\langle \mathbf{vv} \rangle$ and $\langle \mathbf{v}v^2 \rangle$ is required.

9.2.5.1 Electric fields only

In the remainder of this chapter, we consider the case where $B = 0$, so that $\mathbf{a} = q\mathbf{E}/m$ and the field terms simplify accordingly.

9.3 Spatially Homogeneous Systems

9.3.1 Notation

Generally speaking, the mean energy $\bar{\epsilon}$ depends on position; in fact, it is this dependence which leads to the interesting phenomenon of *transport coefficient duality* discussed below. However, for the moment, we consider only spatially homogeneous systems, for which the following notation is adopted

$$\bar{\epsilon} \to \varepsilon; \ \langle \mathbf{v} \rangle \to \mathbf{v}_d^* \tag{9.22}$$

for the mean energy in the CM and average velocity, respectively. Note that $\mathbf{v}_d^*$ is *not* the same as what we normally call the "drift velocity" $\mathbf{v}_d$, unless reactions are absent. The relevant definitions will follow below.

9.3.2 Hot atom chemistry

9.3.2.1 Reaction yield

Consider energetic atoms produced either photochemically or by nuclear processes in a reservoir of neutral molecules. In this case, the particles are uncharged, there is no external forcing term, and the average velocity may be set to zero, $\langle \mathbf{v} \rangle = \mathbf{v}_d^* = 0$. The "hot" atoms are assumed to be distributed uniformly in space, so that all gradient terms vanish. They may be lost in reactions with the background molecules, but there is no production through any process. The relevant balance equations are then the equation of continuity,

$$\frac{\partial n}{\partial t} = -n\nu_*(\varepsilon) \tag{9.23}$$

and the energy balance equation,

$$\frac{\partial \left\langle \frac{1}{2}mv^2 \right\rangle}{\partial t} = -\frac{2\mu}{m+m_0}\nu_m(\varepsilon)\left[\left\langle \frac{1}{2}mv^2 \right\rangle - \frac{3}{2}k_B T_0\right]$$
$$-M_0 \sum_i \left[\langle \overline{\nu}_i \rangle - \langle \overline{\nu}_i \rangle\right] \epsilon_i^* - \nu_*'(\varepsilon)\xi \frac{1}{2} m \langle v^2 \rangle,$$

which follows from Equations 9.18 and 9.20. The latter can be written entirely in terms of the mean energy in the CM (Equation 9.21) as

$$\frac{\partial \varepsilon}{\partial t} + \nu_\epsilon \left(\varepsilon - \frac{3}{2}k_B T_0\right) + M_0^2 \sum_i \left[\langle \overline{\nu}_i \rangle - \langle \overline{\nu}_i \rangle\right] \epsilon_i^*$$
$$= -\frac{2}{3}\nu_*'(\varepsilon)\left(\varepsilon - M\frac{3}{2}k_B T_0\right)^2, \tag{9.24}$$

where $\nu_\epsilon \equiv \frac{2\mu}{m+m_0}\nu_m = 2MM_0\nu_m$ is the collision frequency for energy transfer.

The solution of Equation 9.23 is

$$n(t) = n(0)\exp\left\{-\int_0^t \nu_*[\varepsilon(t')]\,dt'\right\}$$

from which one can obtain the fractional *yield*,

$$Y = \frac{n(0) - n(\infty)}{n(0)} = 1 - \exp\left\{-\int_0^\infty \nu_*[\varepsilon(t')]\,dt'\right\}. \tag{9.25}$$

The right hand side can be evaluated once the mean energy is known as a function of time from the solution of Equation 9.24.

9.3.2.2 Model calculation

Consider now hot atoms undergoing elastic and reactive collisions in a cold gas $T_0 = 0$, characterized by cross sections, σ and σ_*, respectively, as shown in Figure 6.2.

In this case, in the absence of inelastic processes, Equation 9.24 becomes

$$\frac{\partial \varepsilon}{\partial t} + 2MM_0 \nu_m \varepsilon + \frac{2}{3} \nu'_* \varepsilon^2 = 0 \tag{9.26}$$

from which we find $\varepsilon = \varepsilon(t)$ and hence calculate the yield from Equation 9.25.

It is assumed that the atoms start out hot, with an energy $\varepsilon(0) > \varepsilon_2$. Initially, they cool by elastic collisions until the reaction is ignited, and then, as the energy falls further below ε_1, the reaction is quenched, and cooling takes place again by elastic energy loss. If we ignore any smoothing factors associated with the thresholds, and set $\nu_*(\varepsilon) \approx n_0 \sqrt{\frac{2\varepsilon}{\mu}} \sigma_*(\varepsilon)$, then Equation 9.26 can be solved analytically, and the yield calculated from Equation 9.25 as

$$Y = 1 - \left(\frac{\varepsilon_1}{\varepsilon_2} \right)^p \tag{9.27}$$

where

$$p = \frac{3}{1 + 6MM_0 \sigma / \sigma_*} \tag{9.28}$$

Numerical values obtained for $\sigma/\sigma_* = 4$, $\varepsilon_1/\varepsilon_2 = \frac{1}{2}$ are shown in Table 9.1. These results are in accord with the Boltzmann calculations of Knierem et al. [86], to the extent that the yield is a minimum for equal masses, $m/m_0 = 1$, but otherwise the agreement could best be described as qualitative. Applying smoothing factors to the sharp thresholds of the reactive cross section would be expected to significantly improve this agreement, but then no analytic solution would be possible (e.g. even for the model specified in Figure 9.1).

9.3.3 Reactive heating and cooling

9.3.3.1 Particles of arbitrary mass and zero field

In some circumstances, a non-equilibrium *steady energy state* may be reached as $t \to \infty$, in which a balance is achieved between the various

Table 9.1 Model Calculation for the Hot Atom Chemistry Model Fractional Yield

m/m_0	10	5	3	2	1	1/2	1/3	1/5	1/10
Equation 9.27	50.0	38.1	31.5	28.0	25.7	28.0	31.5	38.1	50.0

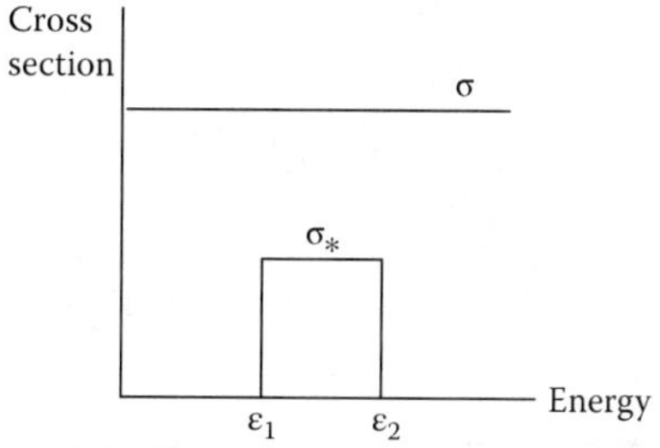

Figure 9.1 Model cross sections [83] for reactive and non-reactive collisions, as a function of CM energy, showing the lower and upper thresholds ε_1 and ε_2 for reactions.

competing collisional processes, such that $\frac{\partial\varepsilon}{\partial t}=0$, and ε =constant. In the absence of inelastic collisions, Equation 9.24 shows that in the steady energy state,

$$\nu_\epsilon(\varepsilon)\left(\varepsilon-\frac{3}{2}k_BT_0\right)+\frac{2}{3}\nu'_*(\varepsilon)\left(\varepsilon-M\frac{3}{2}k_BT_0\right)^2=0. \tag{9.29}$$

First, we observe that if there are no reactive effects, or if reactions are not energy selective, that is, $\nu'_*=0$, then the solution corresponds to thermal equilibrium, $\varepsilon=\frac{3}{2}k_BT_0=\frac{1}{2}m\langle v^2\rangle$, as expected. On the other hand, if $\nu'_*\neq 0$, the situation is quite different, and non-equilibrium steady states may exist where $\varepsilon\neq\frac{3}{2}k_BT_0$.

Consider for simplicity, the situation where ν_ϵ and ν'_* are both constants, so that Equation 9.29 is a quadratic, with two possible roots

$$\varepsilon_\pm=M\frac{3}{2}k_BT_0+\frac{3}{4}\left\{-\frac{\nu_\epsilon}{\nu'_*}\pm\sqrt{\left(\frac{\nu_\epsilon}{\nu'_*}\right)^2+4\frac{\nu_\epsilon}{\nu'_*}M_0k_BT_0}\right\}. \tag{9.30}$$

1. In the case where $\nu'_*>0$, higher energy atoms are lost preferentially, and physical reasoning suggests that on the average, atoms are *cooled* below the thermal equilibrium value. Indeed, there is only one positive root, $\varepsilon_+<\frac{3}{2}k_BT_0$, and it follows from Equation 9.21 that $\frac{1}{2}m\langle v^2\rangle$ is also less than $\frac{3}{2}k_BT_0$. One might refer to this effect as "reaction cooling."

2. On the other hand, if $\nu'_*<0$, there is selective loss of low energy atoms, promoting overall heating of the atoms. Both roots $\varepsilon_\pm$ are positive, but only ε_- corresponds to a stable equilibrium situation [83], with $\varepsilon_->\frac{3}{2}k_BT_0$. It then follows from Equation 9.21 that $\frac{1}{2}m\langle v^2\rangle$ also exceeds $\frac{3}{2}k_BT_0$, and "reaction heating" is said to prevail.

9.3.3.2 Light charged particles in an electric field

Consider now a spatially uniform swarm of electrons, positrons, or muons in a gas, subject to an electric field E. They interact with the gas through elastic collisions and may be lost reactively through attachment, annihilation, or muonic atom formation, respectively. Since $m \ll m_0$, it follows that $M_0 \approx 1$, $M \approx 0$, and one might expect that that $\langle v^2 \rangle$ would be large compared with both $\langle v_0^2 \rangle_0$ and $\langle \mathbf{v} \rangle^2$. Hence by Equations 9.21 and 13.35,

$$\varepsilon \approx \frac{1}{2} m \langle v^2 \rangle$$

$$\xi \approx \frac{2}{3} \varepsilon.$$

If we further assume that the magnitude of reactive collision frequency is sufficiently low that $\xi \nu'_* \ll \nu_m$, then the steady state momentum and energy balance Equations 9.19 and 9.20 simplify to

$$\langle \mathbf{v} \rangle = \mathbf{a}/\nu_m \equiv \mathbf{v}_d^* \tag{9.31}$$

$$\varepsilon = \varepsilon_W - \frac{2}{3} \varepsilon^2 \frac{\nu'_*}{\nu_\epsilon}, \tag{9.32}$$

respectively, where

$$\varepsilon_W = \frac{3}{2} k_B T_0 + \frac{1}{2} m_0 (v_d^*)^2$$

is the Wannier energy relation in the absence of reactions. As in the previous example, we take ν_m and ν'_* to be constants, so that Equation 9.32 is effectively a quadratic in ε with the zeros

$$\varepsilon_\pm = \frac{3}{4} \left\{ -\frac{\nu_\epsilon}{\nu'_*} \pm \sqrt{\left(\frac{\nu_\epsilon}{\nu'_*} \right)^2 + \frac{8}{3} \frac{\nu_\epsilon}{\nu'_*} \varepsilon_W} \right\} \tag{9.33}$$

$$= \frac{3}{4} \frac{\nu_\epsilon}{\nu'_*} \left\{ \pm \sqrt{1 + \frac{8}{3} \frac{\nu'_*}{\nu_\epsilon} \varepsilon_W} - 1 \right\}.$$

For the sake of being definite, we now consider specifically electron attachment, and set $\nu_* = \nu_a$.

1. If the attachment frequency increases with energy, that is, $\nu'_a > 0$, higher energy electrons are lost preferentially, and Equation 9.33 has only one positive root, $\varepsilon_+ < \varepsilon_W$, confirming the expected "attachment cooling."

2. If on the other hand $\nu_a' < 0$, attachment *heating* is expected; indeed, Equation 9.33 indicates that $\varepsilon_- > \varepsilon_W$. (Only the negative root corresponds to a stable steady state.) Note that a physically tenable solution requires the term under the square root to remain positive, that is

$$\left(\frac{\nu_\epsilon}{\nu_a'}\right)^2 > \frac{8}{3}\frac{\nu_\epsilon}{|\nu_a'|}\varepsilon_W.$$

This effectively places an upper limit on ε_W, and hence upon the field strength itself, for the existence of a steady state.

9.3.3.3 Ionization, attachment, and inelastic processes

Consider now electrons in a gas under conditions where both ionization and attachment may be significant, with rates denoted by ν_I and ν_a, respectively. The number density changes in time according to Equation 9.18, but all other quantities are assumed constant in space and time. The drift velocity follows from the small mass limit of Equation 9.19,

$$\mathbf{v}_d^* = \frac{\mathbf{a}}{m(\nu_m + \nu_I + \xi\nu_a')} \approx \frac{\mathbf{a}}{\nu_m(\varepsilon)},$$

where the last step follows if ν_I and $\xi\nu_a'$ are small compared with ν_m. Under the same conditions, the mean energy follows from Equation 9.20

$$\varepsilon = \frac{\nu_e\varepsilon_W - \left[\epsilon_I\nu_I + \sum_i \left(\langle\overrightarrow{\nu}_i\rangle - \langle\overleftarrow{\nu}_i\rangle\right)\,\epsilon_i^*\right]}{\nu_e + \nu_I + \frac{2}{3}\varepsilon\nu_a'}. \qquad (9.34)$$

Note that ν_I and $\xi\nu_a'$ may be small compared with ν_m, but cannot be neglected in comparison with the collision frequency for energy transfer $\nu_\epsilon = 2m\nu_m/m_0$, which is itself small compared with ν_m. This expression encapsulates several key points regarding ionization and attachment.

First, the *denominator* shows that while attachment tends to either raise or lower ε, depending on whether ν_a' is negative or positive, respectively, ionization *always* produces a cooling effect, since it is ν_I which appears, and not its derivative. That is, unlike the energy selective nature of attachment, which leads to cooling or heating, ionization is always accompanied by a *dilution* of energy, since there is a fixed amount of energy to be spread over an increased number of electrons, independently of the nature of the energy dependence of the ionization cross section σ_I.

Second, it is apparent from the term $\epsilon_I\nu_I$ in the *numerator* of Equation 9.34 that ionization contributes to energy loss in the same way as an inelastic process, in which an electron gives up energy ϵ_I at a rate ν_I. However, it is incorrect to consider this as the *only* effect, and to ignore energy dilution, as discussed above. The cross section model of Lucas and Saelee [87] has been constructed specifically to demonstrate this point: the inelastic collision frequency and ν_I both vary, but combine in such a way to make the numerator of Equation 9.34 constant. However, since the denominator changes as ν_I varies, so does the mean energy, and likewise other transport properties. In summary, the popular assumption that ionization can be treated as just another inelastic process is qualitatively and quantitatively incorrect.

9.4 Reactive Effects and Spatial Variation

9.4.1 Hydrodynamic regime

Using Equation 9.18 to eliminate $\frac{\partial n}{\partial t}$ from Equation 9.19 gives the equivalent form of the momentum balance equation,

$$\delta_{\mathbf{v}} + ne\mathbf{E} - k_B\mathsf{T}\cdot\nabla n = -n\mu\langle\mathbf{v}\rangle\widehat{\nu}_m(\overline{\epsilon}) \tag{9.35}$$

in the absence of ionization and inelastic processes, where

$$\widehat{\nu}_m \equiv \nu_m\left[1 + \frac{\nu_I + \xi\nu'_*}{M_0\nu_m}\right] \tag{9.36}$$

is a reaction-modifed momentum transfer collision frequency, $k_B\mathsf{T}$ is the temperature tensor and $\delta_{\mathbf{v}} \equiv nm\frac{\partial\langle\mathbf{v}\rangle}{\partial t} + n\nabla\cdot k_B\mathsf{T}$.

In the *spatially homogeneous* case, where both ∇n and $\delta_{\mathbf{v}}$ vanish, Equation 9.35 becomes

$$ne\mathbf{E} = n\mu\widehat{\nu}_m(\varepsilon)\mathbf{v}_d^*(\varepsilon), \tag{9.37}$$

where, as in Equation 9.22, average velocity and mean CM energy are written as $\mathbf{v}_d^*$ and ε, respectively.

In the *weak-gradient hydrodynamic regime*, ∇n is a small quantity, and the equations are linearized accordingly. Since $\delta_{\mathbf{v}}$ is of second order, the momentum balance equation (Equation 9.35) to first order in ∇n is

$$ne\mathbf{E} - k_B\mathsf{T}\cdot\nabla n = n\mu\widehat{\nu}_m(\overline{\epsilon})\langle\mathbf{v}\rangle, \tag{9.38}$$

which is of exactly the same mathematical form as the non-reactive case discussed in Chapters 7 and 8, the only difference being that $\widehat{\nu}_m$ replaces ν_m. Similarly, the general energy balance equation (Equation 9.20) reduces to the same mathematical form as for the non-reactive case, to both zero and first order in ∇n, with suitably re-defined collision frequency.

It is clear that Equation 9.38 differs from Equation 9.37 only in that $\mathbf{E}$ is replaced by $\mathbf{E} - \frac{k_B\mathsf{T}}{ne}\cdot\nabla n$. Thus, if $\mathbf{v}_d^*(\mathbf{E})$ is solution of Equation 9.37, then the solution of Equation 9.38 is $\langle \mathbf{v} \rangle = \mathbf{v}_d^*\left(\mathbf{E} - \frac{k_B\mathsf{T}}{ne}\cdot\nabla n\right)$. Similarly, we deduce that the spatial varying energy is $\bar{\epsilon} = \varepsilon\left(\mathbf{E} - \frac{k_B\mathsf{T}}{ne}\cdot\nabla n\right)$ where $\varepsilon(\mathbf{E})$ is the spatially homogeneous energy. Both expressions can be expanded as a series in ∇n, and thus

$$\langle \mathbf{v} \rangle = \mathbf{v}_d^*\left(\mathbf{E} - \frac{k_B\mathsf{T}}{ne}\cdot\nabla n\right) \approx \mathbf{v}_d^*(\mathbf{E}) - \frac{1}{n}\mathsf{D}^*\cdot\nabla n$$

to first order in ∇n, or equivalently

$$\mathbf{\Gamma} = n\langle \mathbf{v} \rangle = n\,\mathbf{v}_d^* - \mathsf{D}^*\cdot\nabla n, \tag{9.39}$$

which is Fick's law, where

$$\mathsf{D}^* = \frac{k_B\mathsf{T}}{e}\cdot\frac{\partial \mathbf{v}_d^*}{\partial \mathbf{E}}$$

is a second rank diffusion tensor. This has longitudinal and transverse components given by the familiar generalized Einstein relations

$$D_{\parallel}^* = \frac{k_B T_{\parallel}}{e}\left[K^*\left(1 + \frac{\partial \ln K^*}{\partial \ln E}\right)\right];\quad D_{\perp}^* = \frac{k_B T_{\perp}}{e}K^*. \tag{9.40}$$

To this point, the relations are *formally* the same as for the non-reactive case, although explicit expressions for the spatially homogeneous quantities $K^* = v_d^*/E$, ε, $k_B T_{\parallel}$ and $k_B T_{\perp}$ are of course different.

9.4.2 Diffusion equation and the two types of transport coefficients

9.4.2.1 The two types of transport coefficients

If Equation 9.39 is substituted into the equation of continuity (Equation 9.18), there results

$$\frac{\partial n}{\partial t} + \mathbf{v}_d^*\cdot\nabla n - \mathsf{D}^* : \nabla\nabla n = S(\mathbf{r}, t), \tag{9.41}$$

where $S(\mathbf{r}, t) = n[\nu_I(\bar{\epsilon}) - \nu_*(\bar{\epsilon})]$ is the net particle gain rate per unit volume. Since the mean energy $\bar{\epsilon}(\mathbf{r}, t)$ is space-time dependent, so is the "source" term, that is, $S(\mathbf{r}, t) = S[\bar{\epsilon}(\mathbf{r}, t)]$ and, like all quantities in the hydrodynamic regime, can be represented in terms of a density gradient expansion. For consistency with the left hand side of Equation 9.41, however, the expansion must be taken to *second order* in ∇n, that is,

$$S(\mathbf{r}, t) = n(\mathbf{r}, t)\, S^{(0)} + \mathbf{S}^{(1)}\cdot\nabla n(\mathbf{r}, t) + \mathsf{S}^{(2)} : \nabla\nabla n(\mathbf{r}, t) + \ldots, \tag{9.42}$$

where $S^{(0)}$, $\mathbf{S}^{(1)}$, and $\mathsf{S}^{(2)}$ are tensors of rank 0 (a scalar), 1 (a vector) and 2, respectively, and dots denote appropriate contraction over tensor indices. While it is clear that

$$S^{(0)} = \nu_I(\varepsilon) - \nu_*(\varepsilon) \tag{9.43}$$

$\mathbf{S}^{(1)}$ and $\mathsf{S}^{(2)}$ arise from the energy selective character of reactive processes, that is, from $\nu_I'(\varepsilon)$ and $\nu_*'(\varepsilon)$. The details can be found in Ref. [83].

Substituting Equation 9.42 into 9.41 and rearranging gives the diffusion equation

$$\frac{\partial n}{\partial t} + \mathbf{v}_d \cdot \nabla n - \mathsf{D} : \nabla\nabla n = n\,\left[\nu_I(\varepsilon) - \nu_*(\varepsilon)\right], \tag{9.44}$$

where

$$\begin{aligned} \mathbf{v}_d &\equiv \mathbf{v}_d^* - \mathbf{S}^{(1)} \\ \mathsf{D} &\equiv \mathsf{D}^* + \mathsf{S}^{(2)}. \end{aligned} \tag{9.45}$$

Swarm experiments are traditionally analyzed by solving the diffusion Equation 9.44 for $n(\mathbf{r}, t)$, which gives the distribution of particles throughout the bulk of the medium. The quantities $\mathbf{v}_d$ and D defined by Equation 9.45 are extracted from the experimental data. These quantities are referred to as "bulk" transport coefficients. The flux-gradient relation (Equation 9.39), called Fick's law is *not* in general, used to analyze experiment, and hence the "flux" quantities $\mathbf{v}_d^*$ and D^* are not generally measured. It is bulk transport data, $\mathbf{v}_d$ and D, *not* the flux properties, which are generally tabulated in the literature. The two sets of transport properties converge only when reactive effects are either negligible or not energy selective.

Before exploring detailed difference, we review solution of the diffusion equation.

9.4.2.2 Solution of the diffusion equation

If, for simplicity, we assume plane-parallel geometry, so that particle properties vary only in one direction, say along the z-axis, and that the only non-conservative process taking place is ionization, then Equation 9.44 simplifies to

$$\frac{\partial n}{\partial t} + v_d \frac{\partial n}{\partial z} - D_{\|} \frac{\partial^2 n}{\partial z^2} = \nu_I\, n. \tag{9.46}$$

This is solved for $n(z, t)$ with initial and boundary conditions appropriate to the experiment under consideration. Detailed solutions are left as an exercise, and in what follows we highlight only a few of the most important results.

In an idealized time-of-flight (ToF) experiment, for example, in which a pulse of N_0 particles is released from a plane source at $z=0$ at time $t=0$ into an unbounded medium, the initial and boundary conditions are

$$n(z,0)=N_0\,\delta(z)$$
$$n(z,t)=0 \quad (z\to\pm\infty, \quad t>0),$$

respectively, and the solution is

$$n(z,t)=\frac{N_0}{\sqrt{4\pi D_{\|}t}}\exp\left\{\nu_I\,t-\frac{[z-z_0-v_d\,t]^2}{4D_{\|}t}\right\}. \tag{9.47}$$

This corresponds to a travelling pulse of particles (see Figure 1.3) whose CM travels with velocity $\mathbf{v}_d$, and whose width increases at a rate determined $D_{\|}$. For that reason, these quantities are sometimes called the "time-of-flight" transport coefficients, but in fact they are defined without reference to any particular experiment and are fundamental, universal properties of the particles and the medium in which they move. The corresponding flux transport properties, v_d^* and $D_{\|}^*$, while also fundamental properties, are not normally measured, however. In the absence of non-conservative processes, that is, $\nu_I\to 0$, the two sets of transport coefficients are the same, $v_d^*\to v_d$ and $D_{\|}^*\to D_{\|}$.

In the steady state Townsend (SST) experiment, the source emits particles continuously into the medium, and a steady state is eventually attained in which $\partial n/\partial t=0$. The corresponding diffusion equation,

$$v_d\frac{\partial n}{\partial z}-D_{\|}\frac{\partial^2 n}{\partial z^2}=\nu_I\,n \tag{9.48}$$

is traditionally solved by assuming a solution of the form

$$n(z)\sim\exp(\alpha_T\,z) \tag{9.49}$$

where α_T is a constant. Physically, this corresponds to an avalanche of electrons, growing in number the further they are accelerated by the field away from the source. The two possible values of α_T are found by substituting Equation 9.49 into 9.48

$$\alpha_T=\frac{v_d}{2D_{\|}}\left\{1\pm\sqrt{1-\frac{4\nu_I D_{\|}}{v_d^2}}\right\}. \tag{9.50}$$

Only the negative solution is physically tenable, and

$$\alpha_T = \frac{v_d}{2D_\parallel}\left\{1 - \sqrt{1 - \frac{4\nu_I D_\parallel}{v_d^2}}\right\}$$
$$\approx \frac{\nu_I}{v_d} \tag{9.51}$$

if the effects of diffusion are sufficiently small. The quantity α_T is called the *Townsend ionization coefficient*.

It should be noted that Equation 9.49, with α_T given by Equation 9.51, is actually only an approximate, asymptotic representation of the density, far from the source. The SST experiment is inherently non-hydrodynamic, and basic assumptions underpinning the validity of the diffusion equation are invalid. The problem with the diffusion equation is most clearly illustrated in the absence of non-conservative collisions (set $\nu_I = 0$ in the above expressions) when there is no physical solution anywhere. Details are left as an exercise.

The density profile for the SST experiment can be obtained only by solving the Boltzmann equation in phase space, without making the hydrodynamic assumption or density gradient expansion. Details are discussed in Chapter 16 in terms of a general eigenvalue problem.

The lesson here is that hydrodynamic transport theory has its limitations, and that one should be cautious in applying the diffusion equation to any particular problem.

9.4.3 Light particles

9.4.3.1 Bulk versus flux drift velocity

For light particles, the temperature tensor is approximately diagonal, that is, $k_B \mathsf{T} \approx \frac{2}{3}\varepsilon\ \mathsf{I}$, and hence $\mathbf{E} - \frac{k_B \mathsf{T}}{ne}\cdot\nabla n \approx \mathbf{E} - \frac{2\varepsilon}{3ne}\nabla n$. Thus, the space-time dependent mean energy can be represented as $\bar{\epsilon} = \varepsilon\left(\mathbf{E} - \frac{2\varepsilon}{3ne}\nabla n\right)$ and hence to *first order* in ∇n

$$\nu_*(\bar{\epsilon}) = \nu_*\left[\varepsilon\left(\mathbf{E} - \frac{2\varepsilon}{3ne}\nabla n\right)\right] \approx \nu_*(\varepsilon) - \frac{2\varepsilon}{3ne}\frac{\partial \nu_*(\varepsilon)}{\partial \mathbf{E}}\cdot\nabla n \tag{9.52}$$

and similarly for $\nu_I(\bar{\epsilon})$. Thus, comparing with the corresponding first order term in Equation 9.42, it is clear that

$$\mathbf{S}^{(1)} = \frac{2\varepsilon}{3e}\frac{\partial}{\partial \mathbf{E}}(\nu_* - \nu_I) = \frac{2\varepsilon}{3e}\frac{\partial \varepsilon}{\partial E}\frac{\mathbf{E}}{E}(\nu_*' - \nu_I').$$

Hence by Equation 9.45, the bulk and flux drift velocities are related by

$$v_d = v_d^* - \frac{2\varepsilon}{3e}\frac{\partial \varepsilon}{\partial E}(\nu_*' - \nu_I'). \tag{9.53}$$

Since in general $\frac{\partial \varepsilon}{\partial E} > 0$, the difference between bulk and flux drift velocities depends on the sign and magnitude of $\nu'_* - \nu'_I$. Clearly, if the net reaction rate decreases with energy, $\nu'_* - \nu'_I < 0$, the bulk drift velocity exceeds the flux drift velocity, and conversely. The physical basis for the difference between v_d and v_d^* can best be in terms of the ToF experiment. Note that for these explicit reactive corrections, ionization and attachment simply act in opposite directions. Thus, if ν_* and ν_I both increase or decrease with energy, ionization creates particles at the leading edge of the pulse in a ToF experiment (Chapter 1), while the latter removes them.

Corrections can also be calculated for the diffusion tensor by evaluating $\mathsf{S}^{(2)}$ in the manner discussed Section V of Ref. [83].

9.4.3.2 An attachment model

Consider a model of electrons in an attaching gas, where both ν_m and ν'_* are constants, and assume that $\nu'_* > 0$. Then $v_d^* = \frac{eE}{m\nu_m}$ is proportional to E, and by Equation 9.33 the mean energy is

$$\varepsilon = \frac{3}{4}\frac{\nu_\epsilon}{\nu'_*}\left\{\sqrt{1 + \frac{8}{3}\frac{\nu'_*}{\nu_\epsilon}\varepsilon_W} - 1\right\}, \tag{9.54}$$

where $\varepsilon_W = \frac{3}{2}k_B T_0 + \frac{1}{2}m_0 (v_d^*)^2$. Substitution in Equation 9.53 and setting $\nu_I = 0$, gives the bulk drift velocity

$$\begin{aligned} v_d &= v_d^* - \frac{2\varepsilon}{3e}\frac{\partial \varepsilon}{\partial E}\nu'_* \\ &= \frac{v_d^*}{\sqrt{1 + \frac{8}{3}\frac{\nu'_*}{\nu_\epsilon}\varepsilon_W}}, \end{aligned} \tag{9.55}$$

where the last step follows after a little algebra.

Clearly, $v_d < v_d^*$, a fact that can be best understood physically in the context of a travelling pulse in the ToF experiment: the more energetic electrons at the leading edge of the pulse are preferentially lost by attachment, effectively retarding the motion of the centroid, and reducing v_d.

These results are now illustrated using a numerical calculation. Consider electrons in an oxygen-like gas with $m_0 = 16$ amu. at temperature $T_0 = 293$ K and subject to a reduced field $E/n_0 = 0.4$ Td. Interaction is governed by model cross sections $\sigma_m(\epsilon) = 10\,\epsilon^{-\frac{1}{2}}\,\text{Å}^2$; $\sigma_* = b\,\epsilon^{\frac{1}{2}}\,\text{Å}^2$ (energy ϵ in units of eV) and b is an adjustable parameter representing the strength of attachment. Substitution in Equation 9.55 gives

$$v_d = \frac{v_d^*}{\sqrt{1 + 6 \times 10^2 b}}.$$

Table 9.2 The Effect of Varying the Strength b of the Attachment Cross Section on Transport Properties of Electrons in a Model Attaching Gas.

$b(\text{Å}^2\,\text{eV}^{-\frac{1}{2}})$	ε(eV)	v_d (10^3 m/s)	v_d^* (10^3 m/s)
0	0.155	1.186	1.186
10^{-4}	0.152	1.15	1.186
10^{-3}	0.136	0.938	1.186
10^{-2}	0.085	0.448	1.186

Table 9.2 shows that as b increases, and higher energy electrons are lost preferentially at a greater rate, the bulk drift velocity decreases, as does the mean energy (attachment cooling). The flux drift velocity v_d^* remains unchanged, however, since ν_m is constant in this model.

CHAPTER 10

Fluid Modelling in Condensed Matter

10.1 Introduction

In Sections 6.4.2 and 6.4.4, we considered charged particle transport in soft-condensed structured matter. We discussed coherent scattering processes operative in such materials, and then determined a collision operator that accounted for the structure of the soft-condensed matter and the nature of the interaction of the charged particle with the individual atoms/molecules within the soft-condensed matter. In this chapter, we develop a set of moment equations that account explicitly for coherent and incoherent scattering. We then solve these moment equations to generalise the well-known relations for gaseous systems and obtain, for example, structure-modified mobility and Wannier energy relations and structure-induced anisotropic diffusion and negative differential conductivity (NDC).

10.2 Moment Equations Including Coherent and Incoherent Scattering Processes

10.2.1 Basic fluid equations

As in the previous chapters, the set of moment/balance equations can be found by multiplying Boltzmann's equation with the collision operators by various functions $\phi(\mathbf{v})$ and integrating over all velocities. For simplicity, we focus on the case where the external force derives from an electric field only. When coherent scattering processes are operative, we have essentially performed this operation in the development of the Legendre projections of the coherent collision operator in Section 6.4.2. Utilising these results, and invoking momentum transfer theory as detailed in Chapters 6–8, we find that the particle, momentum, and energy balance equations for light particles take the following form when coherent scattering effects are considered:

$$\frac{\partial n}{\partial t} + \nabla \cdot n\mathbf{v} = 0\,, \tag{10.1}$$

$$\frac{\partial}{\partial t}(mn\mathbf{v}) + \nabla \cdot \big(mn\langle \mathbf{v}\mathbf{v}\rangle\big) - neE = -mn\mathbf{v}\widetilde{\nu}_m\,, \tag{10.2}$$

$$\frac{\partial\left[n\left\langle\frac{1}{2}mv^2\right\rangle\right]}{\partial t}+\nabla\cdot(n\mathbf{J}_E)-ne\boldsymbol{E}\cdot\mathbf{v}=-n\nu_e\left[\left(\langle\frac{1}{2}mv^2\rangle-\frac{3}{2}k_BT_0\right)+\Omega(\langle\epsilon\rangle)\right], \tag{10.3}$$

where ϵ is the energy in the centre-of-mass frame which is approximately $\frac{1}{2}mv^2$ for light particles, $\mathbf{J}_E=\langle\frac{1}{2}mv^2\mathbf{v}\rangle$ is the average electron energy flux and m_0 is the mass of background medium molecules. It has also been assumed that, for light particles, random motion dominates directed motion, that is, $\langle v^2\rangle \geq \langle\mathbf{v}\rangle^2$. The average momentum and energy transfer collision frequencies take the functional form:

$$\nu_m(\epsilon)=n_0\sqrt{\frac{2\epsilon}{m}}\sigma_m(\epsilon)+\sum_{\substack{j,j'\\ j\neq j'}} n_{0j}\sqrt{\frac{2\epsilon}{m}}\sigma_m(j,j';\epsilon)$$

$$\nu_e(\epsilon)=\frac{2m}{m_0}\nu_m(\epsilon)$$

$$\widetilde{\nu}_m(\epsilon)=n_0\sqrt{\frac{2\epsilon}{m}}\Sigma_m(\epsilon)+\sum_{\substack{j,j'\\ j\neq j'}} n_{0j}\sqrt{\frac{2\epsilon}{m}}\sigma_m(j,j';\epsilon)$$

represent collision frequencies for the phase momentum and energy transfers without structure effects and structured–modified momentum transfer, respectively. In the limiting case of isotropic scattering (with cross section σ_0), the structure-modified "momentum transfer cross section" takes on the form:

$$\Sigma_m=\sigma_0(s_0-s_1), \tag{10.4}$$

where

$$s_l(v)=\frac{1}{2}\int_{-1}^{1} S\left(\frac{2mv}{\hbar}\sin\left(\frac{\chi}{2}\right)\right)P_l(\cos\chi)\mathrm{d}(\cos\chi). \tag{10.5}$$

10.2.2 Structure-modified momentum transfer collision frequency

As an example of how coherent scattering from a structured system modified the momentum transfer collision frequency, we consider the simple analytic model of Percus and Yevick (with the Verlet–Weiss correction) [88,89]. The Percus–Yevick structure factor has been shown to approximate well the structure observed in atomic liquids:

$$S(\Delta k)=\left(1+\frac{24\eta}{\Delta k^2}\left[\frac{2}{\Delta k^2}(\frac{12\zeta}{\Delta k^2}-\beta)+\frac{\sin(\Delta k)}{\Delta k}\left(\alpha+2\beta+4\zeta-\frac{24\zeta}{\Delta k^2}\right)\right.\right.$$
$$\left.\left.+\cos(\Delta k)\left(\frac{2}{\Delta k^2}\left(\beta+6\zeta-\frac{12\zeta}{\Delta k^2}\right)-\alpha-\beta-\zeta\right)\right]\right)^{-1}, \tag{10.6}$$

where $\eta = \Phi - \frac{\Phi^2}{16}$, $\alpha = \frac{(1+2\eta)^2}{(1-\eta)^4}$, $\beta = \frac{-6\eta\left(1+\frac{\eta}{2}\right)^2}{(1-\eta)^4}$ and $\zeta = \frac{\eta\alpha}{2}$. This model considers a system of hard-spherical particles, where the volume fraction, Φ is used as a measure of how tightly the hard spheres are packed in the materials. It can be written in terms of the hard-sphere radius r and the neutral number density n_0 as $\Phi = \frac{4}{3}\pi r^3 n_0$. This structure factor depends only on the magnitude of the momentum exchange during a collision.

We have modeled systems with a range of densities, from $\Phi \approx 0$, which approximates a dilute gas, to a strongly structured system with $\Phi = 0.4$, which states that 40% of the volume is excluded by the hard-sphere potentials of the neutral molecules. Figure 10.1 shows the static structure value, $S(K)$, for different values of Φ. While there is a highly oscillatory nature of the structure factor for low K, in the limit $K \to \infty$, we find $S \to 1$, and the dilute gas case is regained [90].

If we assume a simple system where the elastic collision scattering cross section is constant and isotropic ($\sigma_m = 6$ Å^2 with neutral mass of 4 amu and under cold gas conditions), then the dilute gas and dense gas momentum transfer collision frequencies for various volume fractions are shown in Figure 10.2. For low energies, there is a significant

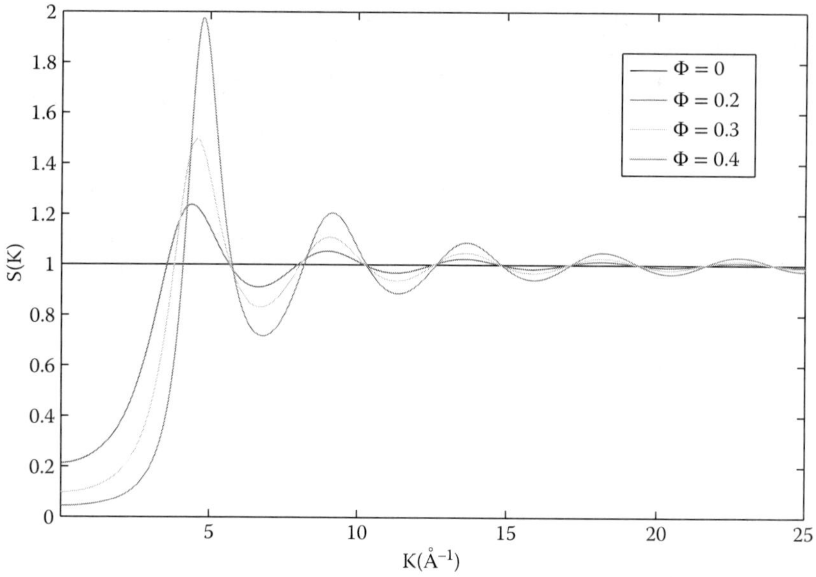

Figure 10.1 The variation of the static structure factor with momentum exchange K for the Percus–Yevick model (with Verlet–Weiss correction) for various volume fractions Φ. (From G. J. Boyle et al., *New Journal of Physics*, 2012 © IOP Publishing and Deutsche Physikalische Gesellschaft. CC BY-NC-S.)

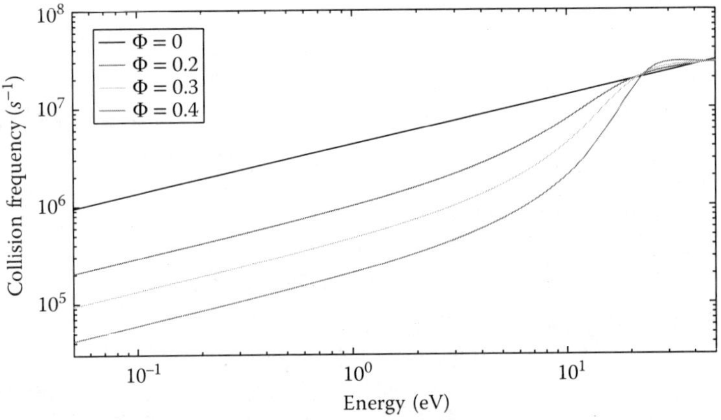

Figure 10.2 The energy variation of elastic collision frequency for various volume fractions Φ for a dense gas of hard spheres. (From G. J. Boyle et al., *New Journal of Physics*, 2012 © IOP Publishing and Deutsche Physikalische Gesellschaft. CC BY-NC-S.)

reduction in the momentum transfer collision frequency arising from coherent scattering. This effect is enhanced as the volume fraction is increased. As the energy is increased, however, the de Broglie wavelength decreases, and hence the effects of coherent scattering are reduced. It then follows that the structure-modified profiles converge on the dilute gas phase profile in this high energy limit as seen in Figure 10.2.

10.3 Structure-Modified Empirical Relationships

Using the techniques implemented in Chapters 7 and 8, the above set of moment equations yield in the steady-state hydrodynamic regime for light particles ($m \ll m_0$), the following hierarchy of coupled equations:

$$\mathbf{v_d} = \frac{q\mathbf{E}}{m\widetilde{\nu}_m}, \tag{10.7}$$

$$\varepsilon = \frac{3}{2}k_B T_0 + \frac{1}{2}m_0 v_d^2 \frac{\widetilde{\nu}_m}{\nu_m} - \Omega \tag{10.8}$$

$$\widetilde{\nu}_m \mathbf{D} = \frac{d\widetilde{\nu}_m}{d\varepsilon}\mathbf{v_d}\gamma + \frac{2}{3}\frac{\varepsilon}{m}\mathbf{I} \tag{10.9}$$

$$\left[1 + \frac{\nu_m'}{\nu_m}\left(\varepsilon - \frac{3}{2}k_B T_0 - \frac{1}{2}m_0 v_d^2 \frac{\widetilde{\nu}_m'}{\nu_m'}\right) + \Omega'\right]\gamma = -m_0 \frac{\widetilde{\nu}_m}{\nu_m}\mathbf{v}_d \cdot \mathbf{D}. \tag{10.10}$$

As we have detailed previously, Equations 10.7 and 10.8 represent coupled non-linear differential equations for the drift velocity and mean energy, which serve as inputs into the linear coupled differential Equations 10.9 and 10.10 for the diffusion coefficients and gradient energy parameter. The heat flux, **Q**, is found from higher order moments, and so an assumption needs to be made to achieve closure. Here, we neglect **Q**, which is often (but not always) a reasonable assumption under hydrodynamic conditions. From this system of equations, further structure-modified generalizations of well-known dilute gas phase results as NDC, generalized Einstein relations (GERs), and others can be made, as detailed below.

10.3.1 Mobility and Wannier energy relations

Equation 10.7 enables us to define a mobility that accounts for coherent scattering in structured systems. Likewise, Equation 10.8 represents a generalization of the Wannier energy relation for dilute gases (Equation 7.32) for such systems.

In Figures 10.3 and 10.4, the variation of swarm mean energy and drift velocity, respectively, with reduced electric field E/n_0 for the above model is displayed for our simple Percus–Yevick model detailed above. The results are compared for various packing volumes Φ and in the limit of the dilute gas phase $\Phi \approx 0$. In the dilute gas case, the mean energy exhibits the expected linear dependence upon field. In the low-field regime, the mean energy increases with increasing volume fraction. The reduction in

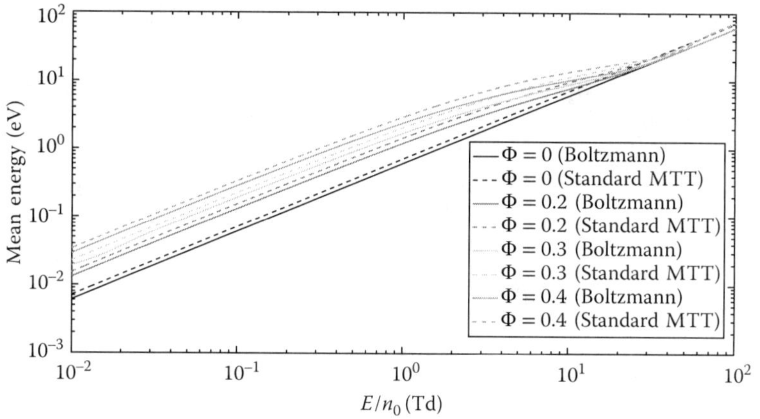

Figure 10.3 Variation of the swarm mean energy with reduced electric field for various volume fractions Φ for a dense gas of hard spheres. Fluid results are compared with those from a multi-term Boltzmann equation solution [91]. (From G. J. Boyle et al., *New Journal of Physics*, 2012 © IOP Publishing and Deutsche Physikalische Gesellschaft. CC BY-NC-S.)

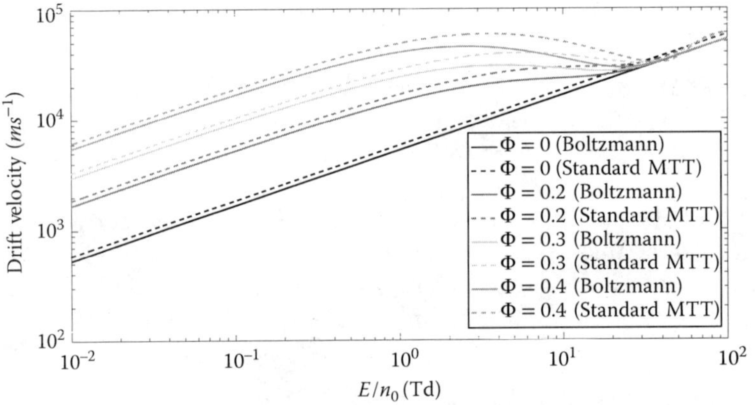

Figure 10.4 Variation of the swarm flux drift velocity with reduced electric field for various volume fractions Φ for a dense gas of hard spheres. Fluid results are compared with those from a multi-term Boltzmann equation solution [91]. (From G. J. Boyle et al., *New Journal of Physics*, 2012 © IOP Publishing and Deutsche Physikalische Gesellschaft. CC BY-NC-S.)

the structure-modified momentum transfer cross section with increasing volume fraction for low energies reduces the randomizing nature of collisions and allows the electric field to pump energy into the swarm efficiently. The coalescence of the various profiles at high fields and hence energies is a reflection of the decrease in the de Broglie wavelength and the suppression of the coherent scattering effects.

In Figure 10.4, the impact of volume fraction on the field dependence of the drift velocity is displayed. At low fields, the impact of the reduction in the randomizing nature of the interactions with Φ is evidenced by the increased directed motion and hence drift velocity. Importantly, we notice, that for volume fractions above a critical value, NDC exists, that is, the reduction in the drift velocity with increasing electric field. This has been discussed previously in Chapter 8 for gaseous systems, where it was shown that certain combinations of elastic and inelastic cross sections can induce NDC and that a non-zero heat flux $\mathbf{Q}$ is important. For a structured media, however, there is a new type of NDC which arises purely as a consequence of coherent scattering from the medium structure. In what follows, we outline the necessary criteria for its existence.

10.3.1.1 Structure-induced NDC

For simplicity, let us assume elastic collisions only and take a cold gas approximation, $T = 0$K. The spatially averaged drift velocity and energy balance Equations 10.7 and 10.8, respectively become

$$v_d = \frac{qE}{m\widetilde{\nu}_m}, \tag{10.11}$$

$$\varepsilon = \frac{1}{2} m_0 v_d^2 \frac{\widetilde{\nu}_m}{\nu_m}. \tag{10.12}$$

From these relations, it can be shown that

$$\frac{d\varepsilon}{d\ln E}\left[1 - \frac{d\ln\left(\frac{\widetilde{\nu}_m}{\nu_m}\right)}{d\ln\varepsilon}\right] = m_0 v_d \frac{\widetilde{\nu}_m}{\nu_m}\frac{d\,v_d}{d\ln E}. \tag{10.13}$$

It follows that NDC arises when the following condition must be met,

$$\frac{d\ln\left(\frac{\widetilde{\nu}_m}{\nu_m}\right)}{d\ln\varepsilon} > 1. \tag{10.14}$$

For isotropic scattering, the condition is given by

$$\frac{d\ln(s_0 - s_1)}{d\ln\varepsilon} > 1,$$

and NDC is induced solely by the structured nature of the material. This condition has no dependence on the interaction between the charged particle and the medium.

Figure 10.5 shows a log-log plot of energy versus $s = s_0 - s_1$ for our Percus–Yevick model, as defined in Equation 10.14, superimposed with straight lines of slope one. It is evident that there are energies for which only the $\Phi = 0.3$ and $\Phi = 0.4$ profiles exhibit a slope of $\log s / \log\varepsilon$ which exceeds one. This coincides with the occurrence of NDC in Figure 10.4 for $\Phi = 0.3$ and $\Phi = 0.4$ and not in the smaller volume fraction profiles.

So what is the origin of this type of NDC? In Figure 10.2, we see that in certain energy regions an increase in energy leads to a sharp increase in the structure-modified momentum-transfer collision frequency $\widetilde{\nu}_m$ (a reflection of sharp increases in the structure factor). Hence, a small increase in the field E in this region is accompanied by a rapid increase in $\widetilde{\nu}_m$ resulting in an overall decrease in the drift velocity. Like the NDC criterion discussed for dilute gases, it is the relative importance of momentum and energy exchange frequencies which control this phenomenon.

10.3.2 Structure-modified GERs

As discussed in Chapter 7 for gaseous systems, the GERs enable the diffusion coefficients parallel to the electric field, $D_\parallel$, and perpendicular, $D_\perp$ to be related to field derivatives of the mobility coefficient. In this section, we generalise the GER to systems where coherent scattering is considered. It is

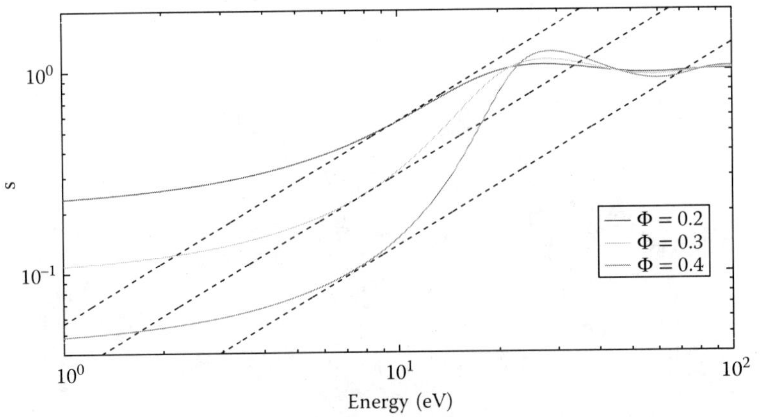

Figure 10.5 Variation of $s = s_0 - s_1$, with reduced electric field, for various volume fractions Φ for a dense gas of hard spheres. (From G. J. Boyle et al., *New Journal of Physics*, 2012 © IOP Publishing and Deutsche Physikalische Gesellschaft. CC BY-NC-S.)

straightforward to show from Equation 10.9 that the diffusion coefficients when coherent scattering is considered are given by

$$D_\perp = \frac{2}{3}\frac{\varepsilon}{m\tilde{\nu}_m} \tag{10.15}$$

$$D_\parallel = \frac{2}{3}\frac{\varepsilon}{m\tilde{\nu}_m} + \frac{\tilde{\nu}'_m}{\tilde{\nu}_m} v_d \gamma \tag{10.16}$$

where dashed quantity represents energy derivatives. Substituting in the longitudinal component of expression Equation 10.10 (neglecting the heat flux) into Equation 10.16 and re-arranging, yields

$$D_\parallel = \frac{2}{3}\frac{\varepsilon}{m\tilde{\nu}_m}\left[1 + \frac{m_0 v_d^2 \frac{\tilde{\nu}'_m}{\nu_m}\frac{\tilde{\nu}_m}{\nu_m}}{1 - \frac{1}{2} m_0 v_d^2 \frac{\partial}{\partial \varepsilon}\left(\frac{\tilde{\nu}_m}{\nu_m}\right)}\right]^{-1}. \tag{10.17}$$

This result can expressed in terms of the mobility, K,

$$K = \frac{v_d}{E} = \frac{q}{m\tilde{\nu}_m}. \tag{10.18}$$

From this definition, in combination with Equation 10.12 the following identity emerges:

$$\frac{\frac{\partial \ln K}{\partial \ln E}}{\left(1 + \frac{\partial \ln K}{\partial \ln E}\right)} = \frac{-m_0 v_d^2 \frac{\tilde{\nu}'_m}{\tilde{\nu}_m}\frac{\tilde{\nu}_m}{\nu_m}}{1 - \frac{1}{2} m_0 v_d^2 \frac{\partial}{\partial \varepsilon}\left(\frac{\tilde{\nu}_m}{\nu_m}\right)}. \tag{10.19}$$

It then follows from Equations 10.15 and 10.17 that the structure-modified GER take the form

$$\frac{D_{\parallel}}{D_{\perp}} = 1 + \frac{\partial \ln K}{\partial \ln E} = \frac{\partial \ln W}{\partial \ln E}. \tag{10.20}$$

This of the same form as the well-known GER in dilute gas transport theory (Equation 7.62).

In Figures 10.6 and 10.7, we display the perpendicular and longitudinal diffusion coefficients with E/n_0 for our model Percus–Yevick system and various volume fractions Φ. For charged particles interacting with a dilute gas of hard spheres with a constant collision cross section, we know that $W \sim E^{1/2}$, and hence

$$\frac{D_{\parallel}}{D_{\perp}} \simeq 0.5 \tag{10.21}$$

for all reduced electric fields. The results shown in Figures 10.6 and 10.7 for $\Phi \approx 0$ are in agreement with this result.

In the first instance, we focus on the Boltzmann equation results in these figures, and discuss the impact of coherent scattering effects on the anisotropic nature of diffusion. For low fields, we observe that diffusion is enhanced by nearly two orders of magnitude with increasing Φ. This is due to the implicit enhanced thermal (or random energy) effects and the explicit effects of a reduction in the momentum transfer cross section.

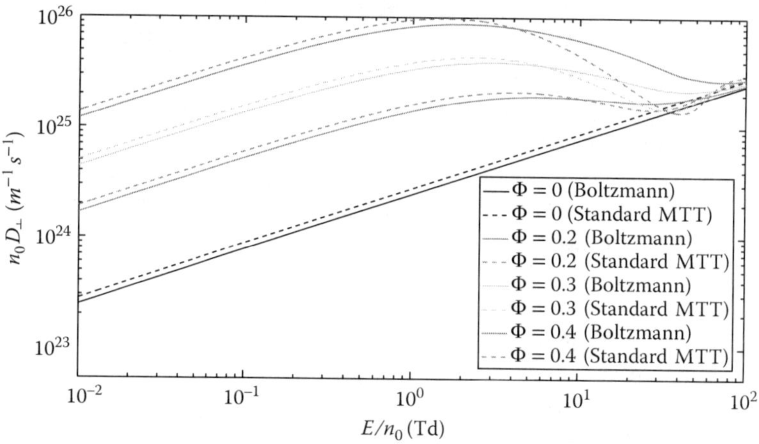

Figure 10.6 Variation of the transverse diffusion coefficient with reduced electric field for various volume fractions Φ for a dense gas of hard spheres. Fluid results are compared with those from a multi-term Boltzmann equation solution [91]. (From G. J. Boyle et al., *New Journal of Physics*, 2012 © IOP Publishing and Deutsche Physikalische Gesellschaft. CC BY-NC-S.)

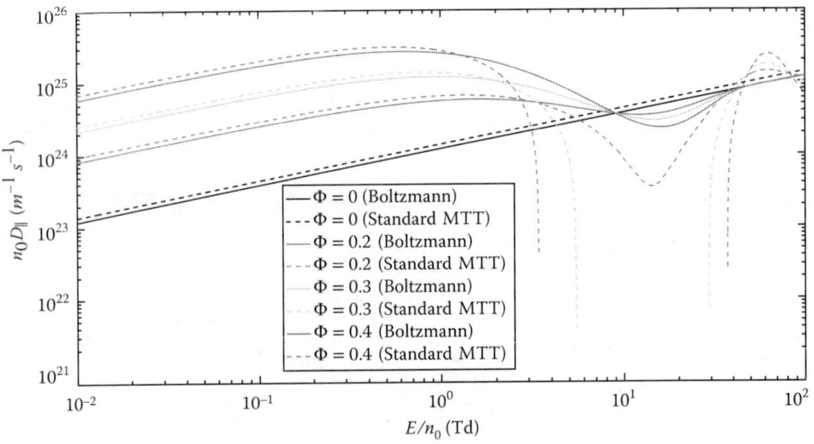

Figure 10.7 Variation of the longitudinal diffusion coefficient with reduced electric field for various volume fractions Φ for a dense gas of hard spheres. Fluid results are compared with those from a multi-term Boltzmann equation solution [91]. (From G. J. Boyle et al., *New Journal of Physics*, 2012 © IOP Publishing and Deutsche Physikalische Gesellschaft. CC BY-NC-S.)

Interestingly, the $D_\parallel/D_\perp$ ratio remains at 0.5 in this field range. Here, the effective momentum transfer collision frequency is approximately constant with energy for all volume fractions, and it follows that the $D_\parallel/D_\perp$ ratio is still fixed at approximately 0.5. For the volume fractions displayed, all diffusion coefficients have a window of field strengths where they are decreasing with increasing field. This is a reflection of the regime where the momentum transfer collision frequency increases sufficiently rapidly with energy as the coherent scattering effects decrease due to the reduced de Broglie wavelength. The decrease is particularly significant for the longitudinal component where it can fall by as much as an order of magnitude. There are two explicit effects which contribute to this modification (1) the thermal (random energy) effects which are also manifest in the transverse diffusion coefficient, and (2) the differential velocity effect [78]. As we have seen in Chapter 7, the differential velocity effect is a combination of a collision frequency which varies with energy and an average energy which varies through the swarm as detailed by the parameter γ. This differential velocity effect plays the key role here, where it is enhanced by virtue of the rapidly increasing momentum transfer collision frequency in this regime.

As detailed in Section 8.5.3, the standard GERs break down in the presence of NDC. This is again evidenced in 10.7 for $\Phi = 0.3$ and 0.4 profiles where longitudinal diffusion coefficient becomes negative in those field regions where NDC occurs. Further terms are required in the moment equations to be retained to adequately model this region.

Part III

Solutions of Kinetic Equations

CHAPTER 11

Strategies and Regimes for Solution of Kinetic Equations

11.1 The Kinetic Theory Program

11.1.1 General statement of the problem

Put most simply, the aim of kinetic theory is to solve the kinetic equation*

$$(\partial_t + \mathbf{v} \cdot \nabla + \mathbf{a} \cdot \partial_{\mathbf{v}}) f = \left(\frac{\partial f}{\partial t} \right)_{\mathrm{col}} \tag{11.1}$$

for the charged phase space distribution function $f(\mathbf{r}, \mathbf{v}, t)$ and hence calculate quantities of physical interest as velocity averages, for example, the particle number, particle current density, and mean energy,

$$n(\mathbf{r}, t) = \int \mathrm{d}^3 v\, f(\mathbf{r}, \mathbf{v}, t) \tag{11.2}$$

$$\Gamma(\mathbf{r}, t) = \int \mathrm{d}^3 v\, v\, f(\mathbf{r}, \mathbf{v}, t) \mathbf{v} \tag{11.3}$$

and

$$\left\langle \frac{1}{2} m v^2 \right\rangle (\mathbf{r}, t) = \frac{1}{n} \int \mathrm{d}^3 v\, v\, \frac{1}{2} m v^2 f(\mathbf{r}, \mathbf{v}, t), \tag{11.4}$$

respectively.

In this book, we consider primarily low density charged particles, where space-charge fields and mutual interactions between particles are negligible, and the force term

$$\mathbf{a} = \frac{q}{m} (\mathbf{E} + \mathbf{v} \times \mathbf{B}) \tag{11.5}$$

---

* In what follows, we shall sometimes refer to Equation 11.1 as the "Boltzmann equation," although this terminology is strictly correct only for the original 1872 collision operator $\left(\frac{\partial f}{\partial t} \right)_{\mathrm{coll}}$ for particles in a gaseous medium. Here, as elsewhere, it is implicitly assumed in writing down Equation 11.1 that we are dealing with *non-relativistic* particles whose energies are small compared with the rest mass, $\epsilon \ll mc^2$. At higher energies, a relativistic kinetic equation would be required, and both the left and right hand sides would have to be modified. However, for most of the problems discussed in this book, Equation 11.1 suffices.

derives from externally applied fields. In this case, there is a major simplification: the collision term $\left(\frac{\partial f}{\partial t}\right)_{\text{col}}$ involves only interactions between the particles and the background medium, and Equation 11.1 is *linear* in f.

11.1.2 Fluid analysis versus rigorous solution

In fluid analysis, as discussed in Chapters 7 through 10, these averages are found without actually calculating f itself, by solving approximate balance equations formed by taking appropriate velocity moments of Equation 11.1. This approach elucidates the physics with a minimum of mathematics and provides reasonable semi-quantitative estimates (accuracy ~10%) of the macroscopic properties of interest. However, in rigorous kinetic theory, we must solve Equation 11.1 for f first, and *then* evaluate the moments, in order to achieve accuracies comparable with experiment, often around 0.1% or better.

Fluid analysis is also important in another respect, in that it can be readily adapted to lay the platform for the required rigorous solution of Equation 11.1, thereby saving a considerable amount of time and effort. Thus, as discussed in Chapter 12, the fluid method is extended and systematized by forming moment equations with respect to a *complete, infinite set* of basis functions $\phi^{[\nu]}(\mathbf{v})$ in velocity space, where $\boldsymbol{\nu}$ is an index enumerating members of the set. The resulting equations for the averages $\langle\phi^{[\nu]}(\mathbf{v})\rangle$ constitute an equivalent representation of Equation 11.1 in the $\{\phi^{[\nu]}(\mathbf{v})\}$ basis.* Thus, the approximate moment equations for a handful of quantities, $\phi = 1,\ m\mathbf{v},\ \frac{1}{2}mv^2$, are replaced by an infinite set of equations, which is to be solved by truncating to finite size systematically until the prescribed accuracy criterion is met. In effect, the method yields *all* the moments $\langle\phi^{[\nu]}(\mathbf{v})\rangle$ of the distribution function, not just a few, as in fluid analysis, and this is, of course, equivalent to knowing the distribution itself. † Importantly,

---

* Here, $\boldsymbol{\nu} = (\nu, l, m)$ stands for a set of three indices characterising the basis functions in three-dimensional velocity space.

† This statement can be best understood by considering the "moments" $\langle x^N\rangle = \int_{-\infty}^{\infty} f(x)\, x^N\, dx$ of some arbitrary function $f(x)$, which is normalized to unity. Its Fourier transform is

$$\bar{f}(k) = \int_{-\infty}^{\infty} f(x)\, e^{-ikx}\, dx = \sum_{N=0}^{\infty} \frac{(-ik)^N}{N!} \int_{-\infty}^{\infty} f(x)\, x^N\, dx \equiv \sum_{N=0}^{\infty} \frac{(-ik)^N}{N!} \langle x^N\rangle.$$

Clearly, knowledge of all the moments furnishes $\bar{f}(k)$, which in turn gives the function itself:

$$f(x) = \frac{1}{2\pi} \int_{-\infty}^{\infty} \bar{f}(k)\, e^{ikx}\, dk.$$

the average collision transfer terms are now calculated precisely, rather than through the momentum transfer theory *Ansatz.*

This "moment method," which is discussed in Chapter 12, provides an accurate and efficient means of numerical solution of Equation 11.1 and is eminently suitable for particles of masses, and all types of interactions. Ideally, the one computer code can cover all possible cases, with minimal adjustment.

11.1.3 Strategies for reducing complexity

In general, whatever the solution technique, the existence of seven independent variables $(\mathbf{r}, \mathbf{v}, t)$ makes the accurate solution of Equation 11.1 an exceedingly challenging proposition, even with modern high speed computers. Although a reduction in dimensionality may be possible by accounting for symmetries (see Section 11.2), the complex mathematical structure of the collision term $\left(\frac{\partial f}{\partial t}\right)_{\text{col}}$ means that the solution of Equation 11.1 must, in general, be carried out numerically, for both gaseous and condensed matter media. One of our main tasks is to establish ways of minimizing the computational effort, while still achieving the high accuracies required. To achieve this, we must limit the number of assumptions and develop solutions capable of dealing with the highly non-equilibrium conditions characterizing many typical systems. In particular, perturbative solutions about equilibrium are to be avoided, for example, the Chapman–Enskog procedure [18], which assumes that fields are weak and deviations from equilibrium are small.

11.1.4 Roadmap to solution of the kinetic equation

Once a numerical method for solving Equation 11.1 is developed, the typical procedure is as follows: cross sections, structure factor, and relaxation function (as appropriate for the medium in question) are input into the computer code, the relevant experimental parameters and boundary conditions specified, and numerical values of the macroscopic properties are calculated, to the specified level of accuracy (see Figure 11.1). This purely numerical procedure provides the required quantitative results rigorously, though at the expense of physical transparency provided by the less accurate fluid analysis. The two approaches are therefore to be viewed as complementary.

11.2 Identifying Symmetries

In order to make the solution of Equation 11.1 more tractable, it is essential to reduce the number of independent coordinates to a minimum, for example, by identifying any symmetries. Examples of systems

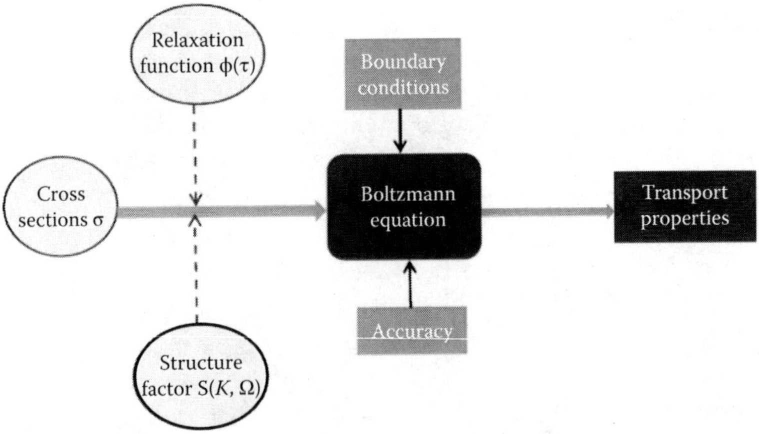

Figure 11.1 "Roadmap" to solution of Equation 11.1, for either gaseous or condensed matter media, proceeding from the required input information to the experimentally measured transport quantities.

where it is possible to reduce the effective dimensionality in this way include

- The steady state Townsend (SST) experiment with plane-parallel electrodes, or the drift tube in the Franck–Hertz experiment (Chapter 16);
- A spherically symmetric system, such as the model of positron emission tomography in Chapter 17; and
- An axially symmetric cylindrical system, of the sort found in plasma processing devices [13].

Further details are given below. As already stated in Chapter 2, it is essential to distinguish carefully between symmetries in configuration and velocity space. To that end, angles in velocity space are distinguished from configuration space through the addition of subscripts "v" in what follows. Complications arise when magnetic fields are applied, and this is discussed further in Chapter 18.

11.2.1 Plane-parallel geometry

The simplest case is the SST experiment in plane-parallel geometry, as shown schematically in Figure 11.2. Particles are emitted at a steady rate from the electrode at $z = 0$ into a gas and driven by a uniform electric field to the electrode at $z = d$. Properties in both configuration and velocity space are invariant under rotations about the the field direction (the z-axis) and furthermore, spatial properties vary along the z-axis, but not in transverse

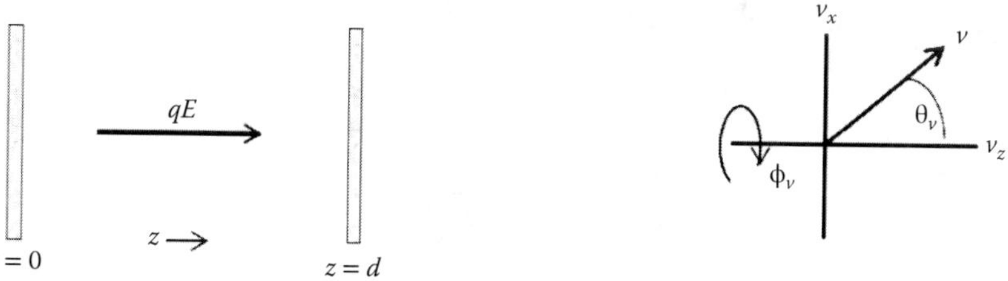

Figure 11.2 A schematic representation of an infinite plane-parallel electrode system in which particles of charge q move under the influence of an electric field E, directed perpendicular to the electrodes, along the z-axis of a system of coordinates. Properties are rotationally symmetric about E in both configuration and velocity space.

directions. On the other hand, while the particles can move in any direction, they do so predominantly in the field direction. With reference to a spherical polar system of coordinates in velocity space (v, θ_v, φ_v), the velocity distribution depends on v and the polar angle θ_v that the velocity vector makes with the z-axis, but not on the azimuthal angle φ_v.

The phase space distribution function, therefore, depends on just three coordinates in this case, z, v, and θ_v only, that is (adding time for the sake of completeness)

$$f = f(z; v, \theta_v; t). \tag{11.6}$$

11.2.2 Spherical geometry

Similar considerations apply to a system confined in a uniform spherical region, or to a spherical source in an infinite medium, in which any fields and spatial gradients are in the radial direction only, as shown in Figure 11.3. Thus, the radial vector $\mathbf{r}$ in configuration space forms an axis of rotational symmetry in velocity space, and therefore as for plane-parallel geometry,

$$f = f(r; v, \theta_v; t) \tag{11.7}$$

where θ_v is the angle between $\mathbf{v}$ and $\mathbf{r}$.

11.2.3 Cylindrical geometry

Consider a cylindrical system described by spatial coordinates $\mathbf{r} = (r_\perp, \varphi, z)$, where the z-axis defines the axis of the cylinder, $r_\perp$ is the distance from the axis, and φ denotes the azimuthal angle. The corresponding cylindrical coordinates in velocity space are $\mathbf{v} = (v_\perp, \varphi_v, v_z)$. Suppose that the system is physically invariant under rotations about the z-axis, that is,

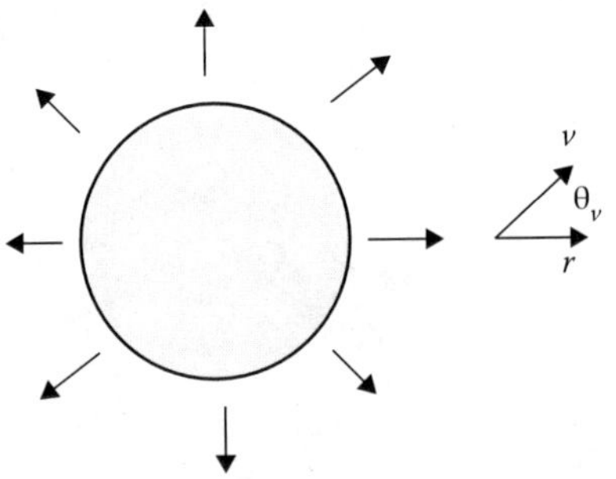

Figure 11.3 Schematic representation of a spherically symmetric source emitting particles with velocities **v** at angles θ_v with respect to the radial direction **r**.

the physical properties should appear exactly the same along one radial direction as any other. In particular, the distribution of velocities should look the same in all radial directions, something which is possible only if $f(\mathbf{r}, \mathbf{v}, t)$ depends on φ_v and φ through their *difference* $\varphi_v - \varphi$. In this case, the phase space distribution function must be of the form [52]

$$f = f(r_\perp, z; v_\perp, v_z; \varphi_v - \varphi; t). \tag{11.8}$$

The cautionary note here is that a symmetry in configuration space is not necessarily reflected in velocity space, and vice versa. Note also that the above examples have been chosen as illustrations of systems which can *not* be described in a hydrodynamic way (see below).

11.3 Kinetic Theory Operators

11.3.1 The collision operator and its adjoint

It is convenient to, now, introduce the kinetic *collision operator J*, defined by

$$J(f) = -\left(\frac{\partial f}{\partial t}\right)_{\text{col}} \tag{11.9}$$

so that the kinetic Equation 11.1 can be written as

$$(\partial_t + L)\, f = 0, \tag{11.10}$$

where

$$L \equiv \mathbf{v} \cdot \nabla + \mathbf{a} \cdot \partial_{\mathbf{v}} + J \tag{11.11}$$

is a linear collision operator in phase space, whose role will become clear in Section 11.4.

In the preceding chapters, expressions for $\left(\frac{\partial f}{\partial t}\right)_{\text{col}}$ and hence J have been derived from first principles for particles in both gaseous and condensed matter media. For example, the Wang Chang–Uhlenbeck–de Boer (WUB) collision operator for point particles in a gas whose molecules have internal states j follows from Equation 5.36 as

$$J(f) = \sum_{j,j'} \int d^3 v_0 \int d\Omega_{\mathbf{g}'} \left[f(\mathbf{r},\mathbf{v},t) f_{0j}(\mathbf{v}_0) - f(\mathbf{r},\mathbf{v}',t) f_{0j}(\mathbf{v}'_0) \right] g\,\sigma(j,j';\mathbf{g},\mathbf{g}'). \tag{11.12}$$

This was obtained in Chapter 5 by first formulating an expression for the collisional rate of change of some arbitrary property $\phi(\mathbf{v})$ of particle velocity, for which the equivalent expression in terms of the collision operator is

$$\begin{aligned} \int d^3 v \phi(\mathbf{v}) J(f) &= \sum_{j,j'} \int d^3 v \int d^3 v_0 f(\mathbf{r},\mathbf{v},t) f_{0j}(\mathbf{v}_0) \\ &\quad \times \int d\Omega_{\mathbf{g}'} g\sigma(j,j';\mathbf{g},\mathbf{g}') \left[\phi(\mathbf{v}) - \phi(\mathbf{v}')\right] \\ &\equiv \int d^3 v f(\mathbf{r},\mathbf{v},t) \left\{ \sum_{j,j'} \int d^3 v_0 f_{0j}(\mathbf{v}_0) \right. \\ &\quad \left. \times \int d\Omega_{\mathbf{g}'} g\sigma(j,j';\mathbf{g},\mathbf{g}') \left[\phi(\mathbf{v}) - \phi(\mathbf{v}')\right] \right\}. \end{aligned} \tag{11.13}$$

The *adjoint collision operator* $J^{\dagger}$ is defined by the relation

$$\int d^3 v\, \phi(\mathbf{v}) J(f) = \int d^3 v\, f J^{\dagger}(\phi) \tag{11.14}$$

for real functions $\phi(\mathbf{v})$ of velocity. By comparison with Equation 11.13 it follows that

$$J^{\dagger}(\phi) = \sum_{j,j'} \int d^3 v_0 f_{0j}(\mathbf{v}_0) \int d\Omega_{\mathbf{g}'} g\sigma(j,j';\mathbf{g},\mathbf{g}') \left[\phi(\mathbf{v}) - \phi(\mathbf{v}')\right] \tag{11.15}$$

is the adjoint of the Wang-Chang et al. collision operator (Equation 11.12) for particles in a gaseous medium.

For particles in soft-condensed matter, as discussed in Chapter 6, similar reasoning shows that the adjoint collision operator is given by

$$J^{\dagger}(\phi) = n_0 \int_0^{\infty} d\omega' \int d\Omega_{\mathbf{k}'} [\phi(\mathbf{v}') - \phi(\mathbf{v})] v \left(\frac{d\sigma}{d\Omega_{\mathbf{k}'}} \right)^{(\text{lab})} S(\Delta\mathbf{k}, \Delta\omega), \tag{11.16}$$

where $S(\Delta\mathbf{k}, \Delta\omega)$ is the dynamic structure function of the medium, and other notation is the same as in Chapter 6.

It is now clear that in the preceding chapters, we have effectively obtained the adjoint collision operator *before* the collision operator itself, but without actually saying so. In some ways, $J^\dagger$ rather than J is the more fundamental operator. This is evident, for example, in fluid analysis, for which the right hand side of the balance or moment equation is of the form

$$\begin{aligned}\int d^3v\, \phi(\mathbf{v}) \left(\frac{\partial f}{\partial t}\right)_{\text{col}} &= -\int d^3v\, \phi(\mathbf{v}) J(f) \\ &= -\int d^3v\, f\, J^\dagger(\phi) = -n \left\langle J^\dagger(\phi) \right\rangle. \end{aligned} \tag{11.17}$$

Thus, the collisional transfer term in the fluid equation for some property ϕ derives from the action of $J^\dagger$ on that property.

Other important results are

- The identity

$$J[w(\alpha, v)\phi] = w(\alpha, v) J^\dagger(\phi), \tag{11.18}$$

where $w(\alpha, v) = \left(\frac{\alpha^2}{2\pi}\right)^{3/2} \exp\left(-\frac{1}{2}\alpha^2 v^2\right)$ is a Maxwellian velocity distribution function at gas temperature T_0, $\alpha^2 = \frac{m}{k_B T_0}$, and

- The property that for any real $\phi(\mathbf{v})$

$$\int d^3v\, w(\alpha, v) \phi J^\dagger(\phi) \geq 0. \tag{11.19}$$

11.3.2 Phase space operator and adjoint

The definition (Equation 11.14) of an adjoint operator in velocity space generalizes to

$$\int_V d^3r \int d^3v\, \phi^* L \psi = \int_V d^3r \int d^3v\, (L^\dagger \phi)^* \psi \tag{11.20}$$

for the adjoint $L^\dagger$ of an operator L acting in phase space $(\mathbf{r}, \mathbf{v})$, for a system confined in a volume V, where the asterisk indicates complex conjugation. The adjoint of the phase space operator L defined by Equation 11.11 is

$$L^\dagger \equiv -\mathbf{v} \cdot \nabla - \mathbf{a} \cdot \partial_{\mathbf{v}} + J^\dagger. \tag{11.21}$$

Like the collision operators J and $J^\dagger$, both L and $L^\dagger$ are *linear*, and *both* operators figure in the eigenvalue formulation of kinetic theory (Section 11.5). Kinetic theory operators are always real, but not self-adjoint, in contrast to operators in quantum mechanics. However, eigenfunctions in phase space calculations may be complex, as indicated by the complex conjugation in Equation 11.20.

11.4 Boundary Conditions and Uniqueness

11.4.1 Uniqueness theorem

In general, boundary and initial conditions must be specified at the outset in order to solve Equation 11.1.

Suppose that the system under investigation is confined in a volume V by a bounding surface A, as shown schematically in Figure 11.4. At time t_0, a source at position $\mathbf{r}_0$ emits a pulse of N particles of charge q, mass m, and velocity $\mathbf{v}_0$, which subsequently move in response to the applied fields, are scattered by the molecules of the medium, and eventually reach the boundary.

The initial condition is thus

$$f(\mathbf{r}, \mathbf{v}, t = t_0) = N\delta(\mathbf{v} - \mathbf{v}_0)\delta(\mathbf{r} - \mathbf{r}_0) \qquad (\mathbf{r} \in V) \tag{11.22}$$

and, if the bounding walls are perfectly absorbing (i.e., particles incident on walls do not return to the medium), the boundary condition is

$$f(\mathbf{r}, \mathbf{v}, t > t_0) = 0 \quad (\mathbf{r} \in A,\ \mathbf{n} \cdot \mathbf{v} > 0), \tag{11.23}$$

where $\mathbf{n}$ denotes a unit vector normal to A, pointing *inwards* (see Figure 11.4). The distribution function also vanishes for large velocities, effectively

$$f(\mathbf{r}, \mathbf{v}, t) \to 0, \quad \mathbf{v} \to \infty. \tag{11.24}$$

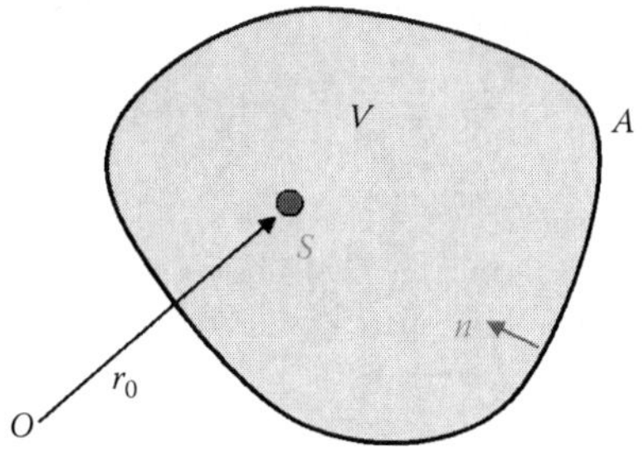

Figure 11.4 At time t_0 a source S located at position $\mathbf{r}_0$ with respect to an arbitrary origin O emits a pulse of N charged particles with velocities $\mathbf{v}_0$ into a uniform gas of number density n_0 in equilibrium at temperature T_0, and confined within a volume V by a boundary surface A. The normal to the surface is denoted by the unit vector $\mathbf{n}$.

Suppose that f_1 and f_2 are two solutions of the Boltzmann Equation 11.1, each satisfying the above initial and boundary conditions, and define

$$\Delta f = f_1 - f_2.$$

This quantity also satisfies an equation similar to Equation 11.1, that is,

$$\left(\partial_t + J + \mathbf{v}\cdot\nabla + \mathbf{a}\cdot\partial_{\mathbf{v}}\right)\ \Delta f = 0, \tag{11.25}$$

with initial and boundary conditions:

$$\Delta f = 0, \quad \mathbf{r}\in V, \quad t = t_0 \tag{11.26}$$

$$\Delta f = 0, \quad \mathbf{r}\in A, \quad \mathbf{n}\cdot\mathbf{v} > 0, t > t_0 \tag{11.27}$$

$$\Delta f \to 0, \quad \mathbf{v}\to\infty. \tag{11.28}$$

We multiply Equation 11.25 by Δf and integrate over time and all $\mathbf{r}$ and $\mathbf{v}$

$$\int d^3v \int_V d^3r \int_{t_0}^{t} dt\ \Delta f\ \left(\partial_t + \mathbf{v}\cdot\nabla + \mathbf{a}\cdot\partial_{\mathbf{v}} + J\right)\Delta f = 0. \tag{11.29}$$

The four terms produced on the left hand side are as follows:

- *Term 1:*

$$\int d^3v \int_V d^3r \int_{t_0}^{t} dt\ \Delta f\ \partial_t(\Delta f) = \frac{1}{2}\int d^3v \int_V d^3r \left[(\Delta f)^2 - (\Delta f)^2_{t=t_0}\right]$$
$$= \frac{1}{2}\int d^3v \int_V d^3r\ (\Delta f)^2 \geq 0,$$

since by (Equation 11.26) $(\Delta f)^2_{t=t_0} = 0$.

- *Term 2:*

$$\int d^3v \int_{t_0}^{t} dt \int_V d^3r\ \Delta f\ \nabla\cdot(\mathbf{v}\Delta f) = \int d^3v \int_{t_0}^{t} dt \int_V d^3r\ \nabla\cdot\left(\frac{1}{2}\mathbf{v}\,(\Delta f)^2\right)$$
$$= \frac{1}{2}\int d^3v \int_{t_0}^{t} dt \int_A dA\,(-\mathbf{n}\cdot\mathbf{v})\ (\Delta f)^2 \geq 0,$$

where Gauss' theorem has been used to transform from a volume to a surface integral, and by (Equation 11.27) the integrand in the last term is non-zero only for $\mathbf{n}\cdot\mathbf{v} \leq \mathbf{0}$.

- *Term 3:*

$$\int_{t_0}^{t} dt \int_V d^3r \int d^3v\ \Delta f\ \left[\mathbf{a}\cdot\partial_{\mathbf{v}}\Delta f\right] = \int_{t_0}^{t} dt \int_V d^3r \int d^3v\ \partial_{\mathbf{v}}\cdot\left[\frac{1}{2}\mathbf{a}\,(\Delta f)^2\right] \equiv 0$$

where the last step follows after evaluating the integral over $\mathbf{v}$, and applying Equation 11.28. Note also that $\partial_{\mathbf{v}}\cdot\mathbf{a} \equiv 0$ for forces of the type (Equation 11.5).

- *Term 4*: In order to evaluate the integral, the collision term needs to be made explicit. For the sake of simplicity, we assume the medium consists of a monatomic gas in equilibrium with distribution function $f_0 = n_0\, w(\alpha_0, v_0)$. If collisions are also assumed to be elastic, the Boltzmann collision operator corresponding to Equation 11.22 of Chapter 3 can be used. Using the notation $d\vartheta = g\sigma d^2 g'\, d^3v\, d^3v_0$, we find

$$\begin{aligned}\int_{t_0}^{t} dt \int_V d^3r \int d^3v \Delta f\, J(\Delta f) &= \int_{t_0}^{t} dt \int_V d^3r \int d^3v\, f_0\, \Delta f\, \left[\Delta f - \Delta f'\right] d\vartheta \\ &\geq n_0 \int_{t_0}^{t} dt \int_V d^3r \int d^3v\, w(\alpha, v) w(\alpha_0, v_0) \\ &\quad \times \Delta f \left[\Delta f - \Delta f'\right] d\vartheta \\ &\geq 0,\end{aligned}$$

  where the second inequality follows from the fact that $0 < w(\alpha, v) < 1$ for all v, and the third line follows from Equation 11.19 with $\phi = \Delta f$.

Thus, all terms in the left hand side of Equation 11.29 can only be positive or zero, whereas the right hand side is zero. This can hold only if each term in the left hand side is zero, which implies $\Delta f = 0$, or in other words, $f_1 = f_2$. Thus, the solution of the Boltzmann equation is indeed uniquely determined by the boundary and initial conditions Equation 11.23 and 11.22, respectively. More precisely, Equation 11.1 is determined *uniquely* by specifying only the distribution function at the boundary for velocities *incident* on the medium, as in Equation 11.23. Specification of the distribution function over the *whole range* of velocities at the boundary, that is, for both $\mathbf{n} \cdot \mathbf{v} > 0$ and $\mathbf{n} \cdot \mathbf{v} < 0$, amounts to an over-specification of boundary conditions, and is incorrect.

We shall assume, without proof, that the theorem applies in general to other interactions and media, and to other types of boundary conditions.

11.4.2 Approximations

In practice, it is very difficult to apply the exact boundary conditions to solve the kinetic equation, and one usually has to settle for some sort of approximation. Whatever compromise one makes, the uniqueness theorem just discussed dictates that, in some sense, only one-half of the distribution function can be specified at a boundary. Thus, for example, one might specify either the odd or the even-l members of the spherical components $f^{(l)}$ of the distribution function, but not both. An example is given in Chapter 13.

As discussed previously, in many experimental situations, the hydrodynamic regime prevails, where the space-time dependence of f separates from its velocity dependence. In effect, all initial and boundary conditions are then projected onto the number density $n(\mathbf{r}, t)$, which is found by solving the diffusion equation, while the velocity dependent components of f are found separately by solving a hierarchy of kinetic equations, without any reference to boundaries or initial conditions.

11.5 Eigenvalue Problems in Kinetic Theory

In many experiments, the primary interest is the behaviour of the system at either long times or large distances from the source and, accordingly, only an asymptotic solution of the kinetic Equation 11.1 is required. In such cases, it is advantageous to discuss solution of the kinetic equation in terms of an eigenvalue problem (actually a dual eigenvalue problem), for then the asymptotic picture may be described by the lowest members of the eigenvalue spectrum.

We return to the situation shown in Figure 11.4, where a pulse of N particles is emitted from a source at time t_0 with uniform velocity $\mathbf{v}_0$ from a point $\mathbf{r}_0$. Following the general theorem developed in Appendix C, the solution of Equation 11.10 may be written as a sum over phase space "modes" p,

$$f(\mathbf{r}, \mathbf{v}, t) = \theta(t - t_0)N \sum_p \phi_p^*(\mathbf{r}_0, \mathbf{v}_0)e^{-\omega_p(t-t_0)}\psi_p(\mathbf{r}, \mathbf{v}), \tag{11.30}$$

where $\theta(t - t_0)$ is the unit step function, which equals unity for $t \geq t_0$ and is zero otherwise, and ψ_p and ϕ_p are eigenfunctions of the dual problems,

$$L\psi_p = \omega_p\psi_p \tag{11.31}$$

and

$$L^\dagger\phi_p = \omega_p^*\phi_p, \tag{11.32}$$

respectively. The eigenfunctions $\psi_p(\mathbf{r}, \mathbf{v})$ are required to satisfy the same boundary conditions as $f(\mathbf{r}, \mathbf{v}, t)$ itself, for example,

$$\psi_p(\mathbf{r}, \mathbf{v}) = 0 \quad (\mathbf{r} \in A,\ \mathbf{n} \cdot \mathbf{v} > 0)$$

for perfect the absorbing boundaries of Figure 11.4.

The two sets of eigenfunctions are orthogonal in phase space, that is

$$\int_V \mathrm{d}^3r \int \mathrm{d}^3v\, \phi_{p'}^*(\mathbf{r}, \mathbf{v})\psi_p(\mathbf{r}, \mathbf{v}) = \delta_{p',p}. \tag{11.33}$$

Otherwise, however, little is known about the general properties of the eigenfunctions or eigenvalues ω_p and numerical calculations are generally required to provide an information about the spectrum in specific cases. In some cases, it is found that the spectrum is real and discrete. An example is the Cavalleri experiment, discussed in Chapter 13. Where there is no production of particles by impact ionization, it is found that $\omega_p > 0$, with $\omega_1 < \omega_2 < \omega_3 < \ldots$.

In reality, the phase space modes p consist of *sets* of two or more indices $p = (n, j, \ldots)$, corresponding to velocity and configuration space modes n and j, respectively, and eigenfunctions and eigenvalues should carry this set of multiple indices. At long times $t \gg t_0$, the fundamental mode $p = 1 \equiv \{1, 1\}$ with smallest eigenvalue $\omega_{\{1,1\}}$ dominates, and Equation 11.30 approximates to

$$f(\mathbf{r}, \mathbf{v}, t) \approx N\phi_1^*(\mathbf{r}_0, \mathbf{v}_0)e^{-\omega_{\{1,1\}}(t-t_0)}\psi_{\{1,1\}}(\mathbf{r}, \mathbf{v}). \tag{11.34}$$

Hence the total number of particles in the container,

$$N(t) = \int_V \mathrm{d}^3 r \int \mathrm{d}^3 v\, f(\mathbf{r}, \mathbf{v}, t) \sim e^{-\omega_{\{1,1\}}(t-t_0)}$$

decays according to a simple exponential law in the asymptotic limit. In the Cavalleri experiment, the number of electrons in a cavity is measured as a function of time in the asymptotic limit, and the decay constant furnishes the eigenvalue $\omega_{\{1,1\}}$ directly, which in turn may be interpreted in terms of a diffusion coefficient.

In other cases, the eigenvalue spectrum may be complex, leading to oscillatory behaviour. This is the case, for example, in the spatially periodic structures characterizing the Franck–Hertz experiment (Chapter 16), where the eigenvalue with the smallest real part controls the asymptotic behaviour downstream from the source. For the relaxation of positrons in a positron emission tomography (PET) environment (Chapter 17), the eigenvalue spectrum is real, and the smallest eigenvalue determines the positron range before being lost to annihilation.

Although the eigenvalue problems Equations 11.31 and 11.32 have been introduced in a formal way in this section, they often arise quite naturally when solving the kinetic equation, typically when using the method of separation of variables. There the eigenvalues are effectively the separation constants, which are determined from either boundary conditions or by imposing physical constraints on the solution of the eigenvalue problem.

11.6 Hydrodynamic Regime

11.6.1 Weak fields and Chapman–Enskog approximation scheme

To begin with, it is useful to look at the simplest possible case where the time and space derivatives, as well as the field, all vanish. A state of *equilibrium* then prevails, with the left hand side of Equation 1.1 identically zero. Hence $\left(\frac{\partial f}{\partial t}\right)_{\text{col}} = 0$, and f is Maxwellian:

$$f_{\text{equil}}(\mathbf{v}) = n\, w(\alpha, v) = n\left(\frac{m}{2\pi k_B T_0}\right)^{\frac{3}{2}} \exp\left(-\frac{mv^2}{2k_B T_0}\right). \tag{11.35}$$

If, on the other hand, the system is slightly perturbed from equilibrium, with the field term, space, and time variations now *all* non-zero, but still weak, then one might consider developing a perturbation procedure for solving Equations 11.1 with 11.35 as the first approximation. This is just the *Chapman–Enskog procedure*, which in the second approximation produces a correction term $\sim\nabla n$, (leading to Fick's law with a constant mobility coefficient), and the third approximation produces a further correction term $\sim\nabla^2 n$, and so on. Full details can be found in the treatise of Chapman and Cowling [18], while a simplified analysis in terms of the relaxation time model (see Section 5.2.2) has been given by Huang [49].

11.6.2 Beyond weak fields

The near-equilibrium analysis, while not useful in most circumstances considered here, nevertheless, provides a guide as to the next step.

- *Role of the Maxwellian:* Although we are generally dealing with strong field, highly non-equilibrium situations, one might still *guess* that a Maxwellian at some non-equilibrium temperature $T_b > T_0$, that is,

$$f(\mathbf{v}) \approx n\left(\frac{m}{2\pi k_B T_b}\right)^{\frac{3}{2}} \exp\left[-\frac{mv^2}{2k_B T_b}\right] \equiv n\, w(\alpha, v) \tag{11.36}$$

  might provide a reasonable first approximation to f. It turns out that while this is not good enough by itself, Equation 11.36 does, nevertheless, provide the "weight function" for an expansion of f in terms of a certain basis set [33].

- *Hydrodynamic regime, density gradient expansion:* If the restriction on weak fields is relaxed, but the first two terms on the left hand side of Equation 11.1 are still to be regarded as small, the space-time dependence of the distribution function may still be represented by an expansion similar in some respects to that which emerges from the Chapman–Enskog theory. Consider, for example, the time-of-flight experiment in a gas or soft-condensed matter medium (Chapter 1) where an initially sharp pulse of particles is released from a point source at time $t=0$ and drifts under the influence of a constant, spatially uniform applied field. At short times, before many collisions have been made, the pulse is still sharp but evolving rapidly in time. This is called the *kinetic* regime. The time and space derivatives on the left hand side of Equation 11.1 are large, and must be retained in any solution. At longer times, however, after many collisions, the pulse spreads out significantly and evolves only slowly. Spatial gradients become weak, time variation is slow, and the first two terms, $\frac{\partial f}{\partial t}$ and $\mathbf{v}\cdot\nabla f$, on the left hand side of Equation 11.1 are small, making a perturbation method of solution possible. The space and time dependence of $f(\mathbf{r},\mathbf{v},t)$ is carried entirely by the density $n(\mathbf{r},t)$ in this *hydrodynamic regime*, and is expressed through a density gradient expansion [31]. The details are given below. Note that some systems, such as those described in Section 11.5, are inherently nonhydrodynamic, and never evolve to the hydrodynamic regime. This means that *all* terms on the left hand side of Equation 11.1 must be considered on the same level, and no perturbation solution is possible.

11.6.3 The hierarchy of velocity space equations

To formalize the perturbation procedure in the hydrodynamic regime, we associate a parameter δ with the small terms on the left hand side of the kinetic equation,

$$\delta\left(\frac{\partial f}{\partial t}+\mathbf{v}\cdot\nabla f\right)+\mathbf{a}\cdot\frac{\partial f}{\partial \mathbf{v}}=\left(\frac{\partial f}{\partial t}\right)_{\text{col}} \tag{11.37}$$

and expand the distribution function in powers of δ:

$$f=f^{(0)}+\delta f^{(1)}+\delta^2 f^{(1)}+\dots. \tag{11.38}$$

This is then substituted into Equation 11.37, and coefficients of powers of δ equated. The zero order (the "unperturbed state") and first order equations are

$$\mathbf{a}\cdot\frac{\partial f^{(0)}}{\partial \mathbf{v}}=\left(\frac{\partial f^{(0)}}{\partial t}\right)_{\text{col}} \tag{11.39}$$

$$\left(\frac{\partial f^{(0)}}{\partial t}+\mathbf{v}\cdot\nabla f^{(0)}\right)+\mathbf{a}\cdot\frac{\partial f^{(1)}}{\partial \mathbf{v}}=\left(\frac{\partial f^{(1)}}{\partial t}\right)_{\text{col}}, \tag{11.40}$$

respectively. Equation 11.39 and the normalization condition (Equation 11.2) can be satisfied by writing

$$f^{(0)}(\mathbf{r},\mathbf{v},t) = n(\mathbf{r},t)F^{(0)}(\mathbf{v}), \tag{11.41}$$

where $F^{(0)}(\mathbf{v})$ satisfies

$$\int \mathrm{d}^3v\, F^{(0)}(\mathbf{v}) = 1$$

and hence

$$\mathbf{a}\cdot\frac{\partial F^{(0)}}{\partial \mathbf{v}} = \left(\frac{\partial F^{(0)}}{\partial t}\right)_{\text{col}}. \tag{11.42}$$

Substitution of Equation 11.41 into 11.40 then gives

$$F^{(0)}(\mathbf{v}-\mathbf{v}^{(0)})\cdot\nabla n + \mathbf{a}\cdot\frac{\partial f^{(1)}}{\partial \mathbf{v}} = \left(\frac{\partial f^{(1)}}{\partial t}\right)_{\text{col}} \tag{11.43}$$

since

$$\begin{aligned}\frac{\partial f^{(0)}}{\partial t} &= F^{(0)}\frac{\partial n}{\partial t}\\ &= -F^{(0)}\nabla\cdot\left[n\langle\mathbf{v}\rangle^{(0)}\right]\\ &\approx -F^{(0)}\mathbf{v}^{(0)}\cdot\nabla n\end{aligned}$$

where

$$\mathbf{v}^{(0)} = \int \mathrm{d}^3v\,\mathbf{v}F^{(0)}(\mathbf{v}) \equiv \mathbf{v}_d \tag{11.44}$$

defines the *drift velocity*.

Finally, by writing

$$f^{(1)} = \mathbf{F}^{(1)}(\mathbf{v})\cdot\nabla n \tag{11.45}$$

substituting into Equation 11.43, and equating the coefficients of ∇n, there follows

$$F^{(0)}(\mathbf{v}-\mathbf{v}^{(0)}) + \left(\mathbf{a}\cdot\frac{\partial}{\partial \mathbf{v}}\right)\mathbf{F}^{(1)} = \left(\frac{\partial \mathbf{F}^{(1)}}{\partial t}\right)_{\text{col}}, \tag{11.46}$$

which can be solved for $\mathbf{F}^{(1)}$ knowing $F^{(0)}$. To first order the perturbation expansion (Equation 11.38) (the arbitrary ordering parameter δ may now be set equal to unity) is thus equivalent to the *density gradient expansion*,

$$f(\mathbf{r},\mathbf{v},t) = F^{(0)}(\mathbf{v})n(\mathbf{r},t) + \mathbf{F}^{(1)}(\mathbf{v})\cdot\nabla n(\mathbf{r},t) + \ldots, \tag{11.47}$$

where $F^{(0)}(\mathbf{v})$ and $\mathbf{F}^{(1)}(\mathbf{v})$ are scalar and vector functions of $\mathbf{v}$ found from the solution of Equations 11.42 and 11.46, respectively.

This shows how the space and time dependence of $f(\mathbf{r}, \mathbf{v}, t)$ is carried entirely by $n(\mathbf{r}, t)$ in this case. Higher order terms, involving higher order tensors, are required for dealing with reactive systems.

In general, in the hydrodynamic regime, the space-time dependence of the distribution function is projected entirely onto the number density. Symbolically, we have in general

$$f(\mathbf{r}, \mathbf{v}, t) = \mathcal{F}\left[\mathbf{v}; n(\mathbf{r}, t)\right], \tag{11.48}$$

where the right hand side is a linear functional of the density, whether ∇n is small or not. If the density gradient is weak, then the functional is the density gradient expansion (Equation 11.47).

The actual kinetic theory problem becomes one in $\mathbf{v}$-space only in the form of a hierarchy of kinetic equations, whose solution is a universal property of the charged particle gas combination. Transport properties obtained in this way are thus *universal functions* of cross sections, gas temperature and pressure, and are the *same* for all experiments operating in the hydrodynamic regime. The geometry, boundary, and initial conditions pertaining to any particular experiment are accounted for entirely through solution of the diffusion equation for $n(\mathbf{r}, t)$.

It is emphasized that the above analysis is specifically for a gas or soft-condensed matter, for which there is no space or time variation associated $\left(\frac{\partial f}{\partial t}\right)_{\text{col}}$ condensed matter. However, for amorphous materials, where trapping and de-trapping introduce an explicit time dependence into $\left(\frac{\partial f}{\partial t}\right)_{\text{col}}$ (see Chapter 6), and large gradients persist for very long times, alternative procedures must be found. Extension of the above treatment to amorphous systems is discussed in Ref. [41].

11.6.4 Diffusion equation and transport coefficients

Physical properties of interest may be found as appropriate velocity moment of (Equation 11.47), for example, the particle flux (Equation 11.3)

$$\boldsymbol{\Gamma} = n\mathbf{v}_d - \mathbf{D} \cdot \nabla n + \ldots, \tag{11.49}$$

where the drift velocity $\mathbf{v}_d$ is defined by Equation 11.44, and

$$\mathbf{D} = \int \mathrm{d}^3 v \, \mathbf{v} \, \mathbf{F}^{(1)}(\mathbf{v}) \tag{11.50}$$

is the *diffusion tensor*. Substituting Equation 11.49 into the equation of continuity (Equation 7.5) gives the *diffusion equation*,

$$\frac{\partial n}{\partial t} + \mathbf{v}_d \cdot \nabla n - \mathbf{D} : \nabla\nabla n = 0, \tag{11.51}$$

which is to be solved for $n(\mathbf{r}, t)$ subject to the initial and boundary conditions of the experiment. The diffusion equation has been established previously in the context of fluid theory (see Section 9.4.2), with transport coefficients obtained from fluid analysis. The above expressions for drift velocity and diffusion coefficients on the other hand are furnished from solution of the Boltzmann equation.

11.6.5 Limitations of the density gradient expansion

The density gradient expansion (Equation 11.47) enables transport coefficients like Equations 11.44 and 11.50 to be defined for gaseous and soft-condensed matter media, but is valid strictly speaking only for the weak-gradient hydrodynamic regime. This prevails in the gaseous time-of-flight experiment, well downstream from the source, but not necessarily for other swarm experiments. The Cavallieri experiment, for example (see Chapter 13), operates in the hydrodynamic regime, but gradients may be large; for that reason, Equation 11.47 is therefore invalid. On the other hand, while the SST and Franck–Hertz experiments (Chapter 16) may have regions where the gradients are small, they are inherently *non-hydrodynamic*, and Equation 11.47 also has no validity.

In summary:

1. Weak gradients are neither necessary nor sufficient for the existence of a hydrodynamic regime;

2. There is no simple rule for prescribing if a system can be characterized by hydrodynamic description—each case has to be treated on its merits; and

3. *Transport coefficients* can be defined *only* in the weak gradient, hydrodynamic regime, when Equation 11.47 holds.

Finally, it is emphasized that the above analysis holds only if the applied fields are constant in time and uniform in space. Extension to consider time dependent fields is developed in Chapter 15.

11.7 Benchmark Models

Cases where the collision term simplifies may facilitate an exact analytic solution of Equation 11.1, which then provides a useful benchmark against which to test the accuracy and integrity of numerical procedures developed for more general situations. We have already encountered a number of such special cases, and go on to discuss them further here.

11.7.1 Constant collision frequency (Maxwell) model

For a spatially uniform, stationary swarm interacting elastically with the neutrals through an inverse fourth power potential (constant collision frequency model) one can find *all* velocity moments of the distribution function exactly, and hence, $f(\mathbf{v})$ itself. For more general interactions, one might then consider developing a scheme of successive approximations to solution of Equation 11.1 using information from the constant collision frequency model to construct the first "guess." This strategy is discussed in some detail by Mason and McDaniel [33] and also provides the basis for fluid modelling using "momentum transfer theory," as discussed in Chapters 8–10.

11.7.2 Light particles (quasi-Lorentz gas)

If $m/m_0 \ll 1$, the Boltzmann collision term can be approximated by the Davydov differential operator (see Chapter 5) and the solution of Equation 11.1 is further simplified by the two-term approximation of the distribution function. *Analytic solutions* can be obtained under stationary, spatially uniform conditions, for example, the Davydov distribution, valid for *any* elastic collision cross section $\sigma_m(v)$. Space and/or time variations limit the possibilities for exact solution, but nevertheless approximate analytic results can be obtained using the variational technique outlined in Chapter 13. Of course, while the two-term approximation (see Chapter 5) provides a very useful way of dealing with situations where elastic collisions dominate, things may change completely when inelastic processes are significant. Then Equation 11.1 must generally be solved numerically from the outset, with an arbitrary number of terms in the spherical harmonic expansion ("multi-term analysis").

11.7.3 Relaxation time model

For the idealized resonant charge transfer model, the collision term in Equation 11.1 can be approximated by a simple expression (see Section 5.2.2) and an *exact analytic solution* is often possible, *including space and time variations*. The complete time development of a travelling pulse of ions in a gas in a time-of-flight experiment is examined in Chapter 14, where there are *both* non-hydrodynamic and hydrodynamic regimes, close to and far downstream from the source, respectively. In this way, we can establish the regime of validity of the Fick's law and the diffusion equation, which underpin the way in which swarm experimental data are unfolded to produce cross sections.

These models and benchmarks are important as a guide to solving the exact kinetic equations, for which numerical procedures are necessary right from at the outset, as discussed in Chapter 12.

CHAPTER 12

Numerical Techniques for Solution of Boltzmann's Equation

12.1 Introduction

Despite the commonality in the fundamental equation describing electron and ion swarms in gases—the Boltzmann equation—the theoretical analyses of electron and ion swarms have, in general, evolved quite separately. This bifurcation is primarily a result of simplifications and approximations, which can be made due to the smallness of the electron to neutral molecule mass ratio, m/m_0. In elastic collisions, the fractional energy transfer is $\sim 2m/m_0$, which means that velocity vectors are essentially rotated in such collisions without any change in the length. The velocity distribution function is, thus, very nearly spherically symmetric and is represented to a good approximation by the first two terms of a spherical harmonic expansion. When inelastic collisions occur, the quasi-spherical symmetry should no longer be expected, and more terms in the spherical harmonic decomposition are generally required. The approximations routinely used in electron swarms do not apply in ion swarms, however, even when collisions are elastic. Consequently, the theoretical development has been on a different path and the mathematical machinery is somewhat different.

In what follows, we highlight a powerful technique for solution of Boltzmann's equation for electrons and ions using a series expansion in Laguerre polynomials and spherical harmonics, commonly referred to as Burnett functions. When dealing with light particles, for example, the Boltzmann collision term can be approximated by a differential operator (see Chapter 5), and standard methods for numerical solution of differential equations can be brought into play. However, if one wishes to develop a *universal* numerical scheme, valid for *both* light and heavy particles, the approach outlined below is possibly the best one to follow.

12.2 The Burnett Function Representation

For any realistic problem, whether hydrodynamic or non-hydrodynamic, with or without any geometrical symmetries, it is generally necessary to solve Equation 11.1 numerically. The following is only an outline; full details can be found in Refs. [29,51,92].

12.2.1 Representation of the directional dependence in velocity space

Since spherical harmonics $Y_m^{(l)}(\theta,\varphi) \equiv Y_m^{(l)}(\hat{\mathbf{v}})$ form a complete orthogonal set of basis functions on the unit sphere in velocity space, that is,

$$\int d^2\Omega_v\, Y_m^{[l]}(\hat{\mathbf{v}}) Y_m^{(l)}(\hat{\mathbf{v}}) = \delta_{l,l'}\delta_{m,m'}. \tag{12.1}$$

The spherical tensor notation is such that $Y_m^{[l]}(\hat{\mathbf{v}}) = Y_m^{(l)}(\hat{\mathbf{v}})^*$. The *directional dependence* of $f(\mathbf{v})$ (suppressing any $\mathbf{r}, t$ dependence for the moment) can be represented through the expansion

$$f(\mathbf{v}) = f(v,\theta,\varphi) = \sum_{l=0}^{\infty}\sum_{m=-l}^{l} f_m^{[l]}(v)\, Y_m^{(l)}(\theta,\varphi), \tag{12.2}$$

where

$$f_m^{[l]}(v) = \int d^2\Omega_v\, Y_m^{[l]}(\hat{\mathbf{v}})\, f(\mathbf{v}).$$

The spherical harmonics are defined by

$$Y_m^{[l]}(\theta,\varphi) = i^l\,(-1)^{(m+|m|)/2}\left[\frac{(2l+1)(l-|m|)!}{4\pi\,(l+|m|)!}\right]^{\frac{1}{2}} P_l^{|m|}(\cos\theta)e^{im\varphi},$$

where

$$P_l^{|m|}(\cos\theta) = \frac{(-1)^l}{2^l\, l!}(\sin\theta)^{|m|}\frac{d^{l+|m|}}{d\cos\theta^{l+|m|}}(1-\cos\theta)^l$$

denotes an associated Legendre function. The upper limit on the l-summation in Equation 12.2 is theoretically infinite, but in practice it is truncated to a finite size $l \le l_{\max}$, for example, in the *two-term approximation* $l_{\max} = 1$. The main thing is that $l_{\max}$ should remain flexible, and its value is ultimately chosen to satisfy some accuracy criterion. It would not be unusual for ions or electrons in molecular gases to have to take $l_{\max} \ge 4$ in order to meet an accuracy criterion of $\le 1\%$ for transport coefficients, for example.

12.2.2 Representation in speed space

Next, we look for a representation of the coefficients $f_m^{[l]}(v)$ in v-space of the form

$$f_m^{[l]}(v) = nw(\alpha, v)\sum_{\nu=0}^{\infty} F_m^{[\nu l]}\, R_{\nu l}(\alpha v), \tag{12.3}$$

where the polynomials

$$R_{\nu l}(\alpha v) = N_{\nu l}\left(\alpha v/\sqrt{2}\right)^{l} L_{\nu}^{(l+\frac{1}{2})}(\alpha^2 v^2/2) \tag{12.4}$$

form a complete set, orthogonal with the Maxwellian weight function $w(\alpha, v)$ defined by Equation 11.36 that is,

$$\int_0^\infty w(\alpha, v) R_{\nu l}(\alpha v) R_{\nu' l}(\alpha v)\, v dv = \delta_{v,\nu'}. \tag{12.5}$$

Here, $N_{\nu l}$ is a normalization coefficient, $L_{\nu}^{(l+\frac{1}{2})}$ is a generalized Laguerre polynomial, and

$$\alpha^2 \equiv m/k_B T_b$$

where T_b is to be understood as some "basis temperature," to be chosen later to optimize convergence of the expansion (Equation 12.3). Like the summation over l, the infinite sum over ν in Equation 12.3 has, in practice, to be truncated at some finite value $\nu_{\max}$, consistent with the imposed accuracy criteria. Further discussion on the various options for weighting functions in this basis set are discussed below.

We emphasize that Equation 12.3 is only one possibility for representation of $f_m^{[l]}(v)$, and one may consider discretization of $f_m^{[l]}(v)$ in v-space, rather than series representation. Formally, although the two approaches are equivalent, practicalities may dictate that one may be preferred over the other [93].

12.2.3 Decomposition in velocity space

If the expansions Equations 12.1 and 12.5 are combined, there follows the complete representation of the distribution function in **v**-space,

$$f(\mathbf{v}) = n\, w(\alpha, v) \sum_{\nu=0}^{\infty} \sum_{l=0}^{\infty} \mathcal{F}_m^{[\nu l]} \phi_m^{(\nu l)}(\alpha \mathbf{v}) \tag{12.6}$$

in terms of the three-dimensional polynomials

$$\phi_m^{(\nu l)}(\alpha \mathbf{v}) \equiv R_{\nu l}(\alpha v)\, Y_m^{(l)}(\hat{\mathbf{v}}) \tag{12.7}$$

called *Burnett functions*. These play a pivotal role throughout kinetic theory and interestingly enough, the quantities $e^{-r^2/2}\phi_m^{(\nu l)}(\mathbf{r})$ are (with appropriate normalization constants) also the eigenfunctions of the three-dimensional

Table 12.1 Low Order Burnett Function Moments

ν	l	m	$\phi_m^{[\nu l]}(\alpha v)$	$\mathcal{F}_m^{[\nu l]}$
0	0	0	1	1
0	1	0	$i\alpha v_z$	$i\alpha\langle v_z\rangle$
1	0	0	$\frac{1}{\sqrt{6}}(3-\alpha^2 v^2)$	$\frac{1}{\sqrt{6}}(3-\alpha^2\langle v^2\rangle)$

harmonic oscillator. Notice that by virtue of Equations 12.1 and 12.5, the Burnett functions have the orthogonality property

$$\int d^3v\, w(\alpha, v)\phi_m^{[\nu l]}(\alpha \mathbf{v})\phi_{m'}^{(\nu' l')}(\alpha \mathbf{v}) = \delta_{v,\nu'}\delta_{l,l'}\delta_{m,m'} \tag{12.8}$$

and hence the expansion coefficients are given by

$$\mathcal{F}_m^{[\nu l]} = \frac{1}{n}\int d^3v\, f(\mathbf{v})\, \phi_m^{[\nu l]}(\alpha \mathbf{v}) \equiv \langle \phi_m^{[\nu l]}(\alpha \mathbf{v})\rangle. \tag{12.9}$$

The connection between low order $\mathcal{F}_m^{[\nu l]}$ and some of the physically meaningful velocity moments of the distribution function is shown in Table 12.1. It is clear that $\mathcal{F}_0^{[00]}$, $\mathcal{F}_0^{[01]}$, and $\mathcal{F}_0^{[10]}$ are related to the normalization condition, average velocity in the z-direction, and mean energy, respectively.

12.2.4 Moments of the Boltzmann equation in the Burnett representation

We now form *moments* of the kinetic equation with respect to the Burnett functions, by multiplying the Boltzmann equation by $\phi_m^{[\nu l]}(\alpha \mathbf{v})$ and integrating over all velocities. Assuming both elastic and inelastic collisions, but neglecting reactive collisions for the moment, we obtain *exactly* as in Part II:

$$\frac{\partial\, [n\langle \phi_m^{[\nu l]}(\alpha \mathbf{v})\rangle]}{\partial t} + \nabla\cdot[n\langle \mathbf{v}\, \phi_m^{[\nu l]}(\alpha v)\rangle] - n\mathbf{a}\cdot\left\langle \frac{\partial \phi_m^{[\nu l]}(\alpha \mathbf{v})}{\partial \mathbf{v}}\right\rangle$$
$$= \sum_{j,j'}\int d^3v \int d^3v_0\, g f(r,v,t) f_{0j}(v_0)\int d^2\Omega_{\mathbf{g}'}\, \sigma(j, j'; g, \chi)$$
$$\times\left[\phi_m^{[\nu l]}(\alpha \mathbf{v}') - \phi_m^{[\nu l]}(\alpha \mathbf{v})\right], \tag{12.10}$$

where it is understood that this now represents an infinite set of equations, $\nu = 0, 1, 2, \ldots, \infty$; $l = 0, 1, 2, \ldots, \infty$; $m = -l, \ldots, l$. In practice, one truncates these equations to a finite set, by imposing upper limits ν_{max} and

Table 12.2 Burnett Function Equations and their Macroscopic Moment Equation

ν	l	m	**Moment Equation**
0	0	0	Equation of continuity
0	1	0	Momentum balance (z-component)
1	0	0	Energy balance

l_{max} on ν and l, respectively. As detailed in Table 12.2, the first few equations are readily recognizable but unlike the fluid description, where we effectively truncate at a very low order, $\nu_{max}=1$, $l_{max}=1$, there are many more equations, as many as are needed to furnish transport coefficients to the required degree of accuracy of ≤1%. In a typical rigorous moment calculation, it would not be at all unusual to have $\nu_{max}\sim 40$, $l_{max}\sim 4$, which together with the fact that m-indices are generally involved as well, implies that several hundred velocity moment equations are to be reckoned with. And this is before any consideration of representation in configuration space and time, which add further indices to the moment equations!

12.2.5 Burnett function representation of Boltzmann's equation

It remains to express the set of Equation 12.10 entirely in terms of the averages $\mathcal{F}_m^{[\nu l]}$ by substituting for $f(\mathbf{v})$ from Equation 12.6:

- *Left hand side:* The time derivative term is exactly $\frac{\partial[n\mathcal{F}_m^{[\nu l]}]}{\partial t}$, while in the next two terms, the quantities $\mathbf{v}\,\phi_m^{[\nu l]}(\alpha\mathbf{v})$ and $\frac{\partial\phi_m^{[\nu l]}(\alpha\mathbf{v})}{\partial\mathbf{v}}$ can be expressed as linear combinations of $\phi_m^{[\nu l]}(\alpha\mathbf{v})$ and $\phi_{m\pm1}^{[\nu\pm1,l\pm1]}(\alpha\mathbf{v})$ using recurrence relations for the Burnett functions (see e.g., Ref. [94]),

$$\mathbf{v}\,\phi_m^{[\nu l]}(\alpha v)=\alpha^{-1}\sum_{\nu'=\nu,\nu\pm1}\ \sum_{l'=l,l\pm1}\ \sum_{m'=-l'}^{l'}\left(\nu lm\left|\mathbf{D}_1\right|\nu'l'm'\right)\phi_{m'}^{[\nu'l']}(\alpha v)$$

$$\frac{\partial\phi_m^{[\nu l]}(\alpha\mathbf{v})}{\partial\mathbf{v}}=\alpha\sum_{\nu'=\nu,\nu\pm1}\ \sum_{l'=l,l\pm1}\ \sum_{m'=-l'}^{l'}\left(\nu lm\left|\mathbf{D}_2\right|\nu'l'm'\right)\phi_{m'}^{[\nu'l']}(\alpha v),$$

  where $\left(\nu lm\left|\mathbf{D}_i\right|\nu'l'm'\right)$ $(i=1,2)$ are "matrix elements" of the operators $\mathbf{v}$ and $\frac{\partial}{\partial\mathbf{v}}$, respectively, in the Burnett function basis. Hence both $\langle\mathbf{v}\,\phi_m^{[\nu l]}(\alpha\mathbf{v})\rangle$ and $\langle\frac{\partial\phi_m^{[\nu l]}(\alpha\mathbf{v})}{\partial\mathbf{v}}\rangle$ are linear combinations of $\mathcal{F}_m^{[\nu l]}$ and $\mathcal{F}_{m\pm1}^{[\nu\pm1,l\pm1]}$, and the left hand side of Equation 12.10 can, therefore, be written as

$$\sum_{\nu'=\nu,\nu\pm1}\ \sum_{l'=l,l\pm1}\ \sum_{m'=-l'}^{l'} (\nu lm\,|D|\,\nu' l' m')\ nF_{m'}^{[\nu' l']}, \tag{12.11}$$

where

$$\begin{aligned}(\nu lm\,|D|\,\nu' l' m') &= \delta_{\nu' l' m';\nu lm}\frac{\partial}{\partial t} + \alpha^{-1}\left(\nu lm\,|\mathbf{D}_1|\,\nu' l' m'\right)\cdot\nabla \\ &\quad - \alpha\left(\nu lm\,|\mathbf{D}_2|\,\nu' l' m'\right)\cdot\mathbf{a}.\end{aligned}$$

Apart from observing that $(\nu lm\,|\mathbf{D}_i|\,\nu' l' m')$ are pure numbers, we do not need any further information about them in the present overview.

- *Right hand side:* The right hand side can be written as

$$-n\sum_{\nu'=0}^{\infty}\sum_{l'=0}^{\infty}\sum_{m'=-l'}^{l'} (\nu lm\,|J|\,\nu' l' m')\ F_{m'}^{[\nu' l']} \tag{12.12}$$

in which the *collision matrix* $(\nu lm\,|J|\,\nu' l' m')$ in the Burnett function representation is defined by

$$\begin{aligned}(\nu lm\,|J|\,\nu' l' m') &= \sum_{j,j'}\int d^3v\int d^3v_0\, g\ \phi_{m'}^{[\nu' l']}(\alpha\mathbf{v})\, f_{0j}(\mathbf{v}_0)\int d^2\Omega_{g'}\ \sigma(j,j';g,\chi) \\ &\quad\times\left[\phi_m^{[\nu l]}(\alpha\mathbf{v}) - \phi_m^{[\nu l]}(\alpha\mathbf{v}')\right].\end{aligned}$$

It can be evaluated by transforming $(\mathbf{v},\mathbf{v}_0)$ to centre of mass and relative velocities $(\mathbf{g},\mathbf{G})$ in *exactly the same way* as in the evaluation of the much simpler collision terms in Part II. The Burnett functions transform according to the *Talmi transformation*, better known in nuclear physics [94]. For differential cross sections which depend only on the angle χ between $\mathbf{g}$ and $\mathbf{g}'$, and not on azimuthal angle φ, it turns out that the collision matrix simplifies considerably:

$$(\nu lm\,|J|\,\nu' l' m') = J_{\nu,\nu'}^{l}\,\delta_{ll'}\,\delta_{m,m'}.$$

Accurate calculation of $J_{\nu,\nu'}^{l}$ is central to any computational procedure in kinetic theory. An explicit expression is not needed for the present discussion, but it is pertinent to point out that the quantities to be ultimately calculated numerically are matrix elements of the partial cross sections detailed earlier

$$\sigma_l(j,j';g) = 2\pi \int_0^{\pi} \sigma(j,j';g,\chi)\,[1 - P_l(\cos\chi)]\sin\chi d\chi$$

with respect to the polynomials (Equation12.4), in what are called "interaction integrals."

- *Burnett function representation:* Equating Equations 12.11 and 12.12 gives the following infinite set of partial differential equations for $\mathcal{F}_{m'}^{[\nu'l']}$,

$$\sum_{\nu'=\nu,\nu\pm1} \sum_{l'=l,l\pm1} \sum_{m'=-l'}^{l'} \left\{ \left(\nu lm\,|D|\,\nu'l'm'\right) + \left(\nu lm\,|J|\,\nu'l'm'\right) \right\} n\, F_{m'}^{[\nu'l']} = 0 \tag{12.13}$$

$(\nu = 0, 1, 2, \ldots, \infty;\ l = 0, 1, 2, \ldots \infty;\ m = -l, \ldots, l)$ which is entirely equivalent to the original, single partial differential–integral Boltzmann kinetic equation.

12.3 Summary of Solution Procedure

1. Choose the basis set of functions, plus the basis temperature T_b.
2. Calculate the matrix elements of the operators, and form the infinite set of equations for $\mathcal{F}_m^{[\nu l]}$ representing the original Boltzmann equation.
3. Truncate this set to finite size, by placing upper limits $(\nu_{\max}, l_{\max})$ on (ν, l) indices, and solve for $\mathcal{F}_m^{[\nu l]}(\nu_{\max}, l_{\max})$.
4. Increment the truncation limits to $(\nu_{\max} + 1, l_{\max} + 1)$ and solve again for $\mathcal{F}_m^{[\nu l]}(\nu_{\max} + 1, l_{\max} + 1)$.
5. Estimate the error $\Delta = \left| \frac{\mathcal{F}_m^{[\nu l]}(\nu_{\max}+1, l_{\max}+1) - \mathcal{F}_m^{[\nu l]}(\nu_{\max}, l_{\max})}{\mathcal{F}_m^{[\nu l]}(\nu_{\max}, l_{\max})} \right|$ for the drift velocity $(\nu = 0, l = 1)$ and/or any other desired transport property.
6. Keep incrementing until the pre-imposed accuracy criterion is satisfied, say $\Delta < 0.1\%$.

Note that this procedure yields *all* the coefficients $\mathcal{F}_m^{[\nu l]}$ up to the truncation limits $(\nu_{\max}, l_{\max})$, and hence by Equation 12.6 furnishes the distribution function also, that is, (now making space–time dependences explicit),

$$f(\mathbf{r}, \mathbf{v}, t) \doteqdot n\, w(\alpha, v) \sum_{\nu=0}^{\nu_{\max}} \sum_{l=0}^{l_{\max}} \mathcal{F}_m^{[\nu l]}(\mathbf{r}, t)\, \phi_m^{(\nu l)}(\alpha \mathbf{v}). \tag{12.14}$$

12.4 Convergence and the Choice of Weighting Function

12.4.1 Convergence in the l-index

As an example, we consider electrons in CH_4, for which Table 12.3 shows convergence of transport coefficients with respect to l_{max} for a fixed value of ν_{max}, for $E/n_0 = 3.0$ Td at a gas temperature $T_0 = 293$ K. Clearly, one must go to $l_{max} = 7$ or 8 before being confident of an accuracy of ~0.1% for *all* transport coefficients. In Chapter 18, we consider the case where there is no longer rotational symmetry due to the presence of a magnetic field. In this case, convergence in the m-index comes under scrutiny.

It must be pointed out that expansion in terms of basis sets, together with calculation of matrix elements of operators, is more or less standard practice for solving the equations in many areas of physics. The procedure for kinetic theory as outlined above naturally differs in detail, but by no means in principle, from any of these other computational schemes. For a discussion, see Ref. [93].

12.4.2 Choice of weighting function

The choice of the weighting function $w(\alpha, c)$ is crucial to the success of the scheme. For a Burnett function basis, there are a variety of different weighting functions that one can consider, all of which have computational benefits in various domains. The parameters are varied in order to optimise convergence in the ν-index:

- Single-temperature method: An expansion about a Maxwellian distribution at the gas temperature. There is no flexibility in this weighting function to optimise convergence.

Table 12.3 Convergence of the Transport Coefficients for Electrons in Methane at 293 K and $E/n_0 = 3.0$ Td

l_{max}	$\mathbf{v}_d$ $(10^5 \mathrm{m\,s^{-1}})$	$n_0 \mathbf{D}_\perp$ $(10^{24} \mathrm{m^{-1}s^{-1}})$	$n_0 \mathbf{D}_\parallel$ $(10^{24} \mathrm{m^{-1}s^{-1}})$	ε (eV)
1	1.16	11.00 ± 0.02	2.38	0.383
2	1.05	$5.96 \pm .02$	3.81	0.364
3	1.07	$7.22 \pm .02$	3.26	0.368
4	1.07	$7.01 \pm .02$	3.37	0.367
5	1.07	$7.00 \pm .02$	3.36	0.367
6	1.07	$7.02 \pm .02$	3.36	0.367
7	1.07	$7.01 \pm .02$	3.36	0.367
8	1.07	$7.01 \pm .02$	3.36	0.367

- Two-temperature method: An expansion about a Maxwellian distribution at some arbitrary temperature T_b, which is varied to optimise convergence [29].
- Bi- and multi-maxwellian treatment: An expansion about a weighted sum of Maxwellian distributions at various temperature T_b^i (see e.g., Ref. [95]).
- Drifted Maxwellian: The basis temperature and drift parameter are simultaneously varied to optimise convergence.
- Gram–Chalier method: Higher order effects including skewness and kurtosis in the weighting function [96].

12.5 Ion Transport in Gases

As detailed above, the most computationally demanding component of this method involves the calculation of the matrix elements of the collision operator from the interaction cross sections. The Talmi transformation methods of Kumar et al. [30] are particularly helpful here, since they enable the separation of mass and interaction effects. To make the former transparent, the collision matrix is written in terms of a mass ratio expansion:

$$J^l_{\nu\nu'} = \sum_{p=0}^{\infty} \left(\frac{m}{m+m_0} \right)^p J^l_{\nu\nu'}(p). \tag{12.15}$$

Truncation at $p_{\max}$ is also required to evaluate the collision matrix (Equation 12.15), though the order of truncation required can usually be estimated a priori [97]. For electrons, truncation at $p_{\max}=1$ is sufficient in the elastic operator while a zeroth order truncation is sufficient for other collision processes. Obviously, the slowest convergence will be for equal mass ratios.

We assess the importance of the p-truncation below, together with truncation of the spherical harmonic expansion. For illustrative purposes, we consider a simple hard-sphere model and vary the ion to neutral mass ratio m/m_0.

12.5.1 Convergence in the l-index

Table 12.4 shows various transport properties are shown for two extreme m/m_0 values for varying $l_{\max}$. For light particles (e.g., $m/m_0=10^{-4}$), the two-term approximation (i.e., $l_{\max}=1$) is sufficient to obtain accuracies in the transport coefficients to within 0.1%. For $m/m_0=1$, however, the two-term approximation is not sufficient and a higher order truncation is required for equivalent accuracies.

Table 12.4 Influence of the Mass Ratio on Convergence in the Spherical Harmonic Expansion for Ions in a Hard Sphere Model Gas ($m_0 = 4$ amu; $T_0 = 293$ K; $E/n_0 = 1$ Td; $\sigma = 6$ Å^2) on the Various Transport Properties/Coefficients

m/m_0	l_{max}	ε (eV)	v_d (10^2ms^{-1})	$n_0 D_\parallel$ 10^{22}(ms)$^{-1}$	$n_0 D_\perp$ 10^{22}(ms)$^{-1}$	$T_\perp$ (10^2K)	$T_\parallel$ (10^2K)	$-n_0\Upsilon$ (10^{-2}kg s^{-2})
10^{-4}	1	0.73326	56.189	158.22	313.31	56.727	56.712	5258.6
	2	0.73324	56.187	158.27	313.20	56.717	56.727	5259.7
	3	0.73324	56.187	158.27	313.20	56.717	56.727	5259.7
10^0	1	0.04284	3.403	0.774	0.977	3.314	2.757	3.430
	2	0.04271	3.366	0.886	0.892	3.072	3.224	3.794
	3	0.04271	3.368	0.884	0.894	3.074	3.219	3.780
	4	0.04271	3.368	0.884	0.894	3.074	3.220	3.781
	5	0.04271	3.368	0.884	0.894	3.074	3.220	3.781
	6	0.04271	3.368	0.884	0.894	3.074	3.220	3.781

Source: R. D. White et al., *Computer Physics Communications*, 142, 349–355, 2001.

As discussed previously, the failure of the two-term approximation is indicative of an anisotropic velocity distribution function. This is highlighted in Figure 12.1, where we display the two-term and converged multi-term approximations to the *velocity* distribution function for the two mass ratios under consideration. For $m/m_0 = 10^{-4}$, the accuracy of the two-term approximation to the velocity distribution function supports the validity of the two-term approximation for the transport coefficients. The enhanced anisotropy in the velocity distribution for increased ion mass is due to the increased energy transfer per elastic collision from the ion to the neutral.* We should highlight, however, that good convergence in l_{max} for the transport coefficients can often mask inadequacies in the accuracy of the *velocity distribution* calculated using the same level of truncation, as shown in Figure 12.1 for $m/m_0 = 1$. The inability to accurately represent the velocity distribution function is clear, yet low order moments of the velocity distribution (e.g., ε and W) are surprisingly accurate. Higher

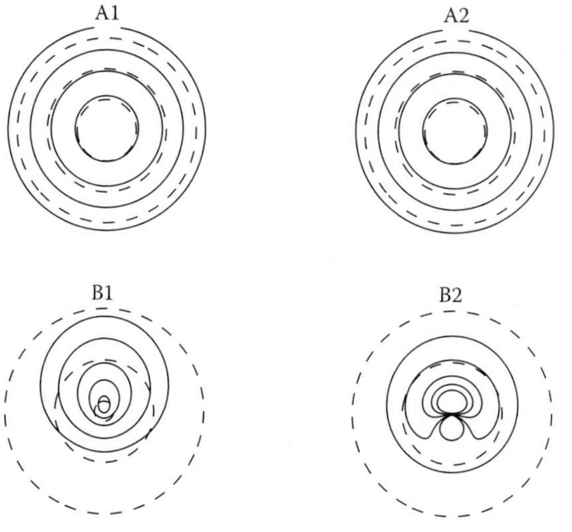

Figure 12.1 Contour plot of f in a plane containing the electric field for ions of masses 4×10^{-4} amu (A1,A2) and 4 amu (B1,B2) in the model gas. Figures labeled "1" refer to converged multi-term results while Figures labeled "2" refer to the two-term approximation. The values of the solid contours from largest to smallest radii in (A1,A2) 0.3,0.6,0.9.1.2,1.5 $(\text{eV})^{-3/2}$; (B1,B2) 25,50,100,150,200 $(\text{eV})^{-3/2}$. The energy scale is denoted by the dashed circular contours. The values of the energy scale contours of increasing radii are respectively for (A1,A2) 0.3,0.6,0.9 eV (B1,B2) 0.01, 0.05, 0.1. (From R. D. White et al., *Computer Physics Communications*, 142, 349–355, 2001.)

* We note that large relative energy exchange in inelastic collisions can also result in an anisotropic velocity distribution function, even for electrons, as reflected in the l-convergence properties in Table 12.3.

Table 12.5 Influence of the Mass Ratio on the Mass Ratio Expansion of the Collision Matrix for Ions in a Hard Sphere Model Gas with Unit Mass Ratio (m_0=4 amu; T_0=293 K; E/n_0=1 Td; σ=6 ÃĚ^2)

m/m_0	p_{max}	ε (eV)	v_d (10^2 ms^{-1})	$n_0 \mathbf{D}_{\parallel}$ 10^{22}(ms)$^{-1}$	$n_0 D_{\perp}$ 10^{22}(ms)$^{-1}$	$T_{\perp}$ (10^2 K)	$T_{\parallel}$ (10^2 K)	$-n_0\gamma$ (10^{-2} kg s^{-2})
10^{-4}	1	0.73318	56.188	158.25	313.20	56.713	56.723	5259.2
	2	0.73324	56.187	158.27	313.20	56.717	56.727	5259.7
	3	0.73324	56.187	158.27	313.20	56.717	56.727	5259.7
10^{0}	2	0.04248	3.382	0.884	0.898	3.051	3.207	3.45
	3	0.04270	3.366	0.887	0.893	3.074	3.218	3.91
	4	0.04271	3.368	0.883	0.894	3.074	3.219	3.79
	5	0.04270	3.365	0.883	0.895	3.073	3.218	3.81
	6	0.04271	3.368	0.884	0.894	3.074	3.220	3.78
	7	0.04271	3.368	0.884	0.894	3.074	3.220	3.78

Source: R. D. White et al., *Computer Physics Communications*, 142, 349–355, 2001.

order moments such as the temperature tensor are much more sensitive to accuracies in the velocity distribution function as demonstrated.

While one may be able to anticipate with some confidence a reasonable value of l_{max} based on the cross-sections, best scientific practice should leave it as a flexible parameter used to ensure a prescribed accuracy.

12.5.2 Convergence in the mass ratio expansion

In Table 12.5, convergence in the p-index is displayed for the two mass ratios under consideration. For $m/m_0 = 10^{-4}$, expectedly, truncation at $p_{max} = 1$ is sufficient to obtain accuracies within 0.1%. For $m/m_0 = 1$, however, setting $p_{max} = 1$ did not provide fully converged values of the transport properties. A priori estimates of p_{max} (based on equal weighting to $J^l_{\nu\nu'}(p)$ for each order p) may have placed p_{max} as high as 10 in order to achieve such accuracy. The numerical results presented here support the physical discussion above—there is a direct correlation between the values of p_{max} and the value of l_{max}, though the converse is in general not true.

CHAPTER 13

Boundary Conditions, Diffusion Cooling, and a Variational Method

13.1 Influence of Boundaries

13.1.1 Boundary effects, diffusion cooling, and heating

Particles in finite geometry generally have properties that are determined by scattering in the medium as well as interaction with the bounding surface. The situation is portrayed schematically in Figure 13.1.

Particles are effectively in thermal contact with both the gas and the walls and therefore they cannot, in general, come into thermal equilibrium with one or the other separately. Thus, although a steady state may be attained in the limit $t \to \infty$, the average particle thermal energy is generally not equal to $\frac{3}{2}k_B T_0$ where T_0 is the temperature of the medium. Whether the particles are cooler or warmer than the medium is determined by the energy-dependence of the scattering cross section. The gas may be relatively "transparent" in a particular energy region as compared with another, and particles with energies in that range are lost to the walls preferentially.

For the sake of being definite, we suppose that the medium is a monatomic gas and that the particles are electrons. Scattering is assumed to be elastic, and described by an energy-dependent momentum-transfer cross section $\sigma_m(\epsilon)$. The discussion which follows takes the walls to be perfect absorbers, but this is by no means necessary.

On the one hand, we expect more energetic electrons to reach the walls faster and that the loss rate is therefore proportional to the electron energy ϵ. On the other hand, collisions act to suppress movement of electrons, and the loss rate should be inversely proportional to the collision rate $\nu_m(\epsilon)$. Putting these two competing factors together suggests that the overall loss rate to the walls is proportional to

$$\epsilon/\nu_m(\epsilon) \sim \frac{\epsilon^{\frac{1}{2}}}{n_0 \sigma_m(\epsilon)} \tag{13.1}$$

from which we may deduce the following:

1. If the cross section $\sigma_m(\epsilon)$ either decreases with energy, or increases with ϵ no faster than $\epsilon^{\frac{1}{2}}$, the more energetic particles will pass

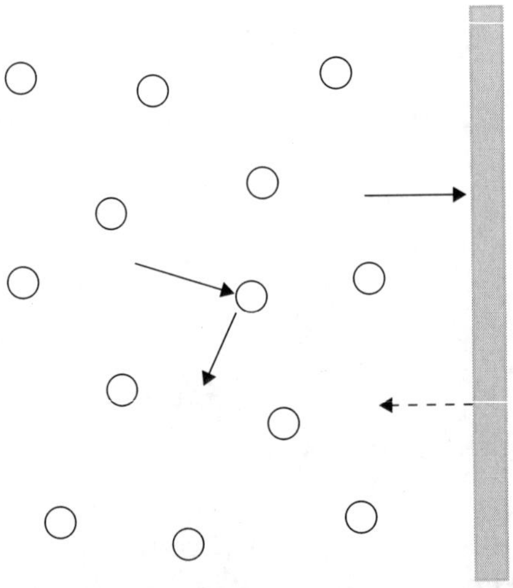

Figure 13.1 Particles diffusing through a gas to the surrounding walls are scattered by the gas atoms according to the cross section $\sigma_m(\epsilon)$. Those particles incident on the walls (solid arrow) may, in general, be absorbed and lost to the system, or reflected back into it (dashed arrow).

preferentially to the walls, and the remaining electrons will be cooled, with an average energy less than $\frac{3}{2}k_BT_0$.

2. If $\sigma_m(\epsilon)$ increases with energy faster than $\epsilon^{\frac{1}{2}}$ lower energy electrons are passed preferentially to the walls, raising the average energy of the remaining electrons to above $\frac{3}{2}k_BT_0$.

3. In the special case where $\sigma_m(\epsilon) \sim \epsilon^{\frac{1}{2}}$ there is neither heating nor cooling, and the particles may come into thermal equilibrium with the gas.

4. In all cases, the "transparency factor" (Equation 13.1) is inversely proportional to gas density n_0.

One speaks of diffusion "cooling" and "heating" in cases 1 and 2, respectively. The implications of this last observation are explored further below.

13.1.2 Pressure variation and practical considerations

We now investigate how the picture in Figure 13.1 changes as gas number density n_0, or equivalently, gas pressure $p_0 = n_0k_BT_0$ varies. Increasing p_0 increases the number of scattering events, and thus reduces the influence of the boundary on electron properties. Put another way, as gas pressure increases, thermal contact between electrons and the gas is enhanced and promotes the establishment of thermal equilibrium, that is, $\langle \frac{1}{2}mv^2 \rangle \to \frac{3}{2}k_B T_0$ in the limit $n_0 \to \infty$. Alternatively, we might think of the influence of the wall as extending through a "boundary layer," whose thickness is, as we shall see, of the order of the mean free path for energy transfer $\lambda_\epsilon \sim \frac{m_0}{m} \left(n_0 \sigma_m\right)^{-1}$. Since this is inversely proportional to gas pressure, the boundary layer expands and shrinks as gas pressure is lowered or raised, respectively.

Pressure-dependent experimental data indicates that boundaries are influencing measurements. While this can, in principle, be suppressed simply by increasing p_0, it may not be feasible to do so. In the case of argon gas, for example, $\sigma_m(\epsilon)$ has a deep Ramsauer minimum at low energies, and a correspondingly large transparency factor (Equation 13.1), which cannot be eliminated using any practically attainable gas pressures. In such cases, traces of a molecular buffer gas like hydrogen may be introduced to promote the required thermal contact with the gas.

In principle, the influence of boundaries might also be reduced by enlarging the container. Note that the limit $p_0 \to \infty$ is formally equivalent to a vessel of infinite size.

These ideas are further discussed below in the context of the Cavalleri experiment [10] which aims at measurement of the thermal equilibrium diffusion coefficient of electrons in a gas. In this case, boundary effects indicated by pressure variation and diffusion cooling are a problem to be eliminated. In other cases, however, such as plasma processing devices [13], interaction of electrons with the bounding surface is of prime interest.

13.1.3 Theoretical considerations

Ideally, the goal would be to develop a general kinetic theory covering all possibilities, but one runs into problems right at the start. Thus, some simplification of boundary conditions is inevitable if one is to make any progress in solution of the kinetic equation. There are generally two steps in this process: (1) idealization of the actual boundary conditions and (2) further approximation to make the problem mathematically tractable. However, even after making approximations, the analysis is generally far from straightforward, requiring new mathematical methods to deal with the additional physics introduced by the boundaries. It is here that the eigenvalue formulation introduced in Chapter 11 plays an important role. For example, the electron population in a Cavalleri experiment decays exponentially, and the measured time constant corresponds to the smallest member of an eigenvalue spectrum. This has an extremal property which allows it to be obtained numerically through a variational procedure.

We commence the theoretical investigation using the simplest possible geometry.

13.2 Plane-Parallel Geometry

We showed in Chapter 11 that the solution of the Boltzmann equation is determined uniquely by specifying the distribution function f at the boundary in the half-space corresponding to inwardly directed velocities only. For the plane-parallel geometry of Figure 13.2, properties are assumed to vary only in the direction normal to the boundaries (the z-axis) and the boundary conditions are

$$\begin{aligned} f(z,\mathbf{v},t) &= 0 \quad \text{for } z=0, \quad v_z>0 \\ f(z,\mathbf{v},t) &= 0 \quad \text{for } z=d, \quad v_z<0. \end{aligned} \tag{13.2}$$

Since the z-axis defines an axis of rotational symmetry in velocity space, $f(z,\mathbf{v},t)=f(z,v,\theta,t)$, where θ is the angle between particle velocity and the z-axis. The boundary conditions can also be written in terms of cylindrical coordinates z and θ, for example, at the left hand wall,

$$f(0,v,\theta,t)=0, \quad (0\leq\theta<\pi/2) \tag{13.3}$$

Solution of the kinetic equation normally commences with decomposition of f in a finite series of spherical harmonics or Legendre polynomials, but then it is impossible to satisfy the true boundary condition (Equation 13.3), and some approximation is essential. For example, the two-term approximation (Chapter 5),

$$f(z,v,\theta,t)\approx f^{(0)}(z,v,t)+f^{(1)}(z,v,t)\cos\theta \tag{13.4}$$

can at best satisfy Equation 13.3 for all v, t at one value of θ only, say $\theta=0$:

$$f(0,v,0,t)\approx f^{(0)}(0,v,t)+f^{(1)}(0,v,t)=0,$$

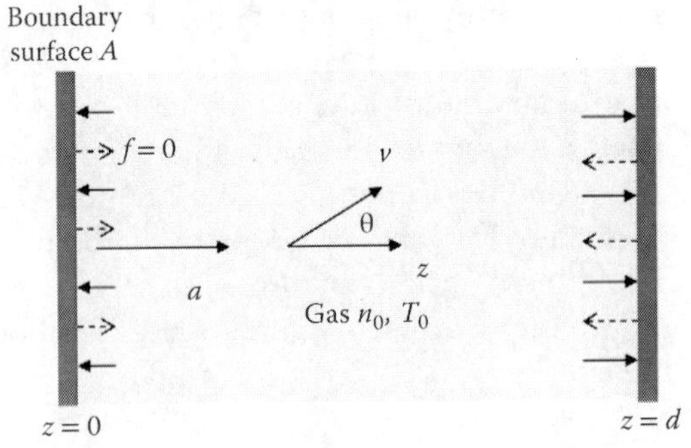

Figure 13.2 Particles in gas confined by absorbing plane-parallel boundaries. Solid and dashed arrows indicate normal components of velocity directed into and out of the bounding surfaces, respectively. In the case of perfect absorption, a particle incident onto the surface does not return to the gas, as expressed by Equation 13.2.

which is a *Mark* type of boundary condition, familiar in neutron transport theory [98]. Since the rationale of the two-term approximation is that $f^{(1)} \ll f^{(0)}$, it has been argued by Parker and Lowke [99] that this implies

$$f^{(0)}(0, v, t) = 0. \tag{13.5}$$

Hence the number density also vanishes on the boundary,

$$n(0, t) = 4\pi \int_0^\infty dv\, v^2 f^{(0)}(0, v, t) = 0, \tag{13.6}$$

which is also a boundary condition often used in solving the diffusion equation.

A similar argument can be made for the right hand boundary at $z = d$, and hence the overall boundary condition can be written as

$$f^{(0)}(z, v, t) = 0, \quad n(z, t) = 0; \quad z \in A.$$

One can relax the assumption of perfect absorption and include partial reflection and absorption, but the fundamental difficulty of dealing with truncated expansions remains, no matter how many spherical harmonics are included. For that reason, we adopt a pragmatic view in the calculations which follow. Before moving onto these details, however, there are some further general observations to be made:

- The presence of boundaries can best be handled in principle through a *half-space* formulation [100], in which the distribution function is considered separately in regions of velocity space for which (say) $v_z > 0$ and $v_z < 0$, respectively, and Boltzmann's equation decomposed accordingly. Boundary conditions like Equation 13.2 can then be dealt with quite readily, even with finite representations of f like Equation 13.4 in each of the respective half-spaces.
- As we have seen in Chapter 11, a *unique* solution of Boltzmann's equation generally requires specification of only the distribution function for velocities *incident into* the medium, as in Equation 13.2. Fixing the total distribution function at the boundaries would amount to an over-specification of boundary conditions. Thus, when representing f by a series of Legendre polynomials, of which Equation 13.4 gives just the first two terms, only *half* the coefficients, for example, $f^{(l)}(\mathbf{r}, v, t)$ with *either* odd *or* even l, need be specified at a particular boundary in order to generate a unique solution of Boltzmann's equation. In the two-term approximation, specification of $f^{(0)}$ as in Equation 13.5 is therefore sufficient to facilitate a solution of Equation 13.8 below, and it would be incorrect to try to specify $f^{(1)}$ as well.

These and other-related matters have for long been studied in the context of neutron transport [100] and many of the results carry over to particles in dilute gases.

13.3 The Cavalleri Experiment

13.3.1 Influence of boundaries

The Cavalleri experiment [10] follows the general arrangement depicted in Figure 13.2, and aims at determining the diffusion coefficient of dilute electrons in a noble gas. It was originally devised with a cylindrical chamber having an alternating electric field applied along the axis, but subsequently interest was focused on the zero field case, and determining the equilibrium diffusion coefficient. In any case, electrons are produced in a cylindrical chamber by ionizing x-rays, and subsequently, diffuse freely through the gas to the walls of the container, where they are absorbed. The number of electrons remaining in the chamber is monitored as a function time and, if the decay is found to be exponential, a decay time constant can be extracted, from which a diffusion coefficient may be inferred. In general, this is an "effective" diffusion coefficient, whose value is pressure and geometry dependent. However, the pressure (and geometry) dependence can be eliminated using the strategies described in Section 13.1.2, and the experiment then furnishes the required equilibrium diffusion coefficient of electrons in the gas.

Note that the measured or "effective" diffusion coefficient is, in all cases, *lower* than the true thermal value, because of the influence of boundaries, and regardless of whether diffusion cooling or heating takes place.

13.3.2 Kinetic theory

13.3.2.1 General two-term equations

For electrons undergoing elastic collisions in a neutral gas, the phase space distribution function $f(\mathbf{r}, \mathbf{v}, t)$ may be represented by the two-term representation (Chapter 6)

$$f(\mathbf{r}, \mathbf{v}, t) = f^{(0)}(\mathbf{r}, v, t) + \mathbf{f}^{(1)}(\mathbf{r}, v, t) \cdot \hat{\mathbf{v}} + \ldots,$$

where the isotropic and vector components $f^{(0)}$ and $\mathbf{f}^{(1)}$ satisfy

$$\frac{\partial f^{(0)}}{\partial t} + \frac{v}{3}\nabla \cdot \mathbf{f}^{(1)} = \frac{m}{m_0}\frac{1}{v^2}\frac{\partial}{\partial v}\left\{v^2 \nu_m(v)\left[v f^{(0)}(v) + \frac{k_B T_0}{m}\frac{\partial f^{(0)}}{\partial v}\right]\right\} \tag{13.7}$$

and

$$\frac{\partial \mathbf{f}^{(1)}}{\partial t} + v\nabla f^{(0)} = -\nu_m(v)\mathbf{f}^{(1)}(v),$$

respectively for zero field. If the timescale for macroscopic variations is long compared with the time ν_m^{-1} between collisions, then $\frac{\partial \mathbf{f}^{(1)}}{\partial t} \ll \nu_m \mathbf{f}^{(1)}$ and

$$\mathbf{f}^{(1)}(v) \approx -\frac{v}{\nu_m(v)}\nabla f^{(0)}.$$

Substituting this in the left hand side of Equation 13.7, and using energy $\epsilon = \frac{1}{2}mv^2$ as the independent variable, we obtain an equation for $f^{(0)}$ alone:

$$\left(\frac{m\epsilon}{2}\right)^{\frac{1}{2}} \frac{\partial f^{(0)}}{\partial t} = \frac{\epsilon}{3n_0\sigma_m(\epsilon)}\nabla^2 f^{(0)} + \frac{2m}{m_0}\frac{\partial}{\partial \epsilon}\left[n_0\epsilon^2\sigma_m(\epsilon)\left(f^{(0)} + k_B T_0 \frac{\partial f^{(0)}}{\partial \epsilon}\right)\right]. \tag{13.8}$$

13.3.2.2 Solution of kinetic equations

Equation 13.8 is solved by separating the variables,

$$f^{(0)}(\epsilon, \mathbf{r}, t) = F(\epsilon)R(\mathbf{r})\mathcal{T}(t) \tag{13.9}$$

where

$$\nabla^2 R = -\Lambda^{-2} R \tag{13.10}$$

$$\frac{\mathrm{d}\mathcal{T}}{\mathrm{d}t} = -\Lambda^{-2} D\mathcal{T} \tag{13.11}$$

and

$$\frac{2m}{m_0}\frac{\mathrm{d}}{\mathrm{d}\epsilon}\left[n_0\epsilon^2\sigma_m(\epsilon)\left(F + k_B T_0\frac{\mathrm{d}F}{\mathrm{d}\epsilon}\right)\right] + \Lambda^{-2}\left[\left(\frac{m\epsilon}{2}\right)^{\frac{1}{2}} D - \frac{\epsilon}{3n_0\sigma_m(\epsilon)}\right] F = 0. \tag{13.12}$$

Here Λ and D are separation constants which, as we shall see, are effectively the eigenvalues of Equations 13.10 and 13.12, respectively.

Note that the number density is given by

$$\begin{aligned} n(\mathbf{r}, t) &= \frac{4\pi}{m}\left(\frac{2}{m}\right)^{\frac{1}{2}} \int_0^\infty f^{(0)}(\epsilon, \mathbf{r}, t)\, \epsilon^{\frac{1}{2}}\, \mathrm{d}\epsilon \\ &= R(\mathbf{r})\mathcal{T}(t) \end{aligned} \tag{13.13}$$

if $F(\epsilon)$ is normalized to unity, that is,

$$\frac{4\pi}{m}\left(\frac{2}{m}\right)^{\frac{1}{2}}\int_0^\infty F(\epsilon)\epsilon^{\frac{1}{2}}\,d\epsilon = 1.$$

From Equations 13.13 through 13.11, it is clear that density satisfies a diffusion equation,

$$\frac{\partial n}{\partial t} = D\nabla^2 n, \tag{13.14}$$

with the eigenvalue D playing the role of a diffusion coefficient.

13.3.3 Diffusion coefficient as an eigenvalue

Equation 13.10 is an eigenvalue equation (see Section 11.5), with eigenvalues Λ_j^{-2} and eigenfunctions $R_j(\mathbf{r})$, ordered by some index (or set of indices) j, which are determined by applying the boundary conditions. For a perfectly absorbing boundary surface A, the boundary condition is taken to be $R(\mathbf{r})=0$ for $\mathbf{r}\in A$. In the one-dimensional, plane-parallel geometry shown in Figure 13.2, where the separation between walls is d, Equation 13.10 becomes

$$d^2R/dz^2 = -\Lambda^{-2}R,$$

whose solution for $R(z)=0$ for $z=0,d$ produces the following spatial "modes" and eigenvalues:

$$R_j(z) \sim \sin\left(\frac{z}{\Lambda_j}\right) \tag{13.15}$$

$$\Lambda_j = \frac{d}{j\pi} \qquad (j=1,2,3,\ldots),$$

respectively. The lowest order spatial mode, $j=1$, corresponds to the smallest value of Λ_j^{-2}, namely $\Lambda_1^{-2}=\left(\frac{\pi}{d}\right)^2$. On the other hand, the energy eigenfunctions $F_n(\epsilon;\Lambda_j)$, and corresponding eigenvalues $D_n(\Lambda_j)$, are determined by solution of the energy space eigenvalue problem (Equation 13.12). Here $n=1,2,\ldots$ is an index ordering the eigenvalues of energy modes from lowest to highest in magnitude. Thus, $D_1(\Lambda_j)$ is the lowest energy eigenvalue for a given spatial mode j.

The solution of Equation 13.11 is

$$\mathcal{T}_{n,j}(t) = \mathcal{T}_{n,j}(0)\exp\left[-\Lambda_j^{-2}D_n(\Lambda_j)t\right].$$

Since the original partial differential equation is linear, its most general solution is a linear combination of all the possible modes (Equation 13.9), that is,

$$f^{(0)}(r,\epsilon,t) = \sum_{n=1}^{\infty}\sum_{j=1}^{\infty} A_{nj}\, F_n(\epsilon;\Lambda_j)\, R_j(\mathbf{r}) \exp\left[-\Lambda_j^{-2} D_n(\Lambda_j) t\right]. \tag{13.16}$$

The coefficients A_{nj} could be found, if desired, from application of the initial conditions and using orthogonality of the dual eigenfunctions (see Section 11.5 and Appendix C), but these are not needed for present purposes. At long times, the lowest mode $n=1, j=1$ dominates, and hence the asymptotic solution of Boltzmann's equation is

$$f^{(0)}(r,\epsilon,t\to\infty) = A_{11}F_1(\epsilon;\Lambda_1)R_1(\mathbf{r})\exp\left[-t/\tau_1\right],$$

where

$$\tau_1 = \frac{\Lambda_1^2}{D_1(\Lambda_1)} \tag{13.17}$$

is the time constant for the fundamental mode. This is the quantity measured in experiment.

We shall focus on the asymptotic solution from now on and, in order to simplify notation, it will be implicit that $F(\epsilon)$ stands for $F_1(\epsilon;\Lambda_1)$, τ for τ_1, D for $D_1(\Lambda_1)$, and Λ for Λ_1.

13.4 Variational Method

13.4.1 Kinetic equation and variational principle

In general, Equation 13.12 must be solved numerically, for example, by discretization in energy space and otherwise employing standard methods for handling second order differential eigenvalue equations. Accuracies of around 0.1% can readily be achieved in this way. An alternative, but somewhat less accurate approach, is to use a *variational method*, based on the extremal (minimum) property of the eigenvalues. This approach is well-known in quantum mechanics (see Appendix C) and atomic and molecular physics, and its accuracy is dependent on the appropriate choice of "trial functions" incorporating variational parameters. In what follows, we sacrifice numerical accuracy in the interests of elucidating the physics and simplifying the mathematics, and opt for a simple trial function, with only one variational parameter.

First, it is generally useful to convert to dimensionless quantities, and thus, we introduce the quantities

$$\begin{aligned} u &= \frac{\epsilon}{k_B T_0} \\ \kappa &= \bar{\lambda}_\epsilon / \left(\sqrt{3}\Lambda\right) \\ \vartheta &= 3D / \left(\bar{v}_{th}\bar{\lambda}_m\right) \\ Q_m &= \frac{\sigma_m}{\bar{Q}}, \end{aligned} \tag{13.18}$$

where $\bar{v}_{th} = \sqrt{\frac{2k_B T_0}{m}}$, representative mean free paths for momentum and energy transfer are defined by

$$\begin{aligned} \bar{\lambda}_m &= \left(n_0 \bar{Q}\right)^{-1} \\ \bar{\lambda}_\epsilon &= \bar{\lambda}_m \left(\frac{m_0}{2m}\right)^{\frac{1}{2}}, \end{aligned} \tag{13.19}$$

respectively, and $\bar{Q} = 10^{-20}\text{m}^2$ is a representative cross section. The time constant (Equation 13.17) is thus

$$\tau = \left(\frac{2m}{m_0}\frac{\bar{v}_{th}}{\bar{\lambda}_m}\kappa^2\vartheta\right)^{-1}. \tag{13.20}$$

If a new unknown $\phi(u)$ is introduced through the substitution

$$F(u) = e^{-u}\phi(u) \tag{13.21}$$

then Equation 13.12 becomes

$$\frac{\mathrm{d}}{\mathrm{d}u}\left(u^2 Q_m e^{-u}\frac{\mathrm{d}\phi}{\mathrm{d}u}\right) + \kappa^2\left(u^{\frac{1}{2}}\vartheta - \frac{u}{Q_m}\right)e^{-u}\phi = 0. \tag{13.22}$$

13.4.2 Minimizing the functional

Equation 13.22 is of standard Sturm–Liouville form, with a self-adjoint differential operator. The calculus of variations can be immediately brought to bear [101], and in particular, we can show that the *minimum value* of the functional,

$$\vartheta = \frac{\int_0^\infty \mathrm{d}u \left[u^2 Q_m e^{-u}(\phi')^2 + \kappa^2 e^{-u}\frac{u}{Q_m}\phi^2\right]}{\kappa^2 \int_0^\infty \mathrm{d}u\, e^{-u} u^{\frac{1}{2}} \phi^2}, \tag{13.23}$$

with respect to all possible variations of the function $\phi(u)$, corresponds to the lowest eigenvalue of Equation 13.22, and the function $\phi(u)$ which brings about this minimum is the corresponding eigenfunction. It is clear that the diffusion coefficient itself $D \sim \vartheta$ also has this extremal property.

The utility of the method is illustrated for model cross sections of the form

$$Q_m = u^{\gamma-1}, \tag{13.24}$$

where γ is a constant. Numerical calculations for real cross sections are left as exercises. Note also that the variational principle can be readily extended to include inelastic processes and gas mixtures [102].

For simplicity, we choose a single parameter trial function,

$$\phi = e^{-\alpha u} \tag{13.25}$$

where α is an adjustable parameter. This together with Equation 13.21 means that we are effectively taking an approximate energy distribution function of the form

$$F(\epsilon) \sim \exp\left[-\frac{\epsilon}{k_B T}\right],$$

where

$$T = \frac{T_0}{1+\alpha} \tag{13.26}$$

is the electron temperature. Substitution of Equation 13.25 in Equation 13.23 yields

$$\vartheta = \frac{2\pi^{-\frac{1}{2}}}{\kappa^2}\left[\frac{\alpha^2\Gamma(\gamma+2)}{(1+2\alpha)^{\gamma+\frac{1}{2}}} + \frac{\kappa^2\Gamma(3-\gamma)}{(1+2\alpha)^{-\gamma+\frac{3}{2}}}\right] \tag{13.27}$$

and setting $\frac{d\vartheta}{d\alpha} = 0$ then gives the value of α which minimizes Equation 13.24. In what follows, we call the latter ϑ_{min}.

13.4.3 Model calculations and diffusion cooling

13.4.3.1 $\gamma = \frac{1}{2}$ (constant collision frequency)

$$\alpha = \frac{1}{2}\left[\left(1+4\kappa^2\right)^{1/2} - 1\right]$$

$$\vartheta_{\text{min}} = \frac{3}{4\kappa^2}\left[\left(1+4\kappa^2\right)^{1/2} - 1\right]. \tag{13.28}$$

In this case, Equation 13.25 is exactly the lowest eigenfunction and Equation 13.28 is the exact lowest eigenvalue, as can be verified by direct substitution into Equation 13.22. In addition, we note that Parker [103] obtained an exact analytic expressions for the nth eigenfunction and eigenvalue,

$$\vartheta_n = \frac{(n-1)}{\kappa^2}\left(1+4\kappa^2\right)^{1/2} + \frac{3}{4\kappa^2}\left[\left(1+4\kappa^2\right)^{1/2} - 1\right] \qquad (n=1,2.,...) \tag{13.29}$$

and setting $n=1$ yields Equation 13.28.

Since $\alpha > 0$, it is obvious from Equation 13.26 that $T < T_0$, that is, *diffusion cooling* occurs, as expected. As the gas pressure (or equivalently, the number density) and/or the size of the containing vessel increase to large enough values, then $\kappa \sim (n_0\Lambda)^{-1} \to 0$, $\alpha \to 0$, and the diffusion cooling effect vanishes. Under these conditions, the electrons come into thermal equilibrium with the gas, $T \to T_0$, and $\vartheta_{min} \to \vartheta_{min}(0) = 3/2$. The equilibrium diffusion coefficient $D_{equil} \sim \vartheta_{min}(0)$ then follows from Equation 13.18. Notice, however, that the effective diffusion coefficient under non-equilibrium conditions is never greater than the equilibrium value, that is,

$$D \leq D_{equil}. \tag{13.30}$$

Although proved in the context of a particular model, this inequality nevertheless expresses a general result, true for any cross section.

13.4.3.2 $\gamma = 1$ (constant cross section)

In this case, the variational method with the very simple trial function (Equation 13.25) produces only approximate results:

$$\alpha = \left(1 - \frac{1}{2}\kappa^2 + \frac{1}{4}\kappa^4\right) - \left(1 - \frac{1}{4}\kappa^2\right)$$

which for $\kappa < 1$, gives $\alpha \approx \frac{1}{4}\kappa^2$, and $\vartheta_{min} \approx 2\pi^{-\frac{1}{2}}\left(1 - \frac{1}{8}\kappa^2\right)$. (Note that the exact value for the coefficient of κ^2 in the expression for ϑ is closer to 1/6 than 1/8.) On reverting to dimensional quantities, we find

$$D \approx D_{equil}\left[1 - \frac{1}{24}\left(\frac{\bar{\lambda}_\epsilon}{\Lambda}\right)^2\right]$$

$$D_{equil} \approx \frac{2\pi^{-\frac{1}{2}}}{3}\bar{v}_{th}\bar{\lambda}_m.$$

13.4.3.3 $\gamma = \frac{3}{2}$ (cross section proportional to speed)

For this case, Equation 13.25 is also an exact eigenfunction, and

$$\alpha = 0, \quad \vartheta = 1, \quad D = D_{\text{equil}} = \frac{1}{3}\bar{v}_{th}\bar{\lambda}_m.$$

There is no diffusion cooling, and the electrons are always in thermal equilibrium with the gas $T = T_0$.

13.4.3.4 $\gamma > \frac{3}{2}$ (cross section increases faster than speed)

Here it can be shown from differentiation of Equation 13.27 that $\alpha < 0$, $T > T_0$, (i.e., diffusion heating) but still $D < D_{\text{equil}}$.

Real cases can be analyzed using the variational principle, with numerical integration over tabulated values of cross sections.

13.5 Diffusion Cooling in an Alternating Electric Field

13.5.1 Variational principle for the time-averaged kinetic equation

Suppose now an alternating field

$$E = E_0 \cos(2\pi\omega t) \tag{13.31}$$

is applied to the system, as was the case in the design of the original Cavalleri experiment. It is assumed that this field not only heats the electrons, but also effectively smooths out any gradients along its direction, leaving only diffusion to the walls in a direction *transverse* to the field. (This approximation would work best for a long, cylindrical diffusion chamber, in which the field is applied along the axis.) The applied frequency ω is assumed to lie in the radio frequency range, and hence

$$\nu_m \ll \omega < \nu_e \tag{13.32}$$

for typical collision frequencies for momentum and energy transfer, respectively. If collisions are predominantly elastic, the electron distribution function may still be well represented by the first two terms of the spherical harmonic expansion (Equation 13.4), even though the field may be strong. For conditions under which Equation 13.32 holds, the energy distribution of the electrons $f^{(0)}$ is only weakly modulated during one cycle of the field, and can be taken to be effectively constant over one period of the oscillating field.

Generalizing Equation 13.8 to include an electric field as in Chapter 5, it is thus found that the (cycle-averaged) number density is again given by an effective diffusion equation (Equation 13.14), and that the experimentally measured relaxation time is again given by Equation 13.20. The effective diffusion coefficient D is the eigenvalue of

$$2\frac{m}{m_0}\frac{\mathrm{d}}{\mathrm{d}\epsilon}\left\{n_0\epsilon^2\sigma_m\,F+\left[n_0\epsilon^2\sigma_m k_B T_0+\frac{m_0}{6m}\left(eE_{\mathrm{rms}}\right)^2\frac{\epsilon}{n_0\sigma_m}\right]\frac{\mathrm{d}F}{\mathrm{d}\epsilon}\right\}$$
$$+\left[\left(\frac{m\epsilon}{2}\right)^{\frac{1}{2}}D-\frac{\epsilon}{3n_0\sigma_m}\right]\Lambda^{-2}F=0, \tag{13.33}$$

which differs from Equation 13.12 through a field term, in which the root-mean-square of Equation 13.31 is $E_{\mathrm{rms}}=E_0/\sqrt{2}$. Equation 13.33 appears to be not greatly different from its zero-field counterpart, but in reality it is much more difficult to solve.

Note, even though a field is present, only one diffusion coefficient appears in the eigenvalue equation. This is because cycle-averaged diffusion in an oscillating field is essentially isotropic if the condition (Equation 13.32) is satisfied, and D is to be interpreted as a *lateral* diffusion coefficient.

As before, dimensionless quantities (Equation 13.18) are introduced, along with a dimensionless field

$$E^*=\frac{eE_{\mathrm{rms}}\bar{\lambda}}{\sqrt{3k_B T_0}}. \tag{13.34}$$

If we further define

$$\xi(u)=\int\frac{\mathrm{d}u}{1+(E^*)^2/uQ_m^2} \tag{13.35}$$

and write instead of Equation 13.21,

$$F=e^{-\xi(u)}\phi(u) \tag{13.36}$$

then Equation 13.33 takes the Sturm–Liouville form

$$\frac{\mathrm{d}}{\mathrm{d}u}\left(p(u)\frac{\mathrm{d}\phi}{\mathrm{d}u}\right)+(\vartheta\, r(u)-s(u))\phi=0, \tag{13.37}$$

where

$$\begin{aligned}p(u)&=Q_m\,u^2e^{-\xi}/\xi'\\ r(u)&=\kappa^2u^{\frac{1}{2}}e^{-\xi}\\ s(u)&=\kappa^2ue^{-\xi}/Q_m.\end{aligned} \tag{13.38}$$

The variational principle (Equation 13.23) for zero field then generalizes to finding the extremal of the functional

$$\vartheta=\frac{\int_0^\infty[p(\phi')^2+s\phi^2]\mathrm{d}u}{\int_0^\infty r\phi^2\mathrm{d}u}. \tag{13.39}$$

Since mainly the asymptotic behaviour of the electron density is of interest, only the lowest eigenvalue needs to be investigated, as explained previously.

13.5.2 Model calculations and diffusion cooling in an alternating field

As before, we take a trial function of the form (Equation 13.25), with variational parameter α, and first, consider application of the variational principle to model cross sections which can be handled analytically:

13.5.2.1 Constant collision frequency $Q_m = u^{-\frac{1}{2}}$, $\nu_m = \text{constant}$

The variational principle gives

$$\alpha_{\min} = \frac{1}{2}\left(\sqrt{1+4\beta\kappa^2} - 1\right) \tag{13.40}$$

$$\vartheta_{\min} = \frac{3}{4\kappa^2}\left(\sqrt{1+4\beta\kappa^2} - 1\right), \tag{13.41}$$

where

$$\beta = 1 + (E^*)^2.$$

This obviously reduces to the field-free results (Equation 13.28) when $E^* \to 0$. Note that while Equation 13.41 is, in fact, the exact lowest eigenvalue of Equation 13.37, the complete spectrum can be calculated for this model.

From Equations 13.18 and 13.41, it then follows that the effective diffusion coefficient is

$$D = \frac{\bar{v}_{th}\,\bar{\lambda}}{4\kappa^2}\left(\sqrt{1+4\beta\kappa^2} - 1\right), \tag{13.42}$$

while the lowest order eigenfunction of Equation 13.33 (again, exact for this model) can be written as (reverting now to dimensional quantities)

$$F(\epsilon) = \exp\left(-\epsilon/k_B T\right). \tag{13.43}$$

The effective electron temperature is given by

$$T = T_e/(1 + \alpha_{\min}) \tag{13.44}$$

where

$$\begin{aligned} k_B T_e &= \beta k_B T_0 \\ &= k_B T_0 + \frac{1}{3} m_0 (eE_{\text{rms}}/m\nu_m)^2 \end{aligned} \tag{13.45}$$

is the free-space electron temperature. (N.B. This is effectively, the Wannier energy relation.) Clearly, since $\alpha_{\min} > 0$,

$$T_{\text{eff}} < T_e \tag{13.46}$$

and the electrons are "cooled" by diffusion to the walls.

In the limit of a large chamber and/or high gas pressure,

$$\kappa \to 0 \tag{13.47}$$

$$\alpha \to 0, \tag{13.48}$$

the effective diffusion coefficient (Equation 13.42) approaches the free-space transverse diffusion coefficient

$$D \to \beta \bar{v}_{th} \bar{\lambda}/2 = k_B T_e / m\nu_m \equiv D_\perp. \tag{13.49}$$

It follows from Equation 13.42 and 13.49 that

$$D \leq D_\perp. \tag{13.50}$$

The zero-field limit is regained by setting $\beta = 1$ in the above formulas. When the field is strong $\beta \gg 1$ and even though the large container/high pressure condition $\kappa \ll 1$ may pertain, the *product* $\beta\kappa^2$ may not be negligible, that is, α and hence the diffusion cooling effect may still be appreciable: the field "pumps" the high-energy tail of the distribution function, the region which suffers the greatest loss to the walls. Thus, diffusion cooling is actually *enhanced* by application of a field.

13.5.2.2 Cross section proportional to speed $q = u^{\frac{1}{2}}$

Direct substitution in Equation 13.37 shows, that for this model, the lowest eigenvalue and eigenfunctions, are respectively

$$\vartheta = 1 \tag{13.51}$$

$$\phi = 1, \tag{13.52}$$

which indicates that diffusion coefficient and temperature retain their free-space values regardless of the presence of boundaries or field. For this model, as explained in Section 13.1.2, electrons diffuse at the same rate to the walls, independently of their energy: There is no preferential loss from any particular energy range to promote any cooling effect, even in the presence of a field.

Apart from these two simple models, analytic expressions, even approximate, are difficult to obtain, and therefore a full numerical treatment is generally required.

13.6 Concluding Remarks

The question of solution of Boltzmann's equation with realistic boundary conditions is a vexed one, with most researchers adopting a pragmatic approach similar to that used in this chapter, for example, taking $n = 0$ on the boundary, or to within the boundary at an "extrapolation length"—mean free path. In the Cavalleri experiment, the aim is to adjust the gas pressure in order to minimize the penetration of the "boundary layer" into the bulk of the system, making detailed representation of the boundary condition not so important. However, in other experiments, such as the Townsend–Huxley experiment [10], where measurements of current are effectively taken at the surface, more accurate representation of the boundary conditions may be required. In industrial plasma processing, where one needs detailed information of the distribution function of charged particles at the surface in order to etch precisely in ever finer detail, there will eventually come a point where crude boundary conditions are no longer satisfactory. From a kinetic theory point of view, the half-space analysis familiar in neutron transport theory [100] offers the best hope of achieving the desired goal.

CHAPTER 14

An Analytically Solvable Model

14.1 Introduction

In this chapter, we examine the way in which a pulse of charged particles evolves in time and space in an idealized time-of-flight experiment [10], from the non-hydrodynamic regime near the source, to the hydrodynamic regime downstream, through exact analytic solution of the relaxation time or Bhatnagar–Gross–Krook (BGK) model kinetic equation introduced in Chapter 5. Full details are given for a gaseous medium, where questions arise concerning the validity of the diffusion equation near the source, where the pulse may be sharp. The results are then extended to an amorphous or dispersive medium, allowing for trapping and de-trapping. In both cases, the kinetic equation is solved using Fourier transformation in space and Laplace transform in time.

14.2 Relaxation Time Model

In this section, we solve a relaxation time collision kinetic model, under initial and boundary conditions appropriate to a time-of-flight experiment. This is done in two ways: first, approximately for weak gradients, from which follows Fick's law and the diffusion equation and second, analytically and exactly, without any limitations on gradients. Exact expressions are obtained for the observables (namely, for the first two spatial moments of the density distribution) and comparison is made with the corresponding expressions obtained from the solution of the diffusion equation, whose limits of validity are thereby established.

In the present analysis, we shall employ a relaxation time collision model,

$$\left(\frac{\partial f}{\partial t}\right)_{\mathrm{col}} = -\nu\left[f - nw(\alpha, v)\right], \tag{14.1}$$

where ν denotes a representative, constant collision frequency, $\alpha^2 = m/k_B T_0$, $n = \int_{-\infty}^{\infty} \mathrm{d}^3v\, f(\mathbf{v})$ is the charge carrier number density, and

$$w(\alpha, v) = \left(\alpha^2/2\pi\right)^{\frac{3}{2}} \exp\left(-\frac{1}{2}\alpha^2 v^2\right) \tag{14.2}$$

is a Maxwellian distribution at the temperature T_0 of the background gas. The relaxation or BGK [63] model kinetic equation

$$\left(\partial_t + \mathbf{v} \cdot \nabla + \mathbf{a} \cdot \partial_{\mathbf{v}}\right) f(\mathbf{r}, \mathbf{v}, t) = -\nu[f(\mathbf{r}, \mathbf{v}, t) - w(\alpha, v)n(\mathbf{r}, t)] \tag{14.3}$$

can be readily justified for ions in their parent gas undergoing resonant charge exchange collisions (see Chapter 5), but it is widely used in both gaseous and condensed matter physics in other circumstances, mainly to simplify the mathematics and help elucidate the underlying physics. It is useful for investigating some of the popular approximations in kinetic theory, as discussed in the next section, and is solved exactly in Section 14.4 for conditions corresponding to the time-of-flight experiment.

Before proceeding, we note that integration of Equation 14.3 yields the equation of continuity,

$$\partial_t n + \nabla \cdot \Gamma = 0, \tag{14.4}$$

where

$$\Gamma = \int_{-\infty}^{\infty} \mathrm{d}^3 v \, \mathbf{v} f(r, \mathbf{v}, t) \tag{14.5}$$

is the particle flux. A further integration over all space gives

$$\partial_t N = 0 \tag{14.6}$$

showing that the total number of particles

$$N = \int_{-\infty}^{\infty} \mathrm{d}^3 r \, n(\mathbf{r}, t) = \int_{-\infty}^{\infty} \mathrm{d}^3 v \int_{-\infty}^{\infty} \mathrm{d}^3 r \, f(\mathbf{r}, \mathbf{v}, t) \tag{14.7}$$

remains constant in time.

14.3 Weak Gradients and the Diffusion Equation

14.3.1 Near-equilibrium case

In the Chapman–Enskog solution procedure (see Chapter 11 and Ref. [18]), the entire left hand side of Equation 14.3, including the field term, is regarded as small, and an iterative scheme of successive approximations to $f(\mathbf{r}, \mathbf{v}, t)$ is established, starting with the Maxwellian $w(\alpha, v)\ n(\mathbf{r}, t)$ as the first approximation. This is substituted in the left hand side to obtain the equation for the second approximation, and so on. At the level of the *second* Chapman–Enskog approximation, the particle flux (Equation 14.5) is given by

$$\Gamma = n\mathbf{v}_d - D\nabla n, \tag{14.8}$$

where

$$\mathbf{v}_d \equiv \mathbf{a}/\nu \quad \text{and} \quad D \equiv \frac{k_B T_0}{m\nu} \tag{14.9}$$

denote the drift velocity and a diffusion coefficient, respectively. Together with the equation of continuity,

$$\partial_t n + \nabla \cdot \Gamma = 0, \tag{14.10}$$

this yields a simplified form of the diffusion equation

$$\partial_t n + \mathbf{v}_d \cdot \nabla n - D\,\nabla n = 0. \tag{14.11}$$

14.3.2 Arbitrary fields, density gradient expansion

We now relax any assumption about the magnitude of the field, and consider only the first two terms on the left hand side of Equation 14.3 to be small. Again an iterative solution is followed and, as explained in Chapter 11, a density gradient expansion for $f(\mathbf{r}, \mathbf{v}, t)$ results:

$$f(\mathbf{r}, \mathbf{v}, t) = n(\mathbf{r}, t)\, f^{(0)}(\mathbf{v}) + f^{(1)}(\mathbf{v}) \cdot \nabla n + \ldots$$

The general structure of the hierarchy of equations for $f^{(i)}(\mathbf{v})$ is outlined in detail by Kumar et al. [31]. For now, we merely observe that to *first order* in ∇n it is found that Equation 14.8 generalizes to

$$\Gamma = n\mathbf{v}_d - \mathsf{D} \cdot \nabla n, \tag{14.12}$$

where I is the unit tensor, and

$$\mathsf{D} = \mathrm{I}\, D + \mathbf{v}_d \mathbf{v}_d / \nu$$

the diffusion tensor. The generalized diffusion equation then follows with Equation 14.45

$$\partial_t n + \mathbf{v}_d \cdot \nabla n - \mathsf{D} : \nabla\nabla n = 0. \tag{14.13}$$

The structure of **D** is as shown below:

$$\mathbf{D} = \begin{bmatrix} D_\perp & 0 & 0 \\ 0 & D_\perp & 0 \\ 0 & 0 & D_\parallel \end{bmatrix},$$

where $D_\perp = D$ and $D_\parallel = D + \frac{v_d^2}{\nu}$.

14.3.3 Solution of the diffusion equation

The solution of the diffusion equation corresponding to the time-of-flight experiment in an infinite medium is a travelling gaussian pulse (see Chapter 1). The solution of Equation 14.12 can be obtained using Laplace and Fourier transformations in time and in configuration space, respectively. For an initial sharp pulse of N particles released at the origin, that is,

$$n(\mathbf{r}, t=0) = N\delta(\mathbf{r}), \tag{14.14}$$

the transformed density is found to be

$$\begin{aligned} \hat{n}_p^{(\mathrm{DE})}(\mathbf{k}) &\equiv \int_0^{\infty} \mathrm{d}t \int_{-\infty}^{\infty} \text{d\^{}\{3\}}r\, n(\mathbf{r}, t) \exp\left\{-pt - i\mathbf{k}\cdot\mathbf{r}\right\} \\ &= \frac{N}{p + i\mathbf{k}\cdot\mathbf{v}_d + \mathbf{kk} : \mathsf{D}}, \end{aligned} \tag{14.15}$$

where a superscript "DE" has been added for future reference to indicate that this is the solution of the diffusion equation and : denotes a double contraction over the tensor indices.

We now obtain an expression for $\hat{n}_p(\mathbf{k})$ by directly solving the kinetic equation (Equation 14.3), without relying on the approximations leading to the diffusion equation. Inversion of the transform (Equation 14.15) yields a travelling pulse, for example, Equation 9.47 for plane-parallel geometry.

14.4 Solution of the Kinetic Equation

14.4.1 Transformed equation

The first step in the exact solution of Equation 14.3 is to take the Laplace transform in time, giving

$$\left[p + \mathbf{v}\cdot\nabla + \mathbf{a}\cdot\partial_{\mathbf{v}}\right] \overline{f}_p = \nu(\overline{f}_p - \overline{n}_p w(\alpha, v)) + f(\mathbf{r}, \mathbf{v}, t=0),$$

where $\overline{f}_p(\mathbf{r}, \mathbf{v}) = \int\limits_0^{\infty} e^{-pt} f(\mathbf{r}, \mathbf{v}, t)\ dt$ and $\overline{n}_p(\mathbf{r}) = \int\limits_0^{\infty} e^{-pt}\ n(\mathbf{r}, t)\ dt$. If N particles of mass m are released from the origin of coordinates at time $t=0$, with a Maxwellian velocity distribution at an arbitrary temperature T', the initial condition is

$$f(\mathbf{r}, \mathbf{v}, t=0) = Nw(\alpha', v)\delta(\mathbf{r}), \tag{14.16}$$

with $(\alpha')^2 = m/k_B T'$. (N.B. This is consistent with the initial condition Equation 14.14 used in solving the diffusion equation, as an integration over all velocities $\mathbf{v}$ shows.) The solution proceeds further through Fourier transformation in *phase space*, with boundary conditions $f(\mathbf{r}, \mathbf{v}, t) \to 0$ as $\mathbf{r} \to \pm\infty$, $\mathbf{v} \to \pm\infty$; thus, the transformed distribution function is found to be

$$\widetilde{f}_p(\mathbf{k}, \mathbf{s}) \equiv \int_0^t \mathrm{d}t \int_{-\infty}^{\infty} \mathrm{d}^3 v \int_{-\infty}^{\infty} \mathrm{d}^3 r f(\mathbf{r}, \mathbf{v}, t)\ \exp\left\{-[i(\mathbf{k}\cdot\mathbf{r} + \mathbf{s}\cdot\mathbf{v}) + pt]\right\}$$

$$= -\exp\left\{\frac{is_{\|}}{k}\left[\frac{1}{2}s_{\|}a_{\|} + \mathbf{s}_{\perp}\cdot\mathbf{a}_{\perp} - \Omega\right]\right\}$$

$$\times \int_{-\infty}^{s_{\|}} \mathrm{d}\sigma\, \frac{1}{k}\left\{\nu\hat{n}_p(\mathbf{k})\exp\left(-\frac{\sigma^2 + s_{\perp}^2}{2\alpha^2}\right) + f_0(\mathbf{k}, \sigma, \mathbf{s}_{\perp})\right\}$$

$$\times \exp\left(-\frac{i\sigma(\frac{1}{2}\sigma a_{\|} + \mathbf{s}_{\perp}\cdot\mathbf{a}_{\perp} - \Omega)}{k}\right), \tag{14.17}$$

where the Fourier transform of the initial condition (Equation 14.16) is

$$f_0(\mathbf{k}, \mathbf{s}) \equiv \int \mathrm{d}^3 r \int \mathrm{d}^3 v\, f(\mathbf{r}, \mathbf{v}, t=0)\ \exp\{-i(\mathbf{k}\cdot\mathbf{r} + \mathbf{s}\cdot\mathbf{v})\} = n_0 \exp\left\{\frac{-s^2}{2(\alpha')^2}\right\} \tag{14.18}$$

and

$$\mathbf{s}_{\|} = \frac{(\mathbf{s}\cdot\mathbf{k})\mathbf{k}}{k^2}, \qquad \mathbf{s}_{\perp} = \mathbf{s} - \mathbf{s}_{\|},$$

$$\mathbf{a}_{\|} = \frac{(\mathbf{a}\cdot\mathbf{k})\mathbf{k}}{k^2}, \qquad \mathbf{a}_{\perp} = \mathbf{a} - \mathbf{a}_{\|}, \tag{14.19}$$

while

$$\Omega = i(p + \nu).$$

It is important to note that the transformed particle density and the initial distribution function are related by

$$\hat{n}_p(\mathbf{k}) = \int \mathrm{d}t \int \mathrm{d}^3 r\ n(r, t)\exp\left\{-pt - i\mathbf{k}\cdot\mathbf{r}\right\} = \widetilde{f}_p(\mathbf{k}, \mathbf{s} = \mathbf{0}). \tag{14.20}$$

Equations 14.17 and 14.20 together furnish the transformed number density

$$\hat{n}_p(\mathbf{k}) = N\ \frac{\frac{\beta'}{i\sqrt{2}k}Z(\zeta')}{1 + \nu\frac{i\beta}{\sqrt{2}k}Z(\zeta)}, \tag{14.21}$$

where $Z(\zeta)$ is the plasma dispersion function [104] defined by

$$Z(\zeta) = \frac{1}{\sqrt{\pi}} \int_{-\infty}^{\infty} dx \, \frac{e^{-x^2}}{x - \zeta} \tag{14.22}$$

for $\mathrm{Im}(\zeta) > 0$, and its analytic continuation for $\mathrm{Im}(\zeta) < 0$, while

$$\beta^{-2} \equiv \alpha^{-2} + \frac{i\mathbf{a} \cdot \mathbf{k}}{k^2}$$

and

$$\zeta \equiv \frac{\Omega \beta}{\sqrt{2}k}, \tag{14.23}$$

with β' and ζ' being similarly defined in terms of α'. Equation 14.21 could now be substituted back into Equation 14.17 to obtain the complete and exact expression for the transformed phase space distribution function, if desired. We can therefore say that the problem has been solved exactly, to this extent.

Note that the exact expression Equation 14.21 for the transformed density appears to be markedly different from Equation 14.20, obtained from the diffusion equation—some reconciliation is obviously required.

If the full, explicit expression for $n(\mathbf{r}, t)$ were desired, it would be necessary to carry out the Fourier–Laplace inversion of Equation 14.21, a difficult task. Instead, we shall concentrate on finding $n(\mathbf{r}, t)$ in various limits, and also upon obtaining its *spatial moments*. In this context, it is useful to note that the inversion of the Laplace transform only leads to the Fourier-transformed number density,

$$\begin{aligned} \tilde{n}(\mathbf{k}, t) &\equiv \int_{-\infty}^{\infty} d^3 r \, e^{-i\mathbf{k}\cdot\mathbf{r}} \, n(\mathbf{r}, t) \\ &= \frac{1}{2\pi i} \int_C dp \, e^{pt} \, \hat{n}_p(\mathbf{k}), \end{aligned} \tag{14.24}$$

which in turn provides all the information necessary to compare with experiment, as explained below. The contour C in the familiar Bromwich integral lies to the right of the singularities of $\hat{n}_p(\mathbf{k})$, which from Equation 14.21 may be seen to include the zeroes p_k of

$$1 + \nu \frac{i\beta}{\sqrt{2}k} Z(\zeta) = 0 \tag{14.25}$$

and this dispersion relation plays a central role in determining transport properties.

14.4.2 Asymptotic expressions

If k is sufficiently small, then from Equation 14.23

$$|\zeta| \equiv \left| \frac{\Omega\beta}{\sqrt{2}k} \right| \gg 1 \tag{14.26}$$

and the asymptotic representation of the plasma dispersion function,

$$Z \approx -\frac{1}{\zeta}\left(1 + \frac{1}{2}\zeta^{-2} + \frac{3}{4}\zeta^{-4} + \ldots\right), \tag{14.27}$$

we can show that the left hand side of Equation 14.25 becomes

$$1 + \nu\frac{i\beta}{\sqrt{2}k}Z(\zeta) \approx \frac{ip}{\Omega}\left\{1 - \frac{\nu}{p\Omega^2}\left(\frac{k^2}{\alpha^2} + i\mathbf{a}\cdot\mathbf{k} - \frac{3(\mathbf{a}\cdot\mathbf{k})^2}{\Omega^2}\right)\right\} + O(k^3). \tag{14.28}$$

Hence by Equation 14.21

$$\begin{aligned}\frac{\hat{n}_p(\mathbf{k})}{N} &= \frac{\frac{i\beta'}{\sqrt{2}k}Z(\zeta')}{1 + \nu\frac{i\beta}{\sqrt{2}k}Z(\zeta)} \\ &\approx \frac{1}{p}\left\{1 + \frac{i\mathbf{a}\cdot\mathbf{k}}{\Omega^2}\left[1 + \frac{\nu}{p}\right] + \frac{k^2}{\Omega^2}\left[\frac{\nu}{\alpha^2 p} + \frac{1}{(\alpha')^2}\right]\right\} - \\ &\quad \frac{(\mathbf{a}\cdot\mathbf{k})^2}{p\Omega^4}\left[\frac{(\nu)^2}{p^2} + \frac{4\nu}{p} + 3\right] + O(k^3)\end{aligned} \tag{14.29}$$

we obtain immediately and *exactly*

$$\frac{i}{n_0}\left\{\frac{\partial\hat{n}_p(\mathbf{k})}{\partial\mathbf{k}}\right\}_{\mathbf{k}=0} = \frac{\mathbf{a}}{p^2(p+\nu)} \tag{14.30}$$

and

$$-\frac{1}{n_0}\left\{\frac{\partial^2\hat{n}_p(\mathbf{k})}{\partial\mathbf{k}\partial\mathbf{k}}\right\}_{\mathbf{k}=0} = \frac{2}{p(p+\nu)^2}\left[\frac{1}{(\alpha')^2} + \frac{\nu}{\alpha^2 p}\right] + \frac{2\mathbf{a}\,\mathbf{a}}{(p+\nu)^3}\left[\frac{\nu}{p^3} + \frac{3}{p^2}\right]. \tag{14.31}$$

Exact expressions for the macroscopically observable quantities, $\langle\mathbf{r}\rangle$ and $\langle\mathbf{r}\,\mathbf{r}\rangle$ follow after inverting the Laplace transforms, as explained in the next section.

14.4.3 Calculation of averages

The quantities inferred in the time-of-flight experiment are not normally the full density distribution $n(\mathbf{r}, t)$ as such, but rather spatial moments, such as the position of the centroid,

$$\langle \mathbf{r} \rangle = \frac{1}{N} \int d^3r \, \mathbf{r} n(\mathbf{r}, t) \tag{14.32}$$

and the dispersion about the centroid

$$\langle \mathbf{RR} \rangle = \frac{1}{N} \int d^3r \, \mathbf{RR} n(\mathbf{r}, t) \equiv \langle \mathbf{rr} \rangle - \langle \mathbf{r} \rangle \langle \mathbf{r} \rangle, \tag{14.33}$$

where

$$\mathbf{R} \equiv \mathbf{r} - \langle \mathbf{r} \rangle.$$

These quantities may be obtained directly from the identities

$$\begin{aligned} \langle \mathbf{r} \rangle &= \frac{i}{N} \left\{ \frac{\partial \tilde{n}(\mathbf{k}, t)}{\partial \mathbf{k}} \right\}_{\mathbf{k}=0} \\ &= \mathbf{a}\nu^{-1} t - \mathbf{a}\nu^{-2} + \mathbf{a}\nu^{-2} e^{-\nu t} \end{aligned} \tag{14.34}$$

and

$$\begin{aligned} \langle \mathbf{rr} \rangle &= -\frac{1}{N} \left\{ \frac{\partial^2 \tilde{n}(\mathbf{k}, t)}{\partial \mathbf{k} \, \partial \mathbf{k}} \right\}_{\mathbf{k}=0} . \\ &= 2\mathrm{I} \left\{ \frac{1}{(\alpha' \nu)^2} \left[1 - (1 + \nu t) e^{-\nu t} \right] + \frac{1}{(\alpha \nu)^2} \left[\nu t - 2 + (2 + \nu t) e^{-\nu t} \right] \right\} \\ &\quad + \frac{\mathbf{aa}}{\nu^4} \left\{ \nu^2 t^2 - 6 + 2(\nu^2 t^2 + 3\nu t + 3) e^{-\nu t} \right\} . \end{aligned} \tag{14.35}$$

At short times $\nu t < 1$, non-hydrodynamic conditions prevail, and the centroid behaves ballistically, $\langle \mathbf{r} \rangle \approx \frac{1}{2} \mathbf{a} t^2$. The hydrodynamic regime is, however, quickly attained after a few collision times, $\nu t \gg 1$, and then

$$\langle \mathbf{r} \rangle \approx \mathbf{v}_d t \tag{14.36}$$

and

$$\langle \mathbf{RR} \rangle \approx 2\mathrm{D} t, \tag{14.37}$$

where

$$\mathbf{v}_d = \frac{\mathbf{a}}{\nu} \quad \text{and} \quad \mathrm{D} = \left(\mathrm{I} \alpha^{-2} \nu^{-1} + \mathbf{v}_d \mathbf{v}_d \nu^{-1} \right).$$

denote the classical drift velocity and the diffusion tensor, respectively. The longtime limit is further explored below.

14.4.4 Validity of the diffusion equation

If in addition to small k, we also consider the small p limit, in the the sense that

$$p \ll \nu,$$

for which it follows from Equation 14.21 that

$$\frac{\hat{n}_p(\mathbf{k})}{N} = \frac{\frac{i\beta'}{\sqrt{2}k} Z(\zeta')}{1 + \nu \frac{i\beta}{\sqrt{2}k} Z(\zeta)} \approx \frac{F_p(\mathbf{k})}{p + p_{\mathbf{k}}^{(0)}} \equiv \frac{\hat{n}_p^{(\infty)}(\mathbf{k})}{N}, \tag{14.38}$$

which is accurate to $O(k^2)$, where

$$F_p(\mathbf{k}) \equiv 1 + \frac{i\mathbf{a}\cdot\mathbf{k}}{(\nu)^2} + \frac{3(\mathbf{a}\cdot\mathbf{k})^2}{(\nu)^4} - \frac{k^2}{(\nu)^2(\alpha')^2}$$

is a factor of the order of unity, and

$$p_{\mathbf{k}}^{(0)} \equiv i\mathbf{v}_d \cdot \mathbf{k} + \nu^{-1}\left(\mathrm{I}\alpha^{-2} + \mathbf{v}_d\mathbf{v}_d\right) : \mathbf{kk}.$$

Comparison of Equation 14.15 with Equation 14.38 shows that the asymptotic and diffusion equation solutions are related by

$$\hat{n}_p^{(\infty)}(\mathbf{k}) = F_p(\mathbf{k})\, \hat{n}_p^{(\mathrm{DE})}(\mathbf{k}) + O(k^3). \tag{14.39}$$

Since $F_p(\mathbf{k}) \longrightarrow 1$ as $k \longrightarrow 0$, it is clear that the exact expression for the small k form of the Fourier transform of density approaches the diffusion equation result asymptotically at long times, that is,

$$\hat{n}(\mathbf{k}, t) \sim \hat{n}^{(\mathrm{DE})}(\mathbf{k}, t) + O(k^3), \tag{14.40}$$

and it follows that the spatial moments $\langle \mathbf{r} \rangle$ and $\langle \mathbf{RR} \rangle$ of $n^{(\mathrm{DE})}(\mathbf{r}, t)$ are just asymptotic expressions for the exact quantities, Equations 14.36 and 14.37. Thus, while the diffusion equation is inadequate for dealing with the short-time, non-hydrodynamic regime, it generates accurate moments in the asymptotic limit.

As for the density distribution itself, Equation 14.40 is to be interpreted as saying that the actual density approaches the solution of the diffusion equation asymptotically,

$$n(\mathbf{r}, t) \sim n^{(\mathrm{DE})}(\mathbf{r}, t) \tag{14.41}$$

for sufficiently long times *and* for distances sufficiently far downstream from the source (the "far field" region), respectively.

Near the source and for short times, non-hydrodynamic conditions prevail, and the diffusion equation is of no use: a complete expression for $n(\mathbf{r}, t)$ can only be found from the full Boltzmann equation solution, which for the present situation means inverting Equation 14.21 numerically, without further approximation. Given that this is a model calculation, and that in any case the observable quantities, the spatial moments of $n(\mathbf{r}, t)$, have been determined exactly for both non-hydrodynamic and hydrodynamic circumstances, such a program may not be worthwhile.

Since the solution of Boltzmann's equation for the pulsed time-of-flight experiment effectively yields the Green's function, other experimental situations can (at least formally) be dealt with in the usual way, by appropriate integration over space and/or time. The hydrodynamic Fick's law/diffusion equation regime can also be identified in these cases, but one should not expect that it should be the same as for the time-of-flight experiment. The situation where charge carriers are emitted from an infinite plane source at a steady rate into an infinite medium, such that a steady state is eventually achieved at long times, is a case in point—it is straightforward to show that the diffusion equation does not yield physically tenable results *anywhere*, except trivially at infinity (see Exercises, Problem 21). There one has an inherently non-hydrodyamic situation and there is no alternative but to solve the phase space kinetic equation without approximation, a task which is undertaken in the next section.

The lesson here is that the diffusion equation should be employed with some care, whatever the context.

14.5 Relaxation Time Model and Diffusion Equation for an Amorphous Medium

14.5.1 Modified BGK kinetic equation with memory

We now generalize the preceding discussion to charge carriers in an amorphous medium, undergoing trapping and de-trapping to and from localized states. From a kinetic theory perspective, the situation is effectively one where collisions have a finite duration, leading to so-called dispersive transport and memory effects.

A schematic portrayal of the model in phase space is shown in Figures 6.1 and 6.2. A charge carrier moving freely in the conduction band is scattered out of a phase space element and is trapped in localized states. The range of trapping times τ is determined by the relaxation function $\phi(\tau)$, defined such that $\phi(\tau)\,d\tau$ is the probability of de-trapping between times τ and $\tau + d\tau$. The corresponding kinetic equation for *free* charge carrier distribution function is

$$\left(\partial_t + \mathbf{v} \cdot \nabla + \mathbf{a} \cdot \partial_{\mathbf{v}}\right) f = -\nu \left[f - w(\alpha, v)\phi * n\right] \tag{14.42}$$

and $\alpha^2 \equiv m/k_B T_0$. Note that only the *second* (de-trapping) term on the right hand side involves a convolution with ϕ,

$$\phi * n = \int_{-\infty}^{\infty} d\tau \, \phi(\tau) n(t-\tau)$$

corresponding to a delayed release from the localized states back to the conduction band. In contrast, scattering out of the conduction band takes place without any delay, and the *first* term on the right hand side is *not* convoluted. The overall result is that a "collision" takes a finite time τ. Zero trapping time corresponds (mathematically) to an instantaneous collision, and thus setting $\phi(\tau) = \delta(\tau)$, gives $\phi * n = n$, and the classical BGK equation (Equation 14.3) for gases and crystalline semiconductors is recovered.

We now define the operator

$$\tilde{\partial}_t \equiv \partial_t + \nu \left[1 - \phi *\right] \tag{14.43}$$

and write Equation 14.42 in the equivalent form

$$\left(\tilde{\partial}_t + \mathbf{v} \cdot \nabla + \mathbf{a} \cdot \partial_{\mathbf{v}}\right) f = -\nu\phi * \left[f - n w(\alpha, v)\right] . \tag{14.44}$$

Integration of Equation 14.44 over velocities yields the equation of continuity,

$$\tilde{\partial}_t n + \nabla \cdot \Gamma = 0 , \tag{14.45}$$

where $\Gamma = \int d^3 v \, \mathbf{v} f(\mathbf{r}, \mathbf{v}, t)$ is the free particle flux. A further integration over $\mathbf{r}$ yields

$$\tilde{\partial}_t N = 0 , \tag{14.46}$$

where $N(t) = \int d^3 r \, d^3 v \, f(\mathbf{r}, \mathbf{v}, t)$ is the total free particle number. Since this implies $\partial_t N = -\nu \left[1 - \phi *\right] N \neq 0$, it is clear that the number $N(t)$ of free carriers is not constant when trapping/de-trapping occurs. Only for classical transport, when collisions are instantaneous, $\phi(t) \to \delta(t)$, $\tilde{\partial}_t \to \partial_t$, and Equation 14.46 reduces to Equation 14.6, is N constant. For that reason, we must now distinguish between the number or particles $N(t)$ at time t and the initial number $N(0)$.

14.5.2 Solution for the time-of-flight experiment

14.5.2.1 Dispersion relation

The exact solution of Equation 14.42 in infinite space, with the initial condition,

$$f(\mathbf{r}, \mathbf{v}, 0) = N(0) \, \delta(\mathbf{r}) \, w(\alpha', v) \tag{14.47}$$

corresponding to release of $N(0)$ particles from the origin of coordinates with a Maxwellian distribution of velocities at some arbitrary temperature T', can be obtained through Fourier and Laplace transformation in space and time, respectively, following the same mathematical procedure as in Section 14.3. The expression (Equation 14.20) for the transformed number density thus generalizes to [105]

$$\hat{n}_p(\mathbf{k}) = \left[\int_{-\infty}^{\infty} \mathrm{d}^3 r \, \exp(-i\mathbf{k}\cdot\mathbf{r})\right]\left[\int_0^{\infty} \mathrm{d}t \exp(-pt)\, n(\mathbf{r},t)\right.$$
$$= N(0)\frac{\frac{\beta'}{i\sqrt{2}k} Z(\zeta')}{1+\nu\,\hat{\phi}(p)\frac{i\beta}{\sqrt{2}k}Z(\zeta)}, \tag{14.48}$$

where $\hat{\phi}(p) = \int_0^{\infty} \mathrm{d}t \exp(-p\tau)\,\phi(\tau)$ is the Laplace transform of the relaxation function, and otherwise all notation is the same as previous, that is, $Z(\zeta)$ is the plasma dispersion function, $\beta^{-2} \equiv \alpha^{-2} + \frac{i\mathbf{a}\cdot\mathbf{k}}{k^2}$, $\zeta \equiv \frac{\Omega\beta}{\sqrt{2}k}$, and $\Omega = i(p+\nu)$, with similar definitions for ζ' and β'in terms of $\alpha'^2 \equiv m/k_B T'$. In passing, we note that if $\phi(\tau)$ were normalized then $\hat{\phi}(0) = 1$ but in general, this is not the case. Neither does the first moment $\int \tau\phi(\tau)\,\mathrm{d}\tau$ necessarily exist—the relaxation function may have strange properties [37].

Inversion of the Laplace transform of Equation 14.48 could be carried out, if desired, through the usual contour integral. This would be evaluated using the residue theorem in terms of the singularities of $\hat{n}_p(\mathbf{k})$, which are given by the zeros of the denominator of Equation 14.48, that is,

$$1+\nu\,\hat{\phi}(p)\frac{i\beta}{\sqrt{2}k}Z(\zeta) = 0. \tag{14.49}$$

The solution of this dispersion relation is not straightforward but, if we focus on the asymptotic, weak gradient, longtime regime, then only the small k solutions of Equation 14.49 need be found. Since in this case $\zeta \equiv \frac{\Omega\beta}{\sqrt{2}k} \gg 1$, we may use the asymptotic representation,

$$Z \approx -\frac{1}{\zeta}\left(1+\frac{1}{2}\zeta^{-2}+\frac{3}{4}\zeta^{-4}+\ldots\right) \tag{14.50}$$

of the plasma dispersion function. Proceeding in this way, the solution of Equation 14.49, valid to second order in k, is found to be approximately

$$\tilde{p} \approx -i\frac{\mathbf{a}\cdot\mathbf{k}}{\nu} - \frac{\mathbf{kk}}{\nu} : \left[\frac{\mathbf{I}}{\alpha^2}+\frac{\mathbf{aa}}{\nu^2}\right], \tag{14.51}$$

where

$$\tilde{p} \equiv p + \nu\left[1-\hat{\phi}(p)\right], \tag{14.52}$$

while I is the unit vector, and as usual : denotes a double contraction over tensor indices.

14.5.2.2 The generalized diffusion equation

A similar result follows if we were to assume a *generalized diffusion equation*

$$\left(\widetilde{\partial}_t + \mathbf{v}_d \cdot \nabla - \mathsf{D} : \nabla\nabla\right) n = 0, \tag{14.53}$$

with $\widetilde{\partial}_t$ specified by Equation 14.43, and take Fourier–Laplace transforms. Singularities of $\overline{n}(p, \mathbf{k})$, in this case, are found from

$$\widetilde{p} = -i\mathbf{v}_d \cdot \mathbf{k} - \mathbf{kk} : \mathsf{D}, \tag{14.54}$$

which, to be consistent with Equation 14.51, requires the drift velocity and diffusion tensor to be given by

$$\mathbf{v}_d = \frac{\mathbf{a}}{\nu}, \quad \mathsf{D} = \frac{1}{\nu}\left[\frac{\mathrm{I}}{\alpha^2} + \frac{\mathbf{aa}}{\nu^2}\right]. \tag{14.55}$$

It is clear that the effect of trapping enters Equation 14.53 only through the operator $\widetilde{\partial}_t$, while the spatial gradient terms are determined by free particle transport. Moreover, the free particle transport coefficients are unaltered by trapping, for example, Equation 14.55 is exactly the same expression obtained in Section 14.4.3 from solution of the classical (non-trapping) BGK model kinetic equation.
Although we have established the credentials of the generalized diffusion equation (Equation 14.53) under idealized time-of-flight conditions in infinite space, with a model kinetic equation collision operator, we will assume that a diffusion equation of the same *mathematical form* is valid in more general circumstances, with appropriate carrier drift velocity and diffusion coefficients, not just those defined by Equation 14.55.

Experiments are normally conducted with a slab of material between two plane-parallel electrodes, the normal direction defining the z-axis of a system of coordinates. Assuming all spatial dependence to be in this direction only, Equations 14.53 and 14.43 together yield the generalized diffusion equation in one dimension,

$$\frac{\partial n}{\partial t} + \nu\left[1 - \phi *\right] n + v_d \frac{\partial n}{\partial z} - D_{\|} \frac{\partial^2 n}{\partial z^2} = 0. \tag{14.56}$$

Solutions of this equation have been obtained by Ref. [105].

Note that a slightly different kinetic equation, in which ϕ is convoluted with both terms on the right hand side of Equation 14.3, was proposed in Ref. [106], and solved using the same mathematical technique as above.

14.6 Concluding Remarks

An exact solution of the relaxation time model kinetic equation for both gases and amorphous matter has been obtained, enabling the regime of validity of the diffusion equation, both standard and fractional forms, to be established.

Part IV

Special Topics

CHAPTER 15

Temporal Non-Locality

15.1 Introduction

In this chapter, we relax this static field assumption and consider temporal non-locality (i.e., transport not a function of the instantaneous field) arising from oscillatory electric fields of the form:

$$\mathbf{E}(t) = \mathbf{E}_0 \cos \omega t \,. \tag{15.1}$$

Once the initial transients have decayed, the system evolves to a state which we will refer to as the "periodic steady state" where all properties oscillate at the applied frequency (or harmonics thereof) of the field. In this section, we shall explore the harmonic dependencies of the distribution function and associated transport properties that arise under these conditions.

Unless otherwise specified, we will consider the case where spatial variations exist, which are sufficiently weak so that the hydrodynamic regime holds and the density gradient expansions considered previously are valid. The spatial dependence of all intensive properties, in that case, arises solely through the number density.

15.2 Symmetries and Harmonics

Under the combined operations

$$\mathbf{r} \rightarrow -\mathbf{r} \qquad \text{and} \qquad \mathbf{v} \rightarrow -\mathbf{v} \tag{15.2}$$

together with phase reversal,

$$t \rightarrow t + \frac{\pi}{\omega} \tag{15.3}$$

the Boltzmann equation is form invariant and $f(-\mathbf{r}, -\mathbf{v}, t + \pi/\omega)$ is therefore also a solution of the Boltzmann equation (provided it satisfies the boundary conditions). In the absence of non-conservative processes, under the combined operations of Equations 15.2 and 15.3, it then follows that

$$f(-\mathbf{r}, -\mathbf{v}, t + \pi/\omega) = f(\mathbf{r}, \mathbf{v}, t) \,. \tag{15.4}$$

However, in the presence of reactive processes, the number density is no longer invariant under the combined operations and it is postulated that

$$n(-\mathbf{r}, t+\pi/\omega) \propto n(\mathbf{r}, t), \tag{15.5}$$

where subsequently

$$f(-\mathbf{r}, -\mathbf{v}, t+\pi/\omega) \propto f(\mathbf{r}, \mathbf{v}, t). \tag{15.6}$$

By definition of the number density, the proportionality constants in Equations 15.5 and 15.6 are equal and it follows by division of Equation 15.6 by Equation 15.5

$$\frac{f(-\mathbf{r}, -\mathbf{v}, t+\pi/\omega)}{n(-\mathbf{r}, t+\pi/\omega)} = \frac{f(\mathbf{r}, \mathbf{v}, t)}{n(\mathbf{r}, t)}. \tag{15.7}$$

In what follows the implications of this symmetry property on the phase-space distribution function and transport coefficients are considered. For simplicity, we will assume that there is only spatial variation in the z-direction, which is parallel to the electric field $\mathbf{E}$.

Under time-dependent hydrodynamic conditions, a sufficient representation of the spatial dependence of the phase-space distribution function is through a density gradient expansion:

$$f(z, \mathbf{v}, t) = \sum_{s=0}^{\infty} f^{(s)}(\mathbf{v}, t) \frac{\partial^s}{\partial z^s} n(z, t). \tag{15.8}$$

Under the combined operations of Equations 15.2 and 15.3, Equation 15.12 then becomes

$$f\left(-z, -\mathbf{v}, t+\frac{\pi}{\omega}\right) = \sum_{s=0}^{\infty} f^{(s)}\left(-\mathbf{v}, +\frac{\pi}{\omega}\right)(-1)^s \frac{\partial^s}{\partial z^s} n\left(-z, t+\frac{\pi}{\omega}\right). \tag{15.9}$$

It then follows that

$$f^{(s)}\left(-\mathbf{v}, t+\frac{\pi}{\omega}\right) = (-1)^s f^{(s)}(\mathbf{v}, t). \tag{15.10}$$

Consider a general moment of the distribution function $\Psi(z, t)$, defined by

$$\Psi(z, t) = \int d\mathbf{v}\, \psi(\mathbf{v}) f(z, \mathbf{v}, t) \tag{15.11}$$

with a parity in velocity space, p, that is, $\psi(-\mathbf{v}) = (-1)^p \psi(\mathbf{v})$. It follows that, the density gradient expansion coefficients of the moment $\Psi(z, t)$:

$$\Psi(z, t) = \sum_{s=0}^{\infty} \Psi^{(s)}(t) \frac{\partial^s}{\partial z^s} n(z, t), \tag{15.12}$$

where

$$\Psi^{(s)}(t) = \int d\mathbf{v}\, \psi(\mathbf{v}) f^{(s)}(\mathbf{v}, t) \tag{15.13}$$

satisfy the following symmetries:

$$\Psi^{(s)}\left(t + \frac{\pi}{\omega}\right) = (-1)^{s+p}\Psi^{(s)}(t). \tag{15.14}$$

For example, if we consider the average velocity $\psi(\mathbf{v}) = \mathbf{v}$ where $p = 1$, then

$$\langle v_z \rangle = v_d(t) - \frac{D_{\|}}{n}\frac{\partial n}{\partial z} + \ldots \tag{15.15}$$

and the following symmetries in time then follow:

$$v_d\left(t + \frac{\pi}{\omega}\right) = -v_d(t) \tag{15.16}$$

$$D_{\|}\left(t + \frac{\pi}{\omega}\right) = D_{\|}(t). \tag{15.17}$$

Likewise, if we consider the average energy $\psi(\mathbf{v}) = \epsilon$, where $p = 0$, then

$$\langle \epsilon \rangle = \varepsilon(t) + \frac{\gamma}{n}\frac{\partial n}{\partial z} + \ldots \tag{15.18}$$

and it follows that

$$\varepsilon\left(t + \frac{\pi}{\omega}\right) = \varepsilon(t) \tag{15.19}$$

$$\gamma\left(t + \frac{\pi}{\omega}\right) = -\gamma(t). \tag{15.20}$$

Suppose we consider a Fourier series representation of the time dependence in the $\Psi^{(s)}(t)$, that is,

$$\Psi^{(s)}(t) = \sum_{k=-\infty}^{\infty} \Psi^{(s,k)} \exp(ik\omega t). \tag{15.21}$$

It then follows from Equation 15.14 together with the reality condition that

$$\Psi^{(s,-k)} = (-1)^{s+k+p}\Psi^{(s,k)}. \tag{15.22}$$

The following harmonic dependencies then emerge in the periodic steady state:

$$\varepsilon(t) = \bar{\varepsilon} + \varepsilon_1 \cos(2\omega t - \phi_e) + \ldots \tag{15.23}$$

$$v_d(t) = \sqrt{2} v_{d\,\mathrm{rms}} \cos\left(\omega t - \phi_m\right) + \ldots \tag{15.24}$$

$$D_{\|}(t) = \bar{D}_{\|} + D_{\|,1} \cos(2\omega t - \phi_{D_{\|}}) + \ldots \tag{15.25}$$

$$\gamma(t) = \sqrt{2}\gamma_{\mathrm{rms}} \cos(\omega t - \phi_\gamma) + \ldots . \tag{15.26}$$

where the various ϕ represent the appropriate phase shifts with respect to the field. Likewise, the transverse diffusion coefficient $D_\perp$ has the harmonic dependency:

$$D_\perp(t) = \bar{D}_\perp + D_{\perp,1}\cos(2\omega t - \phi_{D_\perp}) + \ldots. \tag{15.27}$$

In these expressions, the cycle-averaged properties are defined according to

$$\bar{h} = \frac{\omega}{2\pi}\int_0^{2\pi/\omega} h(t)\,dt.$$

15.3 Solution of Boltzmann's Equation for Electrons in AC Electric Fields

The two-term representation of the Boltzmann equation for an alternating electric field under spatially homogeneous conditions is given by

$$\frac{\partial f^{(0)}}{\partial t} + \frac{q\mathbf{E}\cos\omega t}{3mv^2}\cdot\frac{\partial}{\partial v}[v^2\mathbf{f}^{(1)}] = \frac{m}{m_0}\frac{1}{v^2}\frac{\partial}{\partial v}\left\{v^2\nu_m(v)\left[vf^{(0)}(v) + \frac{k_B T_0}{m}\frac{\partial f^{(0)}}{\partial v}\right]\right\} \tag{15.28}$$

$$\frac{\partial \mathbf{f}^{(1)}}{\partial t} + \frac{q\mathbf{E}\cos\omega t}{m}\frac{\partial f^{(0)}}{\partial v} = -\nu_m(v)\mathbf{f}^{(1)}(v). \tag{15.29}$$

The qualitative nature of the temporal evolution of the spatially homogeneous distribution function components can be described in terms of the relaxation times for energy and momentum transfer. Specifically, the temporal evolution is governed by comparisons of the driving frequency ω to the collision frequencies for momentum ($\nu_m = \tau_m^{-1}$) and energy transfer ($\nu_e = 2m\nu_m/m_0 = \tau_e^{-1}$). From the above equations, it is clear that the anisotropic component $f^{(1)}$ relaxes on a timescale determined by $\tau_m(\epsilon)$, while $f^{(0)}$ relaxes on a timescale $\tau_e(\epsilon)$. Comparison of these time constants with that for the temporal alteration of the field gives rise to four general frequency domains:

1. *Low frequency regime*: $\omega \ll \nu_e(\epsilon) \ll \nu_m(\epsilon)$. In this regime, assuming that the inequality is satisfied over the bulk of the distribution function, then the timescale available for relaxation is sufficiently long that all quantities can relax sufficiently before the field changes. Thus, all quantities follow the field in a quasi-stationary manner with full modulation and no phase delay with respect to the applied field.

2. *Intermediate frequency regime*: $\nu_e(\epsilon) \simeq \omega \ll \nu_m(\epsilon)$. In this regime, on the time scale available for relaxation, quantities which relax with

time constant $\tau_e(\epsilon)$ (e.g., $f^{(0)}$ [for energy domains satisfying the above inequality] and ϵ) can no longer relax sufficiently before the field changes. These quantities undergo a reduction in modulation and a phase lag exists in their temporal profile with respect to the applied field. Quantities which relax on the timescale $\tau_m(\epsilon)$ ($f^{(1)}$ [in the energy range satisfying the above inequality] and v_d) remain fully modulated.

3. *High frequency regime*: $\nu_e(\epsilon) \ll \omega \simeq \nu_m(\epsilon)$. In this regime, on the timescale available for relaxation, quantities which relax according to the time constant $\tau_m(\epsilon)$ can no longer relax sufficiently before the field changes and undergo a reduction in the modulation and an increase in the phase lag with respect to the field. Those properties which relax according to the time constant τ_e are generally time independent in this regime.

4. *Very high frequency (electron-trapping) regime*: $\omega \gg \nu_m(\epsilon)$. Essentially, in this regime, the swarm particles undergo many oscillations per collision, that is, the individual swarm particles and hence the swarm drift velocity oscillate 90° out of phase with the field. Subsequently, on average, the field can no longer pump energy into the system and transport approaches the thermal limit.

These observations allow us to avoid a full time-dependent solution of Equations 15.28 and 15.29 in the low and high frequency regimes. In the high frequency regime, one can appeal to the concept of an effective DC electric field (replacing the time-dependent field), and it can be shown that $f^{(0)}$ is given by the Margenau distribution [107],

$$f^{(0)}(c) = A \exp\left\{ -\frac{3m}{m_0} \int_0^c \frac{c}{\left(\frac{eE_{\text{eff}}}{m\nu_m}\right)^2 + \frac{3k_B T_0}{m_0}} dc \right\}, \tag{15.30}$$

which is the Davydov distribution derived earlier with the DC electric field replaced by an effective DC electric field, defined by

$$E_{\text{eff}}(\epsilon) = \frac{E_0}{\sqrt{2}} \left[\frac{1}{\sqrt{1 + \omega^2/\nu_m^2(\epsilon)}} \right].$$

This allows cycle-averaged scalar properties to be evaluated such as the cycle-averaged mean energy, $\bar{\epsilon}$. In general, however, one must solve the full time-dependent problem.

15.4 Moment Equations for Electrons in AC Electric Fields

If we consider light particles and ignore non-conservative processes for simplicity, and take an electric field parallel to z-direction, the moment equations of Chapter 7 for momentum and energy balance become

$$m\frac{\partial v}{\partial t}+\frac{2}{3}\frac{\epsilon}{n}\frac{\partial n}{\partial z}=eE_0\cos\omega t-\nu_m(\epsilon)mv \tag{15.31}$$

$$\frac{\partial \epsilon}{\partial t}+\frac{Q}{n}\frac{\partial n}{\partial z}=-\nu_e(\epsilon)\left[\epsilon-\frac{3}{2}k_BT_0-\frac{1}{2}m_0v^2+\Omega\right], \tag{15.32}$$

where Q is the heat flux (divided by n) and all other terms have been defined previously in Chapter 7. To solve Equations 15.31 and 15.32 in the time-dependent hydrodynamic regime, we substitute Equations 15.15 and 15.18, and equating coefficients of n and $\frac{1}{n}\partial n/\partial z$, yields the following equations for the transport properties

$$\frac{dv_d}{dt}+\nu_m(\varepsilon)v_d=\frac{eE_0}{m}\cos\omega t \tag{15.33}$$

$$\frac{d\varepsilon}{dt}+\nu_e(\varepsilon)\left[\varepsilon-\frac{3}{2}k_BT_0-\frac{1}{2}m_0v_d^2+\Omega\right]=0 \tag{15.34}$$

$$\frac{dD_\parallel}{dt}+\nu_m(\varepsilon)D_\parallel=\frac{2}{3}\frac{\varepsilon}{m}+\gamma\nu_m'(\varepsilon)v_d \tag{15.35}$$

$$\frac{d\gamma}{dt}+\nu_e(\varepsilon)\left[1+\Omega'-\frac{d\varepsilon}{dt}\frac{\nu_e'(\varepsilon)}{\nu_e^2(\varepsilon)}\right]=m_0\nu_e(\varepsilon)v_dD_\parallel-Q, \tag{15.36}$$

where the dashes refer to energy derivatives. To these equations, we add the following equation for the transverse diffusion coefficient

$$\frac{dD_\perp}{dt}+\nu_m(\varepsilon)D_\perp=\frac{2}{3}\frac{\varepsilon}{m}. \tag{15.37}$$

We should note that since

$$\frac{\partial}{\partial t}\left(\frac{1}{n}\frac{\partial n}{\partial z}\right)=O\left(\frac{\partial^2 n}{\partial z^2}\frac{1}{n}\frac{\partial n}{\partial z}\right)$$

such terms can be regarded as beyond first order in the density gradient.

The equations for drift and energy comprise two first order non-linear differential equations that can be solved simultaneously using standard procedures. Once solved, one can then solve the linear coupled differential equations for the diffusion coefficients and energy gradient parameter.

In what follows, we seek analytic solution by assuming that oscillating time variations in the mean energy, $\delta\varepsilon(t)$ are small relative to the cycle-averaged value $\bar{\varepsilon}$, that is,

$$\varepsilon(t)=\bar{\varepsilon}+\delta\varepsilon(t). \tag{15.38}$$

Linearising functions of energy, for example,

$$\nu_m(\varepsilon) = \nu_m(\bar{\varepsilon}) + \nu'_m(\bar{\varepsilon})\delta\varepsilon(t) + \ldots \tag{15.39}$$

(all collision frequencies henceforth are functions of $\bar{\varepsilon}$) and assuming

$$\left|\frac{\nu'_m}{\nu_m}\delta\varepsilon\right| \ll 1, \tag{15.40}$$

which is equivalent to $\omega/\nu_e > 1$, then it follows that

- *Drift velocity*:

$$v_d(t) = \sqrt{2}v_{d\text{rms}} \cos\left(\omega t - \phi_m\right), \tag{15.41}$$

where

$$v_{d\text{rms}} = \frac{e}{m\nu_m}E_{\text{eff}} \tag{15.42}$$

and

$$\tan\phi_m = \frac{\omega}{\nu_m}. \tag{15.43}$$

- *Mean energy*:

$$\varepsilon(t) = \bar{\varepsilon} + \varepsilon_1 \cos(2\omega t - \phi_e), \tag{15.44}$$

where $\bar{\varepsilon}$ is the solution of the Wannier energy relation (extended to high frequency AC fields)

$$\bar{\varepsilon} = \frac{3}{2}k_B T_0 + \frac{1}{2}m_0 v_{d\text{rms}}^2 - \Omega(\bar{\varepsilon}), \tag{15.45}$$

while

$$\varepsilon_1 = \frac{\frac{1}{2}m_0 v_{d\text{rms}}^2}{\sqrt{(1+\Omega')^2 + (2\omega/\nu_e)^2}} \tag{15.46}$$

and

$$\phi_e = 2\phi_m + \tan^{-1}\left(\frac{2\omega}{\nu_e\left(1+\Omega'\right)}\right). \tag{15.47}$$

- *Longitudinal and transverse diffusion, and gradient energy parameter*:

$$D_{\|}(t) = D_{\perp}(t) - \frac{\nu'_m}{\nu_m}\gamma_{\text{rms}} v_{d\text{rms}}\Big[\cos\tilde{\phi}_e + \cos\Phi_m \times \cos\left(2\omega t - \phi_\gamma - \phi_m - \Phi_m\right)\Big]$$

$$D_\perp(t) = \bar{D}_\perp \left[1 + \frac{\varepsilon_1}{\bar{\varepsilon}} \cos \Phi_m \cos\left(2\omega t - \phi_e - \Phi_m\right)\right]$$

$$\gamma(t) = -\sqrt{2}\gamma_{\text{rms}} \cos\left(\omega t - \phi_\gamma\right),$$

where the cycle-averaged and phase-delay values are given by

$$\bar{D}_\perp = \frac{2}{3}\frac{\bar{\varepsilon}}{m\nu_m}$$

$$\bar{D}_\parallel = \frac{\bar{D}_\perp}{1 + \frac{m_0 v_{\text{drms}}^2}{1+\Omega'}\frac{\nu_m'}{\nu_m}\cos^2\tilde{\phi}_e}$$

$$\gamma_{\text{rms}} = \frac{m_0 v_{\text{drms}} \bar{D}_\parallel}{1+\Omega'}\cos\tilde{\phi}_e$$

$$\tilde{\phi}_e = \tan^{-1}\left(\frac{2\omega}{\nu_m}\right)$$

$$\Phi_m = \tan^{-1}\left(\frac{\omega}{(1+\Omega')\,\nu_e}\right)$$

$$\phi = \phi_m + \tilde{\phi}_e\,.$$

The harmonic dependencies are the same as those obtained through symmetry arguments. We should highlight that the DC formulas derived in Chapter 7 (e.g., Wannier, generalised Einstein relations, etc.), can be straightforwardly adapted to evaluate the cycle-averaged values for high frequency AC electric fields, through the use of the effective field concept in the appropriate limit. The reader is referred to Ref. [108] for further details.

15.5 Transport Properties in AC Electric Fields

In Figure 15.1 [35], we present calculations of mean energy, drift velocity, diffusion coefficients, and the energy gradient parameter for a simple hard sphere model for the full range of applied frequencies ω. We compare results calculated using a multi-term solution of Boltzmann's equation with those from numerical solution of the system of balance Equations 15.33 through 15.36, highlighting typical accuracies of calculations using momentum transfer theory.

The basic phenomenology discussed above is clearly evident in these profiles. Even for the simple hard sphere model, however, we observe that some transport properties violate the prescription detailed above, containing greater structure and variation with frequency than that afforded by the above "traditional" qualitative explanations. To fully understand the

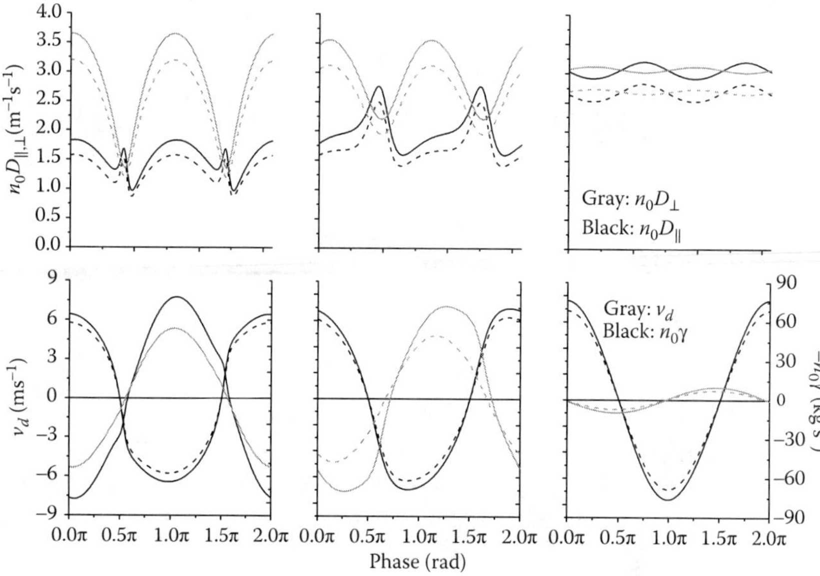

Figure 15.1 A comparison between standard momentum transfer theory (solid lines) and full time-dependent multi-term solution of Boltzmann equation (dashed lines) for various transport coefficients over a range of reduced angular frequencies ω/n_o for a gas of hard spheres. ($E/n_0 = 1\cos\omega t$ Td, $\sigma_m = 5$ Å^2, $T_0 = 0$ K, $m_0 = 4$ amu; Applied frequencies ω/n_0 Column 1 – 1×10^{-18} rad m^{3}s^{-1}, Column 2 – 5×10^{-18}rad m^{3}s^{-1}, Column 3 – 1×10^{-16}rad m^{3}s^{-1}.) (From R. D. White et al., *Journal of Physics D: Applied Physics*, 2009, © IOP Publishing and Deutsche Physikalische Gesellschaft. CC BY-NC-S.)

variation of the properties over the cycle of the field and the variation of the profiles with frequency one must establish:

- The ability of a given transport to relax on the time-scale governed by the frequency of the applied field, *and;*
- The implications associated with an inability to relax.

The reader is referred to Ref. [109] for further details.

15.5.1 Anomalous anisotropic diffusion

One of the more interesting phenomena to come out of studies of temporal non-locality has been that of anomalous anisotropic diffusion [78,110]. For DC electric fields only, we have the following form of the generalised

Einstein relations:

$$\frac{D_{\parallel}}{D_{\perp}} = \frac{T_{\parallel}}{T_{\perp}} \frac{\partial \ln v_d}{\partial \ln E} \tag{15.48}$$

$$= \frac{T_{\parallel}}{T_{\perp}} + \frac{\mu \nu'_m \gamma v_d}{k_B T_{\perp}} . \tag{15.49}$$

Thus, physically, the sources of anisotropic diffusion in DC electric fields only are [111]

1. Thermal anisotropy ($T_{\parallel} \neq T_{\perp}$): dispersion of charged particles associated with random motions is different in the directions parallel and perpendicular to **E**;

2. Differential velocity effect: the spatial variation of local average velocities through the swarm combined with an energy dependent collision frequency act to influence the spread of the swarm parallel to field [78,112].

As discussed in Section 7.6.2, for the simple case of electrons in a gas of hard spheres with $T_{\parallel} \approx T_{\perp}$, we can see from Equation 15.48 that $D_{\parallel}/D_{\perp} \approx 0.5$ (more precise value is 0.491).

When one applies a radiofrequency electric field to the system, interesting anomalous behaviour in the diffusion coefficients emerges as shown in Figure 15.1. The signature effects for "anomalous anisotropic diffusion" are [109]

- At low frequencies, we observe the evolution of a spike in the profile of $D_{\parallel}$ in the low-field phase of the cycle. As the frequency increases, the height and temporal extent of the spike is increased until it becomes the dominant feature in the temporal profile of $D_{\parallel}$;
- There exist phases in the field where $D_{\parallel}/D_{\perp} \neq 0.5$ and indeed cases where $D_{\parallel} > D_{\perp}$ in contrast to the DC steady state case. The fraction of the cycle of the field where the latter relation holds increases with increasing frequency until they are anti-phase. For a given frequency, diffusion is isotropic instantaneously four times per cycle;
- In a cycle-averaged sense, $D_{\parallel}/D_{\perp}$ increases from 0.5 at low frequencies to 1 at high frequencies.

It is interesting in this case that the diffusion is instantaneously isotropic at those phases of the field where drift velocity is zero and the first order spatial variation of the average energy, as characterised by γ, is zero. The phases of the field where $D_{\parallel} > D_{\perp}$ correspond to phases where $\gamma(t)v_d(t) > 0$

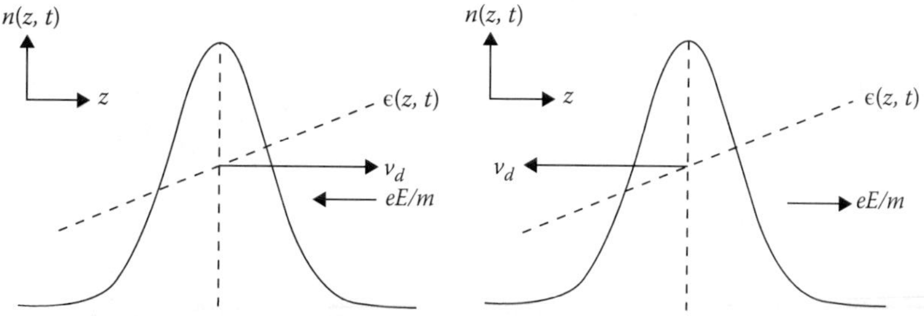

(a) (b)

Figure 15.2 (a) Schematic representation of a travelling pulse showing the density (—) and mean energy (- - -) as functions of position. (b) Schematic representation of a travelling pulse at a phase of the AC field E at which the field and the drift velocity have become negative, but the mean energy continues to increase in the $+z$ direction, that is, $\gamma(t)v_d(t) > 0$. (From R. D. White, Ph. D Thesis, James Cook University, 1997.) See Refs. [78,109] for further details.

(see Figure 15.2, [113]). Due to the differences in the timescales for energy and momentum exchange, γ and v_d do not respond to a change in the field direction on the same timescale [108]. From Equation 15.49, one can see in these phases that the differential velocity effect now acts to enhance longitudinal diffusion [78,111] and consequently $D_{\|} > D_{\perp}$. To fully understand this phenomenon one needs to understand and implement the temporal relaxation profiles (for electric fields) such as those considered above [111].

15.6 Concluding Remarks

This chapter shows how the kinetic and fluid equations are applied to particles in time-varying harmonic electric fields. We showed how general properties could be inferred by considering the role of space–time symmetries. We then looked at the various regimes as characterized by the frequency of the applied field in comparison with the collision frequencies for energy and momentum transfer. The concept of the effective electric field was introduced and it was shown how the Wannier and generalized Einstein relations are modified for time-varying fields. Finally, the anomaly in $D_{\|}/D_{\perp}$ arising from time dependence was investigated.

CHAPTER 16

The Franck–Hertz Experiment

16.1 Introduction

The 1914 experiment of James Franck and Gustav Hertz [46] provided a graphic demonstration of the quantized properties of atoms, and thereby laid the foundations of modern quantum and atomic physics. This chapter deals with its modern interpretation and establishes the link between microscopic processes, which are governed by the laws of quantum mechanics, and macroscopic phenomena as measured in the laboratory. There is also an important connection with the steady state Townsend experiment. We deal with these questions in the framework of kinetic theory outlined in Chapter 5, and go on to show how the Franck–Hertz experimental data may be analysed in terms of an eigenvalue problem, similar to that introduced in Chapter 11.

16.2 The Experiment and Its Interpretation

16.2.1 The original arrangement

Figure 16.1 shows a schematic representation of the experiment in plane-parallel geometry, which is nowadays favoured over the cylindrical arrangement originally used by Franck and Hertz [114].

Electrons are emitted at a steady rate from the cathode into a drift tube containing an atomic gas of number density n_0 and are scattered in collisions with gas atoms (both elastic and inelastic) as they fall through a voltage U over a distance d to the control grid. The retarding voltage U_G applied between grid and anode allows only higher energy electrons to contribute to the measured anode current I_A. An oscillatory $I_A - U$ curve is observed, and the spacing between peaks, designated by ΔU, is supposed to be directly related to some quantized atomic energy level.

The results of the original experiment using mercury vapour published in 1914 are shown in three different $I_A - U$ curves in Ref. [46], with the best known of these reproduced in Figure 16.2, showing $\Delta U \approx 4.9$ V. Franck and Hertz initially interpreted this as the ionisation potential of a mercury atom, but after a subsequent experiment they concluded that their result indicated an electronic excitation level of energy $\epsilon_I = e\Delta U = 4.9$ eV. In addition, they observed that the radiation emitted from excited atoms returning to their ground state had a wavelength $\lambda = 253.6$ nm. This,

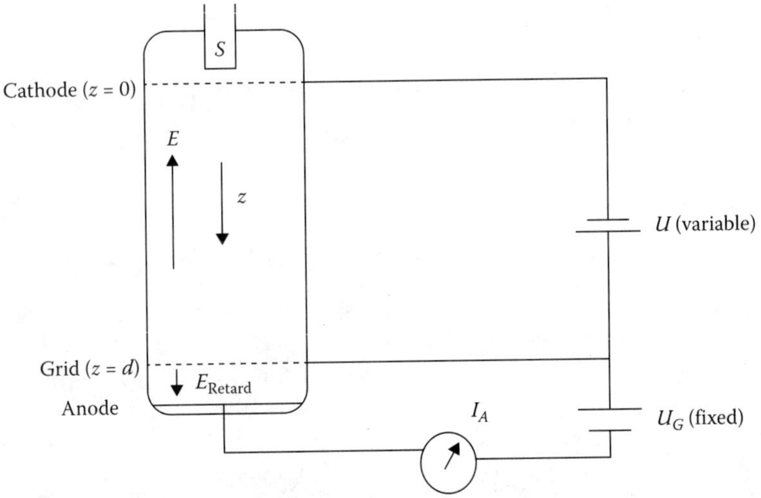

Figure 16.1 Schematic representation of the Franck–Hertz experiment, with the gas under investigation filling the region between plane-parallel electrodes. Electrons emitted from the source S move under the influence of a uniform field $E=U/d$ in the region $0 \leq z \leq d$ between cathode and grid. Beyond the grid at $z=d$ the retarding voltage U_G allows only electrons with energies above eU_G to contribute to the anode current I_A. The voltage U is increased and the resulting $I_A - U$ curve is recorded, as in Figures 16.2 and 16.4 for Hg vapour and Ne gas, respectively. (Reproduced from R. E. Robson, et al., *European Physical Journal D*, 68, 188, 2014. With permission.)

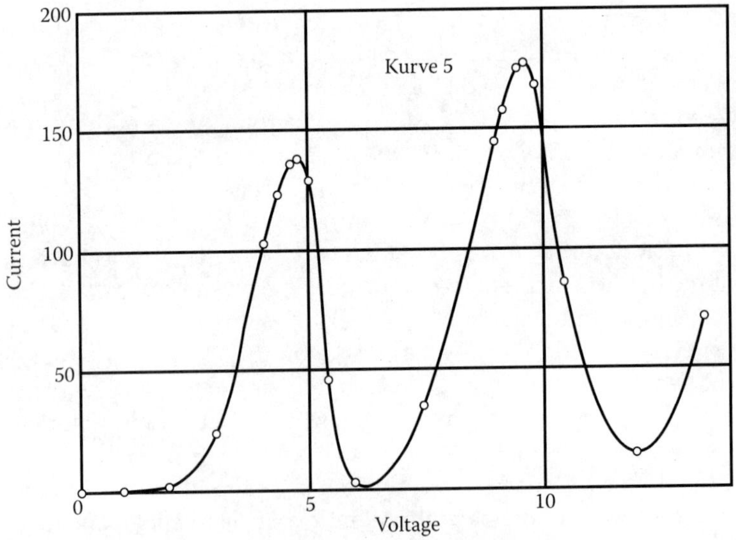

Figure 16.2 The current–voltage characteristic for Hg vapour reported by Franck–Hertz experiment [46] for which the voltage difference between peak currents is $\Delta U = 4.9$ V.

together with $\epsilon_I = hc/\lambda$, could be used to either (1) confirm the measured value of ϵ_I or (2) provide an independent estimate of Planck's constant h.

At first, Franck and Hertz were not aware of the Bohr model, and only appreciated the full implications of their results several years later, in what was, in fact, their last joint publication [115].* On the other hand, Bohr himself realized already that, in 1915, the importance of the Franck–Hertz experiment as confirming his quantized model of the atom.

16.2.2 Traditional model

Within a certain range of voltages and gas pressures, the mean electron energy profile $\bar{\epsilon}(z)$ is oscillatory in the drift region between cathode and grid. Such a spatially periodic structure is a macroscopic reflection of atomic quantisation and, while observed directly in other experiments [47,116], it is usually modelled somewhat crudely in the case of the Franck–Hertz experiment. Thus, the traditional textbook model makes the following assumptions, either explicitly or implicitly:

- Only one excited energy level ϵ_i of the gas atoms is considered.
- Electrons have the same energy and velocity and effectively constitute a monoenergetic beam in the drift tube.
- Each electron starts from rest at the cathode and is accelerated by the electric field $E = U/d$ until it achieves sufficient energy to excite a gas atom to its quantized energy level ϵ_i.
- The electron then gives up all its energy abruptly in an inelastic collision and is again accelerated from rest to energy ϵ_i, when it is once more brought to rest in a collision, and so on.
- There are no elastic collisions.
- The average electron energy $\bar{\epsilon}(z)$, which is the energy of each and every electron in this model, thus fluctuates as a function of the distance z downstream from the source in a sharp, "saw tooth" pattern, with spatial period $\Delta z = \epsilon_i/eE$.
- As U (and therefore E) is increased, the wavelength Δz decreases, that is, the pattern shrinks, and the mean electron energy $\bar{\epsilon}(d)$ at the grid rises and falls accordingly.
- If $\bar{\epsilon}(d) > U_G$, electrons can pass to the anode, and hence the anode current I_A in the external circuit should rise and fall as U increases.
- The "wavelength" of the $I_A - U$ curve (the voltage separation between peaks in I_A) is equal to the increase in voltage $\Delta U =$

* See also: "James Franck - Nobel Lecture" (http:/www.nobelprize.org/nobe prizes/physics/laureates/1925/francklecture.html) and "Gustav Hertz - Nobel Lecture" (http:/www.nobelprize.org/nobel prizes/physics/laureates/1925/hertz-lecture.html).

ϵ_i/e required to produce exactly one more additional internal oscillation.
- In this way, the quantized energy ϵ_I of an atom can be read directly from the $I_A - U$ curve.

Not surprisingly, this simplistic portrayal of electron behaviour in the drift region, and misrepresentation of atoms as having only one possible quantized excited state leads to problems in interpreting of the experiment. As we will see, employing rigorous kinetic theory [117,118] shows that the mean electron energy $\bar{\epsilon}(z)$ in the drift tube oscillates smoothly, not in a sharp saw-tooth manner, with a spatial wavelength Δz corresponding to an energy difference $\Delta\bar{\epsilon} = eE\Delta z$, which is generally influenced by a number of atomic energy levels ϵ_i. The period of the $I_A - U$ curve,

$$\Delta U = \Delta\bar{\epsilon}/e \tag{16.1}$$

must be interpreted accordingly.

16.2.3 Results and interpretation

16.2.3.1 Mercury

The usual textbook interpretation of Figure 16.2 suggests the Hg atom, therefore, has a quantised energy level of 4.9 eV. This is indeed close to 4.89 eV, the energy of the second quantised state (the $6^1S_0 \rightarrow 6^3P_1$ transition), but the first quantized level of Hg, for which $\epsilon_i = 4.67$ eV (the $6^1S_0 \rightarrow 6^3P_0$ transition—see Table 16.1) seems to be bypassed. Since this lowest level is metastable, one would not expect to see any corresponding spectral emission line, but why should it not influence the $I_A - U$ curve? The question was addressed by Hanne [119], who pointed out that the cross section for excitation of the first excited level of Hg is small (see Table 16.1 and Figure 16.3). One must look further, however, to understand why the $6^1S_0 \rightarrow 6^3P_2$ process, with threshold energy 5.46 eV, does not appear to significantly influence the measurement.

Table 16.1 Threshold Energies and Approximate Maximum Cross Sections for Excitation of Hg to the Energy Levels Shown by Inelastic Collisions.

i	Process	Threshold (eV)	Maximum Cross Section (Å^2)
1	$6^1S_0 \rightarrow 6^3P_0$	4.67	0.4/0.5
2	$6^1S_0 \rightarrow 6^3P_1$	4.89	3.5/5.0
3	$6^1S_0 \rightarrow 6^3P_2$	5.46	4.0

Note: Two numbers indicate two peaks of the cross section as seen in Figure 16.3.

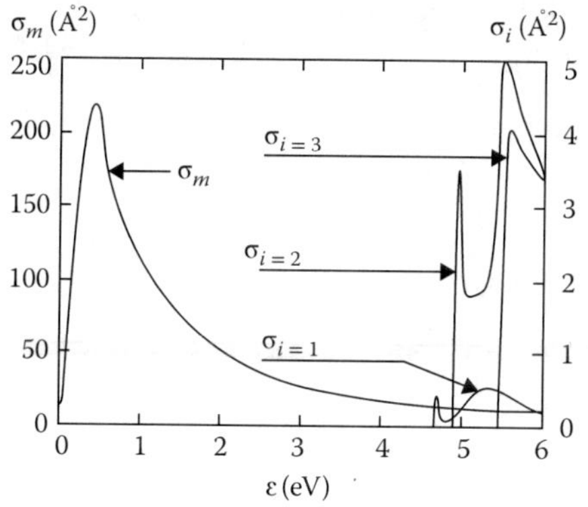

Figure 16.3 The momentum transfer cross section σ_m and three inelastic cross sections σ_i for electron scattering from mercury [117,119]. See Table 16.1 for an explanation of notation. (Reproduced with permission from B. Li. Kinetic theory of charged particle swarms in a.c. fields, PhD Thesis, James Cook University, 1999.)

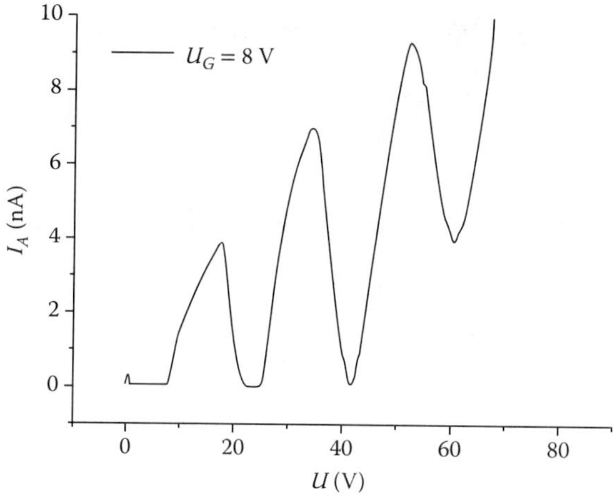

Figure 16.4 Measured Franck–Hertz $I_A - U$ curve for neon using a Leybold Didactic GMBH apparatus (http:/www.ld-didactic.de/literatur/hb/e/p6/p6243_e.pdf).

16.2.3.2 Neon

The Franck–Hertz experimental apparatus for neon is readily available commercially "off the shelf" through Leybold Didactic. The measured current–voltage characteristic is shown in Figure 16.4, from which we might infer an energy wavelength $\Delta\bar{\epsilon} = e\Delta U \approx (18 \pm 1)$ eV. For comparison, the known energy levels of Ne are shown in Table 16.2 and in the energy level diagram of Figure 16.5.

Table 16.2 Threshold Energies and Approximate Maximum Cross Sections for Excitation of Neon to the Energy Levels Shown by Inelastic Collisions.

Process	Threshold energy ϵ_i (eV)	Maximum cross section ($\mathring{A}^2$)
$i=1$	16.62	0.01
2	16.67	0.012
3	16.72	0.0024
4	16.85	0.12
5	18.38	0.033
6	18.97	0.026
7	19.66	0.033

Source: G. F. Hanne, *American Journal of Physics*, 56, 696, 1988.

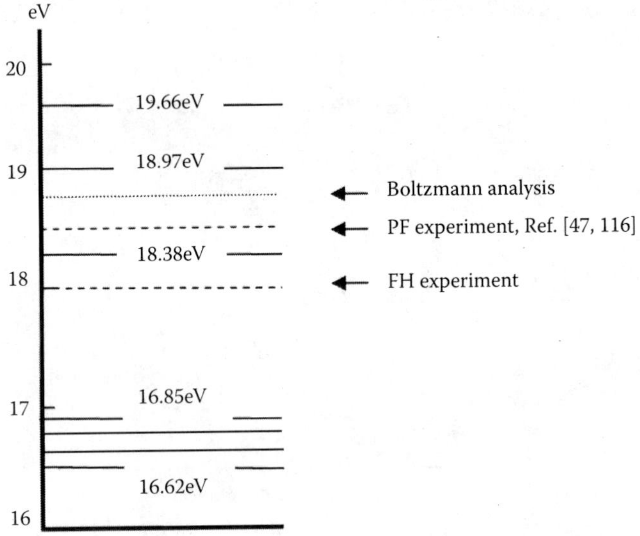

Figure 16.5 Energy level diagram for neon (solid lines), experimentally inferred values of $\Delta\varepsilon$ (dashed lines) from photon flux experiment (PF), Franck–Hertz experiment (FH) and theoretically calculated value of $\Delta\varepsilon$ from Boltzmann equation analysis (dotted line). See Ref. [114] for further details.

In terms of the simplistic text book model, we would say that the experiment samples the level at 18.38 eV to the exclusion of all others, but why should this be so? An inspection of the (*e*, *Ne*) cross section data, shown in abbreviated form in Table 16.2, indicates that, unlike Hg, it is not so easy to dismiss the contribution of any particular channel to the measured value of $\Delta\bar{\epsilon}$. The laboratory manual accompanying the Leybold apparatus at least acknowledges the problem and suggests that an element of "probability" comes into play. In physical terms, this means that the cross sections for each of the possible processes must somehow be accounted for.

16.2.3.3 A critical re-examination

The fundamental question which has emerged is this:

How is the value of ΔU measured in the Franck–Hertz experiment, and hence the value of $\Delta\bar{\epsilon}$ obtained from Equation 16.1 to be interpreted in terms of the quantized energy levels of the atoms comprising the gas?

The answer requires a review of the basic physics of electrons in the drift region.

First, note that elastic scattering is important, for example, the elastic momentum transfer cross section in Hg is enormous (Figure 16.3), and electrons in the drift region, therefore, may make many elastic collisions before exciting an atom in an inelastic collision. This has two implications:

1. Elastic collisions randomise the directions of electron velocities, and hence the electron velocity distribution function $f(z, \mathbf{v})$ may be very nearly isotropic (see Chapter 5) in the drift region; and
2. Although electrons exchange only a small fraction $\sim 2m/m_0$ of energy in elastic collisions with atoms, the net effect after many such collisions is to create a large spread of energies about the mean energy.

Thus, in the drift region of a Franck–Hertz experiment, $f(z, \mathbf{v})$ is broad in energy and nearly isotropic in velocity space. Electrons behave like a swarm, quite the opposite of a unidirectional, monoenergetic beam, as implicit in the textbook model. Most importantly, this means that several inelastic channels may be open simultaneously, and that the measured value of $\Delta\bar{\epsilon}$ is expected to reflect a number atomic energy levels, weighted by cross sections, and averaged over the electron energy distribution function.

16.2.3.4 Direct observation of periodic structures

Periodic electron structures have for long been known in gaseous electronics, although the Franck–Hertz experiment is not normally associated with this literature, in spite of some clear connections. Thus, the steady state Townsend (SST) "swarm" experiment [47,116,118] can be thought of as representing the drift tube region of Figure 16.1, where atoms are excited by electron impact. However, there is no grid and no measurement of currents in an external circuit. Instead, the spatial profile is investigated directly and non-intrusively by measuring the intensity of photons emitted in de-excitation of atoms, in the so-called "photon flux technique" of Fletcher and Purdie [47,116]. Profiles obtained for neon are shown in

Figure 16.6, from which the wavelength Δz can be read, and the energy wavelength estimated as $\Delta\bar{\epsilon} = eE\Delta z \approx 18.5$ eV. This compares reasonably with ~18 eV obtained from the Franck–Hertz experiment (see Figure 16.5). Of course, the wavelengths measured in the two experiments are not expected to coincide exactly, since the photon flux technique does not reflect excitation of atoms to metastable states, whereas the Franck–Hertz measurement includes the effects of all collision processes. In addition, the Franck–Hertz results may also be influenced by the intrusive effect of the grid, something which is quite difficult to account for theoretically. Note that all quantities in both experiments depend on the electric field $E = U/d$ and the gas number density n_0 through the reduced field E/n_0 (unit *townsend* = 1 Td = 10^{-21} V m^2).

Fletcher and Purdie discussed their results in terms of the quantised energy levels shown in Table 16.2 and concluded that some weighting of the respective cross sections would be required in order to explain the measured wavelength of 18.5 eV, just as it is for the Franck–Hertz experiment.

The same conclusion was reached by Fletcher [116] in regard to argon, where the measured $\Delta\bar{\epsilon} \approx 13$ eV does not correspond to any single atomic energy level. This is consistent with the voltage wavelength

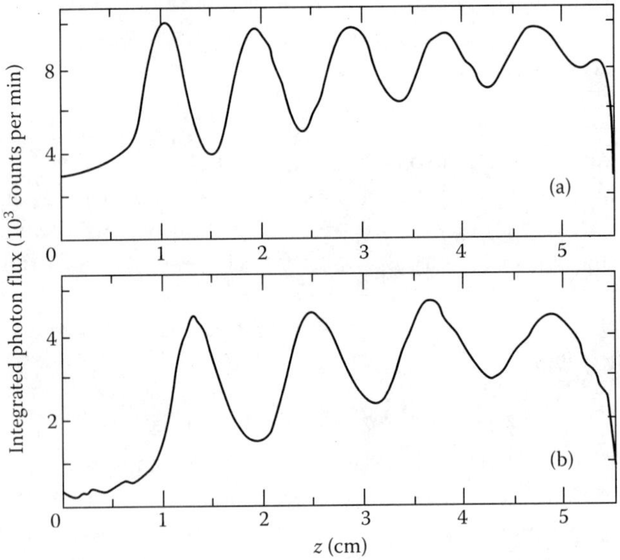

Figure 16.6 Measurements of photon flux as a function of inter-electrode distance z arising from de-excitation of atoms in a steady state Townsend discharge with neon, for a reduced electric field $E/n_0 = 30.4$ Td. Gas pressures are (a) $p = 266$ Pa and (b) $p = 254$ Pa. (From J. Fletcher and P. H. Purdie, *Australian Journal of Physics*, 40, 383, 1987.) (http://www.publish.csiro.au/nid/78/paper/PH870383.htm, reproduced With the permission of CSIRO.)

$\Delta U = (12 \pm 1)$ eV which may be extracted from the argon Franck–Hertz data of Magyar et al. [120], as well as the energy wavelength $\Delta\bar{\epsilon} \approx 12$ eV of their simulated periodic structures in the drift region.

Fletcher [116] also examined the effect of adding trace amounts of a molecular gas, which have quantized rotational and vibrational energy levels which are very closely spaced (~0.1 eV) compared to the large energy gap ($\gtrsim$10 eV) between electronic states in monatomic gases. The effect was to dampen the oscillations, virtually eliminating any periodic structure. This is also the reason that the Franck–Hertz experiment can operate satisfactorily only with monatomic gases.

16.3 Periodic Structures—The Essence of the Experiment

In the detailed discussion on the Franck–Hertz experiment, we focus on the drift region between cathode and grid because

1. This is where quantum effects at the atomic scale become manifest through the formation of macroscopic periodic structures;

2. This is also where the electron physics remains poorly understood, often misrepresented and trivialised;

3. One must fully understand these periodic structures, not only in order to be able to interpret the experimental results, but also to prescribe satisfactory operational conditions, for example, where to place the grid relative to the cathode and the range of operational voltages and gas pressures;

4. The formation of periodic structures lies at the heart of both the Franck–Hertz experiment and the SST experiment of Fletcher and Purdie [47,116]. The method of observation of these structures, either indirectly (and intrusively) through the agency of a grid plus retarding field, and projection onto an external current–voltage characteristic, or directly (and non-intrusively) through the photon flux technique, respectively, is quite another matter;

5. While the experiment furnishes a current–voltage curve over a wide range of voltages, the main interest is confined to its periodicity, the peak to peak voltage difference ΔU. The aim is to infer required atomic properties from Equation 16.1 and

6. The full current–voltage curve may well provide additional information, but its detailed shape is influenced by the properties of the grid and the rather complex fields which develop in its vicinity. Since one wishes to measure the properties of the gas atoms, rather than the apparatus, which is used to make the measurements,

the value of generating the full curve theoretically appears to be limited, especially when the high computational cost is considered [120].

All these considerations aside, we emphasise that the grid, even if it were to operate ideally and non-intrusively, must be placed far enough downstream from the cathode, so that any memory of source properties, specifically, the unknown initial distribution of electron velocities, should be eliminated. Moreover, the field structure in the neighbourhood of the grid is really quite complex [121], so it is something of an idealisation to assume that a uniform field exists over the entire cathode-grid region. In swarm experiments [10], either the distance of the detector from the source or the gas pressure is increased until measurements become length or pressure independent, respectively, indicating that source and end effects are negligible. Apart from Ref. [120], however, Franck–Hertz experiments have not had this flexibility. Thus, there is the possibility that both the internal wavelength Δz and peak-to-peak voltage difference ΔU may be "contaminated" by the source or grid effects.

16.4 Fluid Model Analysis

Although there have been attempts to analyse the Franck–Hertz experiment in terms of mean free paths [122], this approach is too crude to be of much help, as it is with other areas of charged particle transport. If a qualitative or semi-quantitative picture is required, the recommended approach is to use momentum transfer theory to form fluid equations, as in Chapters 7 through 10. The fluid equations for electrons in a Franck–Hertz drift tube or SST experiment can give a good estimate of the energy wavelength $\Delta\bar{\epsilon}(z)$ of the observed periodic structures [85,118]. However, the details are lengthy, and at present, we merely focus one key qualitative aspect that emerges at the fluid level.

In the fluid formalism developed in Chapter 8, the effects of inelastic collisions enter through the quantity*

$$\Omega(\bar{\epsilon}) \approx \frac{m}{2m_0\sigma_m(\bar{\epsilon})} \sum_{i=1} \epsilon_i \sigma_i(\bar{\epsilon}) \exp\left(-\frac{3\epsilon_i}{2\bar{\epsilon}}\right), \tag{16.2}$$

where the threshold energies and cross sections for excitation of atomic levels are denoted by ϵ_i and σ_i (see Tables 16.1 and 16.2 for Hg and Ne, respectively). The influence of any level i for which σ_i is very small may be neglected, while the exponential term acts to suppress the influence of

* Since we are dealing with electronic states, for which $\epsilon_i \gg k_B T_0$, all atoms are effectively initially in the ground state. For that reason, any contributions from superelastic processes can be safely neglected.

higher levels, for which $\epsilon_i > \bar{\epsilon}$. As the applied voltage U (and hence E/n_0) increases, the mean electron energy $\bar{\epsilon}$ also increases, allowing contributions from successively higher threshold inelastic channels to contribute to Equation 16.2. However, the influence of higher levels is suppressed because of a complex interplay between elastic and inelastic processes, and the periodic structure actually disappears above a certain critical value of E/n_0, for example, about 30 Td for Hg [85]. Then the energy profile $\bar{\epsilon}(z)$ becomes monotonic, not oscillatory, and the experiment yields no further useful information.

These two reasons explain why the dominant contribution to the Franck–Hertz experiment with Hg comes from only the second excited level with energy 4.89 eV:

1. The lowest level 4.67 eV is indeed excited, but at a negligibly small rate, since the cross section is so small;
2. Even though the third level 5.46 eV may contribute to Equation 16.2 at sufficiently high voltages U, the oscillatory "window" of operational E/n_0 has by then closed.

For Ne, the situation is even more complicated. No single inelastic process dominates (see Table 16.2) and, while Equation 16.2 offers some qualitative insight into the way these processes might be weighted, a quantitative assessment is best left to a more rigorous analysis.

16.5 Kinetic Theory

16.5.1 The kinetic equation

The essential physics of drift region in Figure 16.1 can be illustrated using idealized plane-parallel model geometry, as in Figure 11.2. Electrons are assumed to be emitted at a steady rate from the source at $z = 0$ into the gas, with the field and all spatial variation directed along the z-axis. In the steady state, the electron distribution function $f = f(z, \mathbf{v})$ satisfies the kinetic equation

$$\left(v_z \partial_z + a \partial_{v_z}\right) f = \left(\frac{\partial f}{\partial t}\right)_{\text{col}}, \tag{16.3}$$

where $a = eE/m$ and $\left(\frac{\partial f}{\partial t}\right)_{\text{col}}$ is the Wang-Chang et al. collision term (see Equation 5.36). The quantity of most interest is the mean electron energy,

$$\bar{\epsilon}(z) = \frac{1}{n(z)} \int \mathrm{d}^3 v\, f(z, \mathbf{v}) \frac{1}{2} m v^2, \tag{16.4}$$

where $n(z) = \int \mathrm{d}^3 v\, f(z, \mathbf{v})$ is the electron number density.

16.5.2 Eigenvalue analysis

16.5.2.1 The importance of the eigenvalue analysis?

Since the physical situation portrayed in Figure 16.1 has rotational symmetry about the z-axis, the corresponding phase space can be described by just three independent variables (z, v, θ). Therefore, we might consider solving Equation 16.3 by discretization and/or orthogonal polynomial expansion techniques (see Chapter 12). While this offers useful information [117], it does have several disadvantages:

1. In spite of some reduction in complexity, it is still numerically challenging.
2. It requires information about the electron distribution function at the source, which is not usually known.
3. It provides more information than is necessary to interpret the current–voltage curve.

For these and other reasons as explained below, we seek an alternative procedure through eigenvalue analysis.

In Chapter 13, we explained how the solution of the general time-dependent kinetic equation could be expressed in terms of a sum over the eigenfunctions of a certain kinetic eigenvalue problem. The procedure could be carried over formally to the present steady state problem by replacing the time coordinate t with z, and redefining key operators, but we prefer to proceed as in earlier chapters, and set up the eigenvalue problem from first principles.

In the situation considered in Chapters 13, the prime interest was in the asymptotic regime, where the quantities of physical interest could be described by a single fundamental eigenvalue. This is also the case for the Franck–Hertz experiment of Figure 16.1, where the grid is (or should be) placed in the asymptotic region downstream from the source, providing a physically meaningful current–voltage wavelength ΔU.

Before proceeding, we recall that unlike quantum mechanics, kinetic theory operators are generally not self-adjoint, and hence eigenvalues can be complex. In fact, the Franck–Hertz experiment effectively measures the imaginary part of a certain fundamental eigenvalue.

16.5.2.2 The eigenvalue problem

The simplest way of understanding how the eigenvalue problem arises is as follows: the steady state Boltzmann equation (Equation 16.3) is separable in z and $\mathbf{v}$, with possible elementary solutions of the form $f(z, \mathbf{v}) \sim \psi(\mathbf{v})\ \exp(Kz)$, where K is a separation constant, and the function $\psi(\mathbf{v})$ satisfies $\left(Kv_z + a\,\partial_{v_z}\right)\psi = \left(\frac{\partial \psi}{\partial t}\right)_{\text{col}}$. Requiring $\psi(\mathbf{v})$ to behave physically restricts the values of K to belong to an eigenvalue spectrum, found from solution of the eigenvalue problem

$$\left(K_j v_z + a\,\partial_{v_z} + J\right)\psi_j = 0, \tag{16.5}$$

where $J(\psi_i) \equiv -\left(\frac{\partial\psi_j}{\partial t}\right)_{\text{col}}$. In general, as explained in Chapter 11, one also needs to solve a dual eigenvalue problem involving the adjoint collision operator $J^\dagger$, but this is not needed for present purposes.

16.5.2.3 The nature of the eigenvalue spectrum

The eigenvalue spectrum $K_0, K_1, K_2, \ldots$ is generally found to be discrete and the index $j = 0, 1, 2, \ldots$ orders the allowed "modes" in the following way:

- The lowest mode $j = 0$ has a real eigenvalue K_0, which characterizes the asymptotic behaviour far downstream from the source. If the field were high enough so that ionization was possible, then number of electrons would increase exponentially in the asymptotic region as governed by the magnitude of K_0, which is the Townsend ionization coefficient α_T introduced in Chapter 9. However, the present discussion assumes that collisions conserve particle number, and in that case $K_0 = 0$. The asymptotic region then corresponds to spatially homogeneous conditions, for which all properties are constant. The corresponding eigenvalue problem is

$$\left(a\,\partial_{v_z} + J\right)\psi_0 = 0, \tag{16.6}$$

  where $\psi_0(\mathbf{v})$ is the distribution function which would, at least in principle, be attained very large distances downstream from the source.
- The eigenvalues K_j for the higher order modes $j > 0$ occur in complex conjugate pairs. Only those with negative real part, that is,

$$Re\{K_j\} = -k_j$$

  are taken to ensure the correct asymptotic behaviour at large z, where it is implicit that

$$k_1 < k_2 < k_3 < \ldots .$$

- The imaginary part of the eigenvalue is written as

$$Im\{K_j\} = 2\pi/\Delta z_j\,,$$

  where Δz_j is the "wavelength" of the jth mode.

16.5.2.4 Distribution function

The most general solution of the kinetic equation (Equation 16.3) is then a linear superposition of all possible eigenmodes $\psi_j(\mathbf{v})\ \exp(K_j z)$, that is,

$$f(z,\mathbf{v}) = \text{Re}\left\{\sum_{j=0}^{\infty} S_j \psi_j(\mathbf{v})\ \exp(K_j z)\right\}$$

$$= f_\infty(\mathbf{v}) + \text{Re}\left\{\sum_{j=1}^{\infty} S_j \psi_j(\mathbf{v})\ \exp\left(-k_j\, z + 2\pi i z/\Delta z_j\right)\right\}, \tag{16.7}$$

where

$$f_\infty(\mathbf{v}) \equiv S_0 \psi_0(\mathbf{v}).$$

The expansion coefficients S_j could, in principle, be found from the distribution function at the source. However, this is usually not known. Fortunately, we do not need values of S_j to understand the Franck–Hertz experiment measurements.

16.5.2.5 Asymptotic region

Near the source, many terms in the summation (Equation 16.7) are generally needed, but sufficiently far downstream, at distances z such that $k_1 z > 1$, the fundamental mode $j = 1$ dominates, and the asymptotic distribution function,

$$f(z,\mathbf{v}) = f_\infty(\mathbf{v}) + \text{Re}\left\{S_1 \psi_1(\mathbf{v}) \exp\left(-k_1 z + 2\pi i z/\Delta z_1\right)\right\} \tag{16.8}$$

is characterised by a single decay constant k_1 and a single wavelength Δz_1. For simplicity, these fundamental properties are henceforth written as just k and Δz, respectively.

Substitution of Equation 16.8 into 16.4 then gives the mean energy in the asymptotic region

$$\bar{\varepsilon}(z) = \varepsilon_\infty + \varepsilon_1 \exp(-kz)\cos(2\pi z/\Delta z + \phi) + \ldots,$$

where the constants $\varepsilon_\infty, \varepsilon_1$ and the phase difference ϕ can be found as integrals over $\psi_0(\mathbf{v})$and $\psi_1(\mathbf{v})$. It is clear that electron properties in the asymptotic region may be described in terms a single, pure harmonic of wavelength Δz, which is a property of the gas atoms alone.

The position $z = d$ of the grid in Figure 16.1 should be such that $k_1 d > 1$, for then it lies in the asymptotic region, and the voltage wavelength (Equation 16.1) in the external circuit reflects only intrinsic atomic properties, as required.

16.6 Numerical Results

16.6.1 Numerical procedure

First, we note that as elsewhere, the quantities of interest in the Franck–Hertz experiment depend on the value of the field $E = U/d$ and gas number density n_0 through the reduced field E/n_0. Distances and wavenumbers scale as $n_0 z$ and k/n_0, respectively.

Since there is axial symmetry about the field direction (the z-axis), we may decompose the eigenfunctions in velocity space through an expansion in a finite number of Legendre polynomials,

$$\Psi_j(\mathbf{v}) \approx \sum_{l=0}^{l_{max}} \Psi_{j,l}(v) P_l(\cos\theta),$$

where θ is the angle between $\mathbf{v}$ and the z-axis. The upper limit l_{max} on the l-summation is incremented successively until some accuracy requirement is met. It turns out that $l_{max} = 2$ is generally sufficient to furnish quantities of interest to accuracies of around 1% or so. The coefficients in $\Psi_{j,l}(v)$ are subsequently decomposed in speed (or energy) space, for example, through an expansion in Sonine (generalised Laguerre) polynomials about a Maxwellian weight function with an arbitrary, adjustable temperature, as in the "two-temperature" method outlined in Chapter 12. It is found that up to ten Sonines are required to achieve the designated accuracy.

In this way, the integro-differential equation (Equation 16.3) is reduced to an equivalent set of algebraic equations, and the eigenvalues can be found by standard linear algebra techniques.

16.6.2 Mercury

We now employ the eigenvalue technique for electrons in mercury gas, as originally used by Franck and Hertz, with the cross sections as shown in Figure 16.3. Figure 16.7 shows the theoretically calculated mean energy $\bar{\epsilon}(z)$ for Hg, in a range E/n_0 corresponding to experimental conditions. For simplicity of presentation, we show the whole region downstream from the source, not just the asymptotic regime, for which these results are strictly valid. Distance is scaled to a representative mean free path, $\lambda = 1/(\sqrt{2} n_0 \sigma_0)$, where $\sigma_0 = 10^{-20}\, m^2$. The relation $\Delta\bar{\epsilon} = eE\Delta z$ between energy and spatial wavelengths can then be written as

$$\Delta\bar{\epsilon}\ (eV) = 0.0707 \left(\frac{E}{n_0}\right)_{Td} \frac{\Delta z}{\lambda}.$$

Thus at $E/n_0 = 5$ Td, for example, the dimensionless wavelength is $\Delta z/\lambda \approx 14$ yielding $\Delta\bar{\epsilon} \approx 4.9$ eV, in accord with the experimental result.

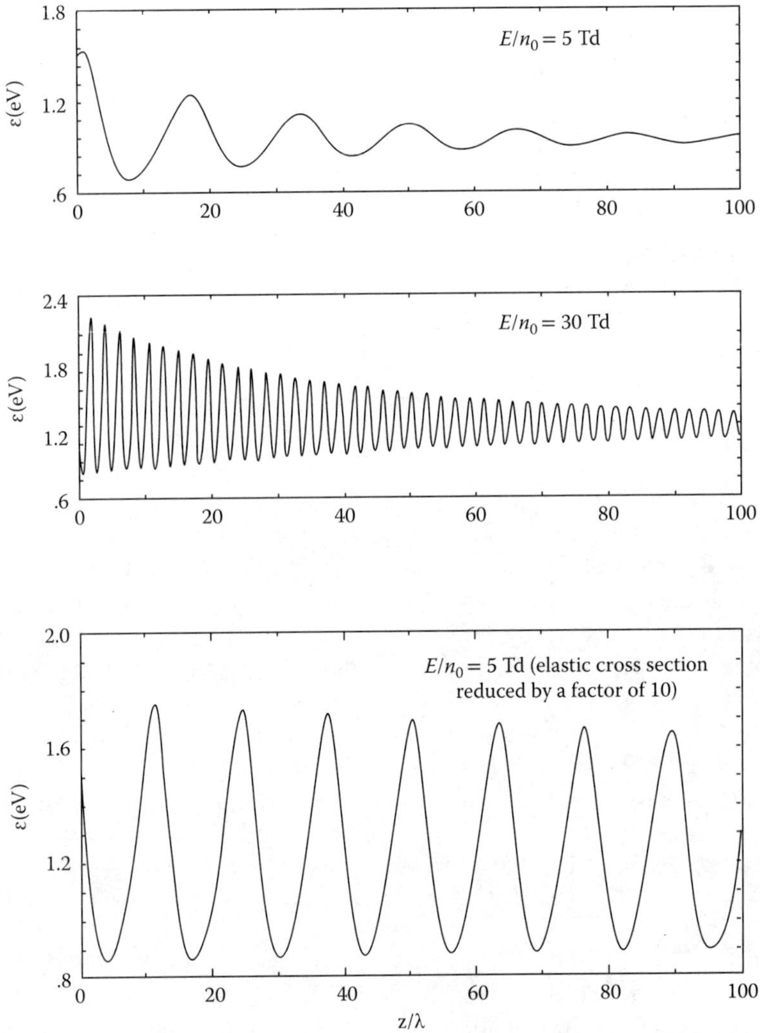

Figure 16.7 Spatial relaxation of the mean energy for electrons in mercury vapour for a range of reduced fields E/n_0 realized in the Franck–Hertz experiment. Distance from the source is scaled by a representative mean free path, $\lambda = 1/(\sqrt{2}n_0\sigma_0)$, where $\sigma_0 = 10^{-20}$ m^2 is a representative cross section. (Reproduced From B. Li, PhD Thesis, James Cook University, 1999, with permission.)

These calculations include the effect of elastic collisions through the momentum transfer cross section σ_m. While this is clearly enormous (note the different scales in Figure 16.3), the traditional discussion ignores such collisions entirely, that is, $\sigma_m = 0$, but still gets the right result, if the other inelastic processes are ignored. How can this be?

The last curve in Figure 16.7 shows that reducing σ_m by a factor of 10 from the real value shown in Figure 16.3 has the following effects:

1. The amplitude of oscillations is markedly increased, but
2. Damping is also significantly reduced, and
3. Most significantly, the wavelength is reduced below 4.9 eV.

It is clear that the experimental result $\Delta U = \Delta\bar{\epsilon}/e \approx 4.9$ V for Hg depends on a combination of very special circumstances:

- Although inelastic collisions are responsible for the oscillations *per se*, elastic collisions of the magnitude shown in Figure 16.3 play a significant role in "tuning" the energy wavelength to the particular value of 4.9 eV.
- In addition, as discussed in Section 16.3, it is somewhat fortuitous that the only inelastic channel which influences the outcome of the experiment is the one with excitation energy 4.9 eV.

16.6.3 Neon

Similar oscillatory mean energy profiles exist for Ne over a range of reduced fields E/n_0 [118]. The estimated energy wavelength from eigenvalue analysis is found to be $\Delta\bar{\epsilon} \approx (18.8 \pm 0.1)$ eV. The Franck–Hertz experimental curve of Figure 16.4 gives $\Delta\bar{\epsilon} \approx (18 \pm 1)$ eV, while Fletcher and Purdie [47] obtained a value of 18.5 eV from their photon flux technique. None of these values of $\Delta\bar{\epsilon}$ corresponds to the threshold energy of any of the Ne inelastic channels shown Figure 16.5. Consequently, no particular energy level of Ne can be inferred from the Franck–Hertz experiment.

16.7 Concluding Remarks

In this chapter, we have investigated the kinetic theory behind the Franck–Hertz experiment and have shown that

- The experiment provides a graphic illustration of the quantization of atomic energy levels through the oscillatory anode current, even though the distance between current peaks is generally determined by several energy levels, not just one;
- Electrons in the drift region behave as a swarm, with a broad distribution of energies, not as a beam as in the traditional textbook model;

- The periodic structures which develop in the drift region are the natural oscillations of a system of electrons undergoing inelastic collisions in a gas subject to a uniform electric field; and
- From a rigorous theoretical point of view, the Franck–Hertz experiment measures the imaginary part of an eigenvalue problem associated with the Boltzmann kinetic equation.

CHAPTER 17

Positron Transport in Soft-Condensed Matter with Application to PET

17.1 Why Anti-Matter Matters

While kinetic modelling of positrons in gases is well established in the literature [124,125], the corresponding analysis for positrons in soft-condensed matter is still in its infancy [126–129]. This chapter investigates the problem in the context of an important application in medicine, positron emission tomography (PET [130]), where positrons annihilate with the electrons in the constituent molecules of human tissue, giving off gamma radiation. Before embarking on this task, we begin by looking at the positron simply as the anti-particle of the electron. A good review of positron physics is given by Charlton [124].

When an anti-particle of mass m meets its corresponding particle (also of mass m), they annihilate, and the total mass of $2m$ is converted completely into radiant energy E which, according to Einstein's famous mass–energy equivalence equation, is given by $E=2mc^2$, where $c=3\times10^8$ m/sec is the speed of light in vacuum. In the case of a positron e^+ and electron e^-, for which each has $mc^2=511$ keV, there are two radiation quanta produced in the annihilation process and, in order to conserve momentum they are emitted, to a first approximation at least, "back-to-back, that is, in opposite directions. This satisfies Newton's third law of motion, that for every reaction there is an equal and opposite reaction (see Figure 17.1). Moreover, the available energy is divided equally between the two, so that each photon has an energy of 511 keV, or equivalently a wavelength of 0.0024 nm. This corresponds to the gamma radiation region of the electromagnetic spectrum.

The two factors outlined above, that is, Einstein's mass–energy equivalence and Newton's laws of motion, underpin the operation of PET [124], the focus of this chapter.

We digress briefly to consider the reverse process, called pair production, that is, the production of matter from radiation, which also follows Einstein's mass-energy equivalence. In a state of equilibrium, annihilation and pair-production proceed at an equal rate, maintaining overall equal populations of matter and anti-matter particles. According to the standard model of cosmology, the early universe contained equal numbers of particles and anti-particles. Given this initial symmetry, we have to ask why matter should dominate over anti-matter in the *present* universe, over

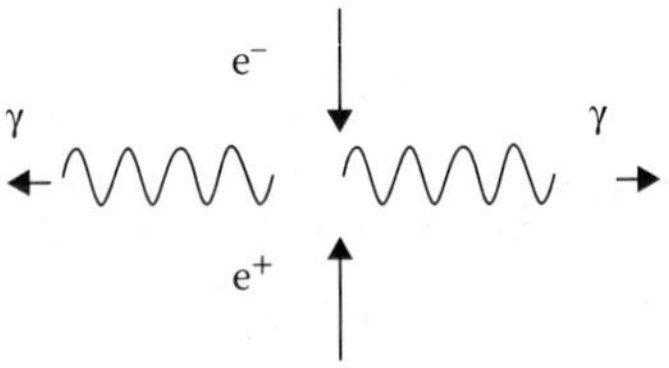

Figure 17.1 Schematic representation of the annihilation of a positron and electron, resulting in creation of two back-to-back gamma rays, γ (in centre-of-mass frame) each of energy equal to the rest energy $E = mc^2 = 511$ keV of a positron and an electron.

13 billion years later? Some sort of subtle asymmetry must exist in the laws of physics, the search for which occupies a prominent place in modern day fundamental physics. Thus, for example, the international ALPHA project based at CERN aims at producing and trapping anti-hydrogen atoms (consisting of an anti-proton and a positron) for a sufficient length of time to enable examination of the putative symmetries of matter and anti-matter [131]. Recent reports of a confinement time of about 1,000 seconds would be sufficient for most of the trapped anti-atoms to reach the ground state. These advances open up a range of experimental possibilities, including precision studies of possible violations of charge–parity–time-reversal symmetry.

These fundamental questions are, however, well beyond the scope of this book and, for present purposes, the practical implication of this asymmetry in the current universe is that positrons must be produced artificially by the radioactive decay of manufactured radioisotopes, for example, ^{22}Na or ^{18}F, for use in either positron beam lines in the physics laboratory, or for radio-pharmaceuticals in PET in medicine, respectively.

17.2 Positron Emission Tomography

17.2.1 The nature of PET

PET is an established technology for pinpointing abnormalities in living tissue and is commonly used as a cancer and brain function diagnostic [130]. A radiopharmaceutical such as fluoro-deoxyglucose, incorporating the positron emitting atom, ^{18}F (half-life 110 minutes), is injected into the patient, and seeks out regions of high metabolic activity, for example, a cancerous tumour. Positrons are emitted at energies of typically several hundred keV from the site of the tumour into the surrounding body tissue, slow down through collisions, and may also annihilate directly with an electron bound to molecule of the medium. However, for energies below about 100 eV, the cross section σ_a for such *direct* annihilation is much smaller than the cross section σ_{Ps} for formation of positronium Ps, in which e^+ and e^- are bound together to form an atom. They subsequently annihilate to produce the two (approximately) back-to-back gamma ray photons, which are detected by coincidence measuring techniques in the external imaging apparatus.

The point to note is that the image reflects the point of annihilation of the positron, *not* the point of emission from the abnormality, which is the information actually desired. In other words, each point in the external image is displaced from the point in the source at which the positron originates and, since there is a distribution of such displacements and corresponding recorded points due to the randomising nature of the intervening scattering processes, the overall image is blurred.

The dislocation is characterised in part by the *positron range*, which we define as the distance a typical positron moves from the point of emission from the source to the point of Ps formation. Subsequently, the Ps atom moves a further distance before annihilation of the constituent electron and positron occurs. This distance is thought to be small, and in any case would require an additional rigorous transport analysis of Ps in soft matter to be developed. What is more important about the subsequent motion of the Ps atom is the momentum and energy that it carries which, by application of the usual conservation laws, means that the annihilation process does not result in two gamma rays moving exactly in opposite directions. This "non-collinearity" is considered to be the main limitation on the accuracy of PET.

For the present, however, we focus exclusively on the calculation of the positron range, using the kinetic theory developed in this book.

17.2.2 Calculation of positron range

The first point to note is that positrons in a PET environment are in a highly non-equilibrium state, since they are emitted at high energies (typically several hundred keV) by the radioactive source into tissue at room temperature and must slow down to much lower energies (well below 100 eV) before Ps formation is appreciable. At these low energies, positrons may also undergo coherent multiple scattering due to the structured nature of the soft-condensed medium, raising the interesting possibility of a synergism between Ps formation and structure effects. Treating the medium as a structureless dense gas is obviously questionable, and yet this has been common in PET modelling.

While structure factors for various types of soft matter media have been available for some time, it is also only recently that cross sections for many of the key low energy scattering processes have become available [132,133] and have not been included in the previous modelling.

Any serious physical model of positrons in a PET environment has to combine all these elements and, whether it is calculation of positron range or estimation of non-collinearity, kinetic theory is an ideal vehicle to carry out the investigation. Monte Carlo investigations are also underway [127,134].

17.3 Kinetic Theory for Light Particles in Soft Matter

17.3.1 Structure-modified cross sections

In Chapter 6, we investigated how coherent, elastic scattering modifies the collision term of the Boltzmann equation for light particles in a soft matter medium.The scalar part of the collision operator, remains intact, but in the vector and higher order tensor components the elastic single-scattering differential cross section $\sigma(v,\theta)$ is replaced by

$$\widetilde{\sigma}(v,\theta) = \sigma(v,\theta) S\left(2mv \sin\frac{\theta}{2}\right), \tag{17.1}$$

where S is the static structure factor of the medium and θ the scattering angle. The elastic momentum-transfer cross section and collision frequency,

$$\widetilde{\sigma}_m(v) = 2\pi \int_0^{2\pi} (1-\cos\theta)\,\widetilde{\sigma}(v,\theta)\sin\theta d\theta$$

and

$$\widetilde{\nu}_m(v) = n_0 v \widetilde{\sigma}_m(v),$$

respectively, are similarly modified. For all other processes, the scattering may be assumed to be coherent, and the corresponding cross sections and collision frequencies are the same as for a gaseous medium.

17.3.2 Two-term analysis

To simplify matters as much as possible, we assume that scattering by the medium quickly slows positrons to non-relativistic energies and randomizes their directions. Thus, we assume that the positron distribution function has only weak anisotropy in velocity space and can be adequately represented through the two-term approximation (see Chapter 5),

$$f(\mathbf{r},\mathbf{v},t) \approx f^{(0)}(\mathbf{r},v,t) + \mathbf{f}^{(1)}(\mathbf{r},v,t)\cdot\hat{\mathbf{v}} + \ldots. \tag{17.2}$$

The scalar component of Boltzmann's equation remains the same as for a gaseous medium (see Chapter 7),

$$\frac{\partial f^{(0)}}{\partial t} + \frac{v}{3}\nabla\cdot\mathbf{f}^{(1)} + \frac{\mathbf{a}}{3v^2}\cdot\frac{\partial}{\partial v}\left[v^2\mathbf{f}^{(1)}\right] = \left(\frac{\partial f^{(0)}}{\partial t}\right)_{\text{col}} \equiv -J_0\left(f^{(0)}\right), \tag{17.3}$$

where the collision term includes contributions from elastic and inelastic collisions, ionization, annihilation, and *Ps* formation, as discussed in Chapter 6. On the other hand, the collision term in the vector component,

$$\frac{\partial \mathbf{f}^{(1)}}{\partial t} + v\nabla f^{(0)} + \mathbf{a}\frac{\partial f^{(0)}}{\partial v} = \left(\frac{\partial \mathbf{f}^{(1)}}{\partial t}\right)_{\text{col}} \approx -\widetilde{\nu}_m(v)\mathbf{f}^{(1)}(v) \tag{17.4}$$

includes, to a first approximation, only the effects of elastic scattering, and that is modified to the extent that $\nu_m \to \widetilde{\nu}_m$.

17.3.3 Multi-term analysis

In "multi-term" analysis, higher order tensor contributions are added to the right hand side of Equation 17.2, and the hierarchy of Equations 17.3 and 17.4 is extended accordingly. The right hand side of the lth member of the extended hierarchy is of a similar form to Equation 17.4, with $\widetilde{\nu}_m(v)$ replaced by $\widetilde{\nu}_l(v) = nv2\pi \int_0^{2\pi} \left(1 - P_l(\cos\theta)\right) \widetilde{\sigma}(v,\theta) \sin\theta \, d\theta$.

17.3.4 Fluid analysis

Fluid equations for light particles in soft matter have been formulated in Chapter 10 on the basis of momentum transfer theory. The energy balance equation is the same as for a gaseous medium, while in the momentum balance equation, the average momentum transfer collision frequency is replaced by its structure modified counterpart, that is $\nu_m(\bar{\epsilon}) \to \widetilde{\nu}_m(\bar{\epsilon})$. Structure-modified expressions for mobility and the Wannier energy formula were obtained.

The PET problem is inherently non-hydrodynamic and therefore cannot be analyzed using a density gradient expansion or the diffusion equation. A fluid analysis using the full moment equations is required, along with an *Ansatz* for the heat flux (see Chapter 8 and Ref. [85]).

17.4 Kinetic Theory of Positrons in a PET Environment

17.4.1 The model

At the low densities involved, mutual interaction between positrons can be neglected, and hence the same analytical and numerical techniques which have been developed for low density electron swarms (see Chapter 12) may be applied to the present problem, with inclusion of structural effects. The main difference is that in a PET environment, positrons have a much wider range of energies than in electron swarm experiments [10], specifically hundreds of keV as compared with a few eV respectively, and the analysis has to be modified accordingly. Although strictly speaking, a relativistic treatment is required *near* the source, it is reasonable to use the non-relativistic Boltzmann equation in the bulk of the medium, with spherical components given by Equations 17.3 and 17.4.

To simplify the analysis and elucidate the essential physics, we take an idealized, spherical situation in which high energy positrons of mass m and charge e are emitted isotropically at a steady rate from a spherical source into an infinite, spatially homogeneous soft matter medium at temperature (see Figure 17.2). As mentioned in Chapter 11, such a situation is inherently non-hydrodynamic, and a density gradient expansion cannot be used—the Boltzmann equation has to be solved in phase space.

Positrons are quickly slowed to lower energies by elastic, inelastic, and ionizing collisions with the constituent molecules of mass m_0. The corresponding cross sections σ_m, σ_{inel}, and σ_{ion}, respectively, together with the cross sections for annihilation and positronium formation σ_a and σ_{Ps}, respectively, and the static structure function $S(Q)$ of the medium, are incorporated in the respective collision operator. Note that as discussed in Chapter 5, ionization by positron impact may be considered as just another an inelastic process, as distinct from electron impact ionization.

The model also assumes that a steady state has been attained whereby there is a balance between the rate at which positrons are produced at the source and the rate at which they are lost (by annihilation and *Ps* formation) in the medium. There is no electric field in this problem.

17.4.2 Two-term equations

17.4.2.1 Statement of the problem

For a steady state, with no external field, Equations 17.3 and 17.4 simplify to

$$\frac{v}{3}\nabla\cdot\mathbf{f}^{(1)}=\left(\frac{\partial f^{(0)}}{\partial t}\right)_{\text{col}}\equiv -J_0(f^{(0)}) \tag{17.5}$$

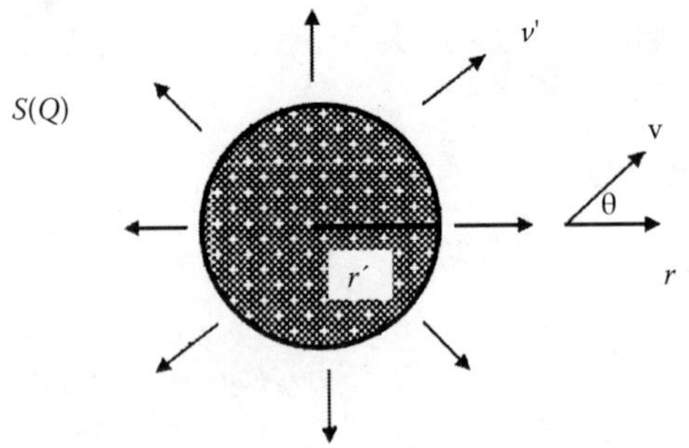

Figure 17.2 An idealized spherical source of positrons emitted into an unbounded soft matter medium with structure factor, density, and temperature prescribed.

and

$$\mathbf{f}^{(1)} \approx -\frac{v}{\widetilde{\nu}_m} \nabla f^{(0)}, \tag{17.6}$$

respectively, where the scalar collision operator

$$J_0(f^{(0)}) \equiv -\frac{m}{m_0 v^2} \frac{\partial}{\partial v} \left[\nu_m v^2 \left(v f^{(0)} + \frac{k_B T}{m} \frac{\partial f^{(0)}}{\partial v} \right) \right] + J_{\text{inel}}(f^{(0)}) + [\nu_{Ps}(v) + \nu_a(v)] f^{(0)} \tag{17.7}$$

includes contributions from all processes. The collision frequencies for direct annihilation and *Ps* formation are related to the corresponding cross sections by $\nu_a(v) = n_0 v \sigma_a$ and $\nu_{Ps}(v) = n_0 v \sigma_{Ps}$, respectively, and $J_{\text{inel}}(f^{(0)})$ is the inelastic collision term, including ionisation by positron impact (Chapter 5).

If Equation 17.6 is substituted into Equation 17.5, there follows

$$\frac{v^2}{3\widetilde{\nu}_m} \nabla^2 f^{(0)} = J_0(f^{(0)}) \tag{17.8}$$

respectively. The task is to solve this equation for $f^{(0)}(\mathbf{r}, v)$ with appropriate boundary conditions, following the prescription of Chapter 11, which stipulates that the solution of the kinetic equation is determined uniquely by the distribution of velocities directed *into* the system.

17.4.2.2 Moments of physical interest

Once $f^{(0)}$ is found, the moments of physical interest follow:

- *Positron number density*

$$n(\mathbf{r}) = 4\pi \int_0^\infty dv\, v^2 f^{(0)}(\mathbf{r}, v) \tag{17.9}$$

- *Mean energy*

$$\bar{\epsilon}(\mathbf{r}) \equiv \langle \frac{1}{2} m v^2 \rangle = \frac{1}{n} 4\pi \int_0^\infty dv\, v^2 \, \frac{1}{2} m v^2 f^{(0)}(\mathbf{r}, v) \tag{17.10}$$

- *Ps formation rate per unit volume*

$$R_{Ps}(\mathbf{r}) = 4\pi \int_0^\infty dv\, v^2 \nu_{Ps}(v) f^{(0)}(\mathbf{r}, v). \tag{17.11}$$

17.4.2.3 Adjoint collision operator

For later reference, we note that the *adjoint* $J_0^\dagger$ of the collision operator (Equation 17.7) is defined such that

$$\int_0^\infty \Phi(v)\, J_0[\Psi(v)] v^2 dv = \int_0^\infty \Psi(v)\, J_0^\dagger[\Phi(v)]\, v^2 dv$$

for any functions $\Psi(v)$ and $\Phi(v)$ of positron speed v. It turns out that an explicit expression is required only for the adjoint of the elastic collision operator, which can be shown

$$J_{0,\text{elas}}^\dagger \Phi = v\nu_m \frac{\partial \Phi}{\partial v} - \frac{k_B T}{mv^2}\frac{\partial}{\partial v}\left(v^2 \nu_m \frac{\partial \Phi}{\partial v}\right). \tag{17.12}$$

17.4.3 Solution for spherical symmetry

17.4.3.1 The eigenvalue problem

For the isotropic source of Figure 17.2 in a uniform medium, all quantities may be assumed to be spherically symmetric, that is, $f^{(0)}(\mathbf{r}, v) = f^{(0)}(r, v)$ and similarly for the moments. In this case, we can set

$$\nabla^2 f^{(0)} \equiv \frac{1}{r}\frac{\partial^2}{\partial r^2}(r f^{(0)})$$

in Equation 17.8. By the method of separation of variables, this has the solution $f^{(0)}(v) \sim \frac{e^{-K_n r}}{r}\Psi(v)$, where K is a constant, and $\Psi(v)$ satisfies $\frac{v^2}{3\tilde{\nu}_m} K^2 \Psi = J_0(\Psi)$. From a purely mathematical perspective, we cannot say anything about K. It is after requiring $\Psi(v)$ to be physically meaningful, for example, well behaved as $v \to \infty$, that values of K are effectively prescribed. Formally, we may represent the allowed K and Ψ as members of the sets $\{K_n\}$ and $\{\Psi_n\}$ which satisfy the eigenvalue problem,

$$\frac{v^2}{3\tilde{\nu}_m} K_n^2\, \Psi_n = J_0(\Psi_n), \tag{17.13}$$

where $n = 0, 1, 2, \ldots$ is an index characterizing the eigenvalue spectrum. There are no general theorems for kinetic theory eigenvalue problems, but numerical and model calculations for the present problem indicate that $\{K_n\}$ consists of a discrete, real set of values, occurring in pairs of equal magnitude and opposite sign. The most general solution of Equation 17.8 valid at infinity is then a linear combination of all possible eigenmodes,

$$f^{(0)}(r, v) = \sum_{n=0}^{\infty} A_n \frac{e^{-K_n r}}{r} \Psi_n(v), \tag{17.14}$$

where A_n are constants to be found from boundary conditions. Equation 17.6 then furnishes the vector part of the distribution function,

$$f^{(1)}(r,v) = -\frac{v}{\tilde{\nu}_m}\frac{\partial}{\partial r}f^{(0)}$$
$$= \frac{v}{\tilde{\nu}_m}\frac{1}{r^2}\sum_{n=0}^{\infty} A_n e^{-K_n r}\left[rK_n + 1\right]\Psi_n(v) \quad (17.15)$$

directed in the radial direction.

Equation 17.13 is actually an example of a more general eigenvalue problem which arises naturally in kinetic theory, as discussed in Chapter 11. We have already met other examples of eigenvalue problems in Chapters 13 and 16 in which only the lowest eigenvalue was required and the expansion coefficients played no role. In the present problem, however, several members of the spectrum are necessary, and it is necessary to find A_n. In that case, as outlined in Appendix C and Section 11.5, it is also necessary to solve the *dual* eigenvalue problem,

$$\frac{v^2}{3\tilde{\nu}_m}K_n^2\Phi_n = J_0^{\dagger}(\Phi_n) \quad (17.16)$$

for the corresponding dual eigenfunctions Φ_n. As we will see, it turns out that only the solution of the equation with the adjoint elastic collision operator (Equation 17.12) is required.

17.4.3.2 Eigenvalue problems for gaseous media

Note that eigenvalue problems also arise in investigations of time-dependent properties of positrons undergoing loss by annihilation in gases [124]. In that case, in contrast to the series of spatially relaxing modes represented by Equation 17.14, one has a series of terms decaying exponentially in time, with time constants determined by the eigenvalue spectrum. A useful comparison can be made with the eigenvalue problem describing electrons lost by diffusion to the walls of a Cavalleri cell, as discussed in Chapter 13.

17.4.3.3 Constant collision frequency model

In general, numerical solution of the two eigenvalue problems is required. However, in the special model of a swarm undergoing elastic collisions only, with constant $\tilde{\nu}_m$, an exact, analytic solution can be found, either directly or by adapting the result of Parker [103] to a steady state. The eigenfunctions may be written in terms of Laguerre polynomials and the eigenvalues are found to be real and occur in pairs of equal magnitude, that is (apart from a constant)

$$K_n \sim \pm\frac{\sqrt{2n(2n+3)}}{4n+3}, \quad n = 0, 1, 2, \ldots \quad (17.17)$$

Only non-negative values are physically meaningful in the context of Figure 17.2.

This result is useful in several respects, for it

- Is an example indicating the reality of K_n;
- Provides a benchmark for testing the accuracy of numerical solutions of Equation 17.13 for more realistic cases; and
- Illustrates an important and seemingly general property of the spectrum, namely that it becomes *dense* at larger values of n, $K_1 \sim 0.451, K_2 \sim 0.481, K_3 \sim 0.489, \ldots, K_\infty \sim 0.5$.

17.4.3.4 Properties of eigenfunctions and eigenvalues and an identity

There are a number of general results which can readily be established from Equations 17.13 and 17.16:

- *Orthogonality relation*

$$4\pi \int_0^\infty \frac{v^2}{3\tilde{\nu}_m} \Psi_n(v)\Phi_{n'}(v)v^2 dv = \delta_{nn'}. \tag{17.18}$$

- *Completeness in speed space, closure relation*

$$\frac{v^2}{3\tilde{\nu}_m} \sum_{n=0}^{\infty} \Psi_n(v)\Phi_n(v') = \frac{\delta(v'-v)}{4\pi v^2}. \tag{17.19}$$

- *Reality of spectrum*: It can be shown that $K_n^2 \geq 0$ and thus, the eigenvalue spectrum K_n consists of pairs of *real* numbers of the same magnitude, but of opposite sign. However, in order that the solution (Equation 17.14) remains physical as $r \to \infty$, only the non-negative part of the spectrum $K_n \geq 0$ contributes.
- *An identity:* Since elastic and inelastic collisions conserve positron number, integration of the first two terms of Equation 17.7 describing elastic and inelastic collisions over all speeds yields zero identically. Thus, integration of Equation 17.13 over all speeds yields

$$\begin{aligned}
&\frac{4\pi}{3} K_n^2 \int_0^\infty \frac{v^4}{3\tilde{\nu}_m} \Psi_n \, dv \\
&= 4\pi \int_0^\infty v^2 J_0(\Psi_n) \, dv \\
&\equiv 4\pi \int_0^\infty v^2 \left[\nu_{Ps}(v) + \nu_a(v)\right]\Psi_n \, dv \approx 4\pi \int_0^\infty v^2 \, \nu_{Ps}(v)\Psi_n \, dv.
\end{aligned} \tag{17.20}$$

The last approximation follows from the fact that annihilation is generally dominated by *Ps* formation. This identity proves useful for evaluating the *Ps* formation rate.

17.4.3.5 Boundary conditions

The constants A_n are found from boundary conditions at the source as follows. Let $4\pi S(v')\, v'^2 dv'$ be the number of positrons emitted by the source per unit time, from its entire surface of area $4\pi r'^2$, with speeds in the range v' to $v' + dv'$, into the surrounding medium. In the present isotropic model, positron velocities v' at the source are assumed to be everywhere directed radially outward from the source at $r = r'$. The *exact* boundary condition is a statement that the radial flux of positrons away from the surface is equal to the number of positrons produced by unit area of the surface per unit time, at all speeds v', that is,

$$\mathbf{v}' \cdot \hat{\mathbf{r}}\, f(r', \mathbf{v}') = \frac{S(v')}{r'^2}\, \delta(\hat{\mathbf{v}}' \cdot \hat{\mathbf{r}} - 1) \tag{17.21}$$

for all velocities directed outwards, that is, $\hat{\mathbf{v}}' \cdot \hat{\mathbf{r}} > 0$. However, as we already have observed in Chapter 11, it is impossible to apply such an exact condition to a full range spherical harmonic representation of the distribution function, like Equation 17.2, and an some approximation is inevitable. To that end, we integrate Equation 17.21 over all directions and obtain

$$\frac{2}{3} v' f_1(r', v') \approx \frac{S(v')}{r'^2}. \tag{17.22}$$

Equation 17.15 is substituted into the left hand side of this equation, which is then multiplied by $v' \Phi_{n'}(v')\, 4\pi v'^2$, and integrated over all v'. The orthogonality relation (Equation 17.18) is then applied to yield

$$A_n = S_n \frac{e^{K_n r'}}{1 + K_n r'}, \tag{17.23}$$

where

$$S_n \equiv 2\pi \int_0^\infty dv'\, \Phi_n(v') S(v')\, v'^2. \tag{17.24}$$

17.4.4 Complete solution

Substitution of Equation 17.23 into 17.14 then gives the solution to the Boltzmann equation

$$f^{(0)}(r, v) = \frac{1}{r} \sum_{n=0}^{\infty} \frac{S_n}{1 + K_n\, r'} \Psi_n(v) e^{-K_n (r - r')} \tag{17.25}$$

from which follows the Ps formation rate per unit volume:

$$R_{Ps}(r) = 4\pi \int_0^\infty dv\, v^2\, \nu_{Ps}(v)\, f^{(0)}(r,v)$$
$$= \frac{1}{rr'} \sum_{n=0}^{\infty} \rho_n \frac{S_n}{1+K_n r'} e^{-K_n(r-r')}, \tag{17.26}$$

where

$$\rho_n \equiv 4\pi \int_0^\infty dv\, v^2\, \nu_{Ps}(v)\, \Psi_n(v) \approx \frac{4\pi}{3} K_n^2 \int_0^\infty \frac{v^4}{3\widetilde{\nu}_m} \Psi_n\, dv, \tag{17.27}$$

the last step following from the identity Equation 17.20.

17.5 Calculation of the Positron Range

17.5.1 Definition of positron range

The *positron range* is defined for practical purposes as the distance from the source at which the average Ps formation rate reaches a maximum, that is, when the derivative of the Ps formation rate Equation 17.26,

$$\frac{dR_{Ps}}{dr} = -\frac{1}{r^2} \sum_{n=0}^{\infty} \rho_n \frac{S_n(1+K_n r)}{1+K_n r'} e^{-K_n(r-r')} \tag{17.28}$$

is zero.

17.5.2 Evaluation of the summation

First, note that far from the source, $r \gg r'$, all terms in the summation Equation 17.26 are exponentially small, and hence $R_{Ps}(r) \to 0$ asymptotically, as would be expected on physical grounds. To evaluate the summation in other cases, it is useful to note the following identity,

$$\sum_{n=0}^{\infty} S_n \Psi_n(v) = \frac{3\widetilde{\nu}_m S(v)}{2v^2}, \tag{17.29}$$

which follows from the closure relation Equation 17.19, and its corollary,

$$\sum_{n=0}^{\infty} \rho_n\, S_n = 6\pi \int_0^\infty \widetilde{\nu}_m\, S(v) \nu_{Ps}(v)\, dv \approx 0. \tag{17.30}$$

The last approximation follows from the fact that $S(v)$ is small when $\nu_{Ps}(v)$ is appreciable and vice versa, and hence the integrand is small for all v. At the source $r = r'$ it then follows with Equation 17.28 that

$$\frac{dR_{Ps}}{dr}(r') = \frac{1}{r'^2} \sum_{n=0}^{\infty} \rho_n \, S_n \approx 0. \tag{17.31}$$

Between these two limiting cases, the *Ps* formation rate rises to a maximum, and then decays to zero. In this intermediate region, one has to approximate the infinite sums in Equations 17.26 and 17.28 by truncating them to finite size N. In a rigorous investigation of the complete spatial profile $R_{Ps}(r)$, N would be incremented, and the whole process repeated until some convergence criterion is satisfied. However, a less stringent approach suffices for present purposes, where the aim is limited to finding the position of the maximum in R_{Ps}. Note also that the contribution from the $n = 0$ mode is negligible, since K_0 is very small (see Table 17.1), and hence by Equation 17.27 $\rho_0 \approx 0$. Thus, in what follows, we retain only the $n = 1, 2$ modes in the summation, and write

$$\sum_{n=0}^{\infty} \rho_n \frac{S_n(1 + K_n\, r)}{1 + K_n\, r'} e^{-K_n(r-r')} \approx \rho_1 S_1 \frac{(1 + K_1\, r)}{1 + K_1\, r'} e^{-K_1(r-r')} + \rho_2 S_2 \frac{(1 + K_2\, r)}{1 + K_2\, r'} e^{-K_2(r-r')}. \tag{17.32}$$

The respective terms can be found from solutions $\left\{K_n, \Psi_n\right\}_{n=\,0,1,2,\ldots,N}$ of the eigenvalue problem Equation 17.13.

Table 17.1 Dimensionless Row Order Eigenvalues

Order	K_n^* (Gas Phase)	K_n^* (Liquid Phase)
0	$1.32\ 10^{-8}$	$1.88\ 10^{-7}$
1	0.68	2.68
2	1.33	2.91

Note: Dimensionless low order eigenvalues $K_n^* = K_n / (\sqrt{2}N\ \sigma_0)$, with $\sigma_0 = 10^{-20} m^2$ for positrons in water using the cross section set from (*Source:* W. Tattersall, et al., *Journal of Chemical Physics*, 140, 044320, 2014; A. Bankovic, et al., *New Journal of Physics*, 14, 035003, 2012), and the static structure factor from (*Source:* Y. S. Badyal, et al., *Journal of Chemical Physics*, 112, 9206, 2000). Since $N \approx 3 \times 10^{28}\ m^3$ for liquid water, $\sqrt{2}N\sigma_0 \approx 4.2 \times 10^8\ m^{-1}$.

The lowest eigenvalues for a liquid water medium (a surrogate for human tissue), as characterised by the cross sections shown in Refs. [133,135], and the structure function of Ref. [136], are shown in Table 17.1. Since a typical positron source has dimension $r' \approx 10^{-2}m$, it is clear from Table 17.1 that $K_n r > K_n r' \gg 1$. Hence by Equations 17.28 and 17.32, the *Ps* formation rate attains a maximum at a position r determined by

$$\rho_1 S_1 e^{-K_1(r-r')} + \rho_2 S_2\, e^{-K_2(r-r')} = 0. \tag{17.33}$$

17.5.3 Numerical example

The example here uses the eigenvalue data of Table 17.1, which derive from the low-energy cross section set of Refs. [133,135]. However, the same procedure is to be used when a more complete set of cross sections becomes available. In the same spirit, a monoenergetic, non-relativistic source of positrons is assumed, where all emitted positrons have the same speed v' and energy $\epsilon' = 1/2mv'^2$ when $r = r'$. Furthermore, near the source is assumed that elastic collisions dominate all other processes. The asymptotic solution of Equation 17.16 at high v', together with Equation 17.24, yields

$$S_n \sim \Phi_n(v') \sim (-)^n \exp\left\{\frac{MK_n^2}{3m}\int^{v'} \frac{v}{\nu_m^2}\mathrm{d}v\right\}. \tag{17.34}$$

Hence, neglecting the contribution from the comparatively slowly varying constants ρ_n the solution of Equation 17.33 is

$$(r - r')_{\max} \approx \frac{M}{3m}(K_2 + K_1)\int^{v'} \frac{v}{\nu_m^2}\mathrm{d}v. \tag{17.35}$$

The right hand side may be conveniently written in terms of the dimensionless eigenvalues shown in Table 17.1, and the integral in the right hand side evaluated approximately using the high energy momentum transfer cross section data of Refs. [133,135]. Thus, for liquid water of density $n_0 \approx 3 \times 10^{28} m^{-3}$ we find

$$(r - r')_{\max} \approx 2 \times 10^{-4}(K_2^* + K_1^*)\, \varepsilon'_{keV}, \tag{17.36}$$

where ε'_{keV} is the energy of the positrons at the source in units of *keV*. For a mono-energetic source with $\varepsilon' \sim$ a few keV, this formula predicts a positron range of the order of a few millimeters. If the structure of the matter medium is ignored, then the eigenvalues are smaller, as shown in Table 17.1, and the estimate of the range as provided by Equation 17.36 is reduced by more than a factor of 2.

17.5.4 Concluding remarks

Refining the estimate of positron range given in this chapter depends on

- Improved modelling of high energy positrons near the source, in particular;
- Extension of the cross section set to include higher energy processes, which will modify the eigenvalues of Table 17.1 and the estimate of the range; and
- The development of a relativistic kinetic theory for positrons in soft-condensed matter.

In any case, this takes us only to the point of *Ps* formation, and questions remain about processes which occur subsequently. How do we estimate, for example, how much further *Ps* drifts in soft matter, before the constituent positron and electron annihilate? What is momentum and energy of the *Ps* atom at this point, given that this information is required to determine the directions of the two emitted gamma rays after annihilation, and thus address the problem of non-collinearity? These fundamental issues can be resolved by building on the kinetic theory presented in this chapter.

CHAPTER 18

Transport in Electric and Magnetic Fields and Particle Detectors

18.1 Introduction

An accurate transport theory for charged particles in gases subject to both electric and magnetic fields is essential for analyzing laboratory experiments, understanding naturally occurring phenomena and optimizing technological processes. Examples include swarm experiments, in which a magnetic field is introduced with the aim of refining the accuracy of extracted low energy scattering cross section data, the dispersion of meteor trains in the Earth's atmosphere, and low temperature magnetron plasma discharges used in the plasma processing industry, respectively. The theory finds another important application in the operation of gaseous radiation detectors, the subject of the second part of this chapter.

In all the applications mentioned above, transport processes are significantly influenced by short-range particle–neutral collisions. Magnetic fields are also fundamental to the operation of hot, fully ionized plasmas in toroidal fusion devices. Here, the constituent electrons and ions interact through the long-range Coulomb force and, since the Rutherford cross section Equation 4.12 decreases rapidly with energy, collisions are not so important. Thus, the early chapters of plasma physics texts are typically devoted to studying the collisionless motion of a single particle in electric and magnetic fields. Even in the present context, where collisions play a more significant role, it is instructive to say a few words about the single particle picture.

18.2 Single, Free Particle Motion in Electric and Magnetic Fields

Solution of Newton's equation of motion for a particle of charge q mass m in a uniform, static magnetic field **B** gives a spiral trajectory, consisting of a combination of uniform circular motion at constant speed $v_\perp$, with angular frequency

$$\Omega_L = \frac{qB}{m} \tag{18.1}$$

(the Larmour frequency) in a plane perpendicular to the field, plus constant motion with velocity $v_\parallel$ along the field direction. The (Larmour) radius of the orbit is given by $r = \frac{v_\perp}{\Omega_L}$. In the jargon of plasma physics, one speaks of a particle confined in an orbit about a "guiding centre," which moves freely along the lines of **B**.

The picture is that the particle sticks to the line of **B**, unless knocked off it by a collision. Transport perpendicular to **B** is inhibited by the field, especially in the *strong field* regime,

$$\Omega_L/\nu_m \gg 1, \tag{18.2}$$

where ν_m is the momentum transfer collision frequency. Here, the particle orbits the field lines many times between collisions, effectively corresponding to the collisionless limit. In the opposite extreme of *weak fields*, where $\Omega_L/\nu_m \ll 1$, a particle does not have time to complete an orbit before undergoing a collision. The ratio Ω_L/ν_m emerges in natural way here and in the transport analysis which follows.

Plasma physics texts go on to show that in the presence of an electric field **E**, and in the absence of collisions, the guiding centre drifts across the magnetic field with velocity

$$\mathbf{v}_{\mathrm{GC}} = \frac{\mathbf{E}\times\mathbf{B}}{B^2}. \tag{18.3}$$

This, as we shall see, is the collisionless limit Equation 18.2 of a more general formula for drift velocity in the presence of both electric and magnetic fields.

These elementary considerations suggest that transport theory in the presence of a magnetic field can be expected to be more complex as compared with the *E* only situation studied previously.

18.3 Transport Theory in E and B Fields

The task at hand can be simply stated: it is, as before, to solve the kinetic equation

$$\frac{\partial f}{\partial t} + \mathbf{v}\cdot\nabla f + \mathbf{a}\cdot\frac{\partial f}{\partial \mathbf{v}} = \left(\frac{\partial f}{\partial t}\right)_{\mathrm{col}}, \tag{18.4}$$

for the charged particle distribution function $f(\mathbf{r}, \mathbf{v}, t)$, using an appropriate form of $\left(\frac{\partial f}{\partial t}\right)_{\mathrm{col}}$, and then form velocity moments to obtain the quantities of physical interest. Here, $\mathbf{a} = \frac{q}{m}(\mathbf{E} + \mathbf{v}\times\mathbf{B})$ is the acceleration suffered by a particle of mass m, charge q, due to electric and magnetic fields **E** and **B**, respectively, both of which are assumed to be uniform and static.

The corresponding fluid equation is formed, as in Chapters 7 through 9 by multiplying Equation 18.4 with some function $\phi(\mathbf{v})$ of velocity and integrating over all $\mathbf{v}$. The most interesting cases are the momentum and energy balance equations, which are formed by taking $\phi(\mathbf{v}) = m\mathbf{v}$ and $mv^2/2$, respectively, and are given by Equations 7.6 and 7.9. Note that $\mathbf{B}$ does not appear explicitly in the latter, since

$$\langle \mathbf{a} \cdot \mathbf{v} \rangle = \left\langle \frac{q}{m} (\mathbf{E} + \mathbf{v} \times \mathbf{B}) \cdot \mathbf{v} \right\rangle = \frac{q\mathbf{E} \cdot \langle \mathbf{v} \rangle}{m}$$

reflecting the fact that the magnetic field does no work. The collision terms in the fluid equations can again be approximated by momentum transfer theory. The fluid approach is further discussed in Section 18.5.

In general, $\mathbf{E}$ and $\mathbf{B}$ may be oriented at any arbitrary angle ψ with respect to each other. Figure 18.1 shows the coordinate system used in this discussion.

Furthermore, there is no longer a single preferred direction (unless the fields are parallel, i.e., $\psi = 0$) and thus, any symmetries applying to the electric field only case are lost; there is, for example, no rotational invariance about the z-axis in Figure 18.2. When the fields are at right angles, that is, $\psi = \pi/2$, the fields are said to be "crossed."

In general, any representation of f in polar coordinates in velocity space (see Figure 2.3), using the spherical harmonic decomposition, for example,

$$f(\mathbf{v}) = f(v, \theta, \varphi) \approx \sum_{l=0}^{l_{\max}} \sum_{m=-l}^{l} f_m^{[l]}(v)\, Y_m^{(l)}(\theta, \varphi) \tag{18.5}$$

must involve *both* polar and azimuthal angles θ and φ, respectively. Representation of f in terms of an expansion in Legendre polynomials $P_l(\cos\theta)$ is not possible. A multi-term solution with $l_{\max} > 1$ is generally needed to give accuracies consistent with swarm experiments (around 1% or so), and this requires a lengthy mathematical analysis and irreducible tensor algebra. The reader is referred to the original research articles for further details [95]. The problem is complicated by the fact that *three* angles, θ, ϕ, and ψ,

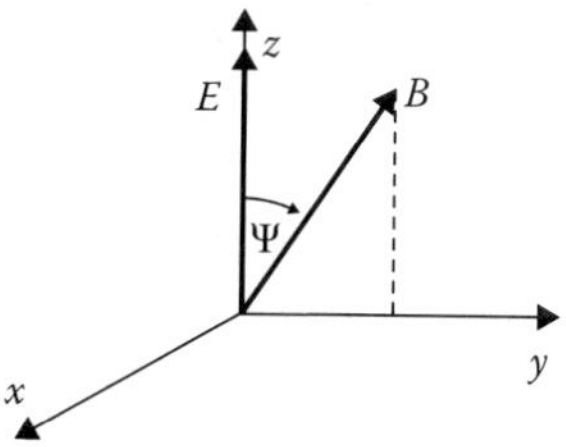

Figure 18.1 Configuration space coordinate system employed in this study.

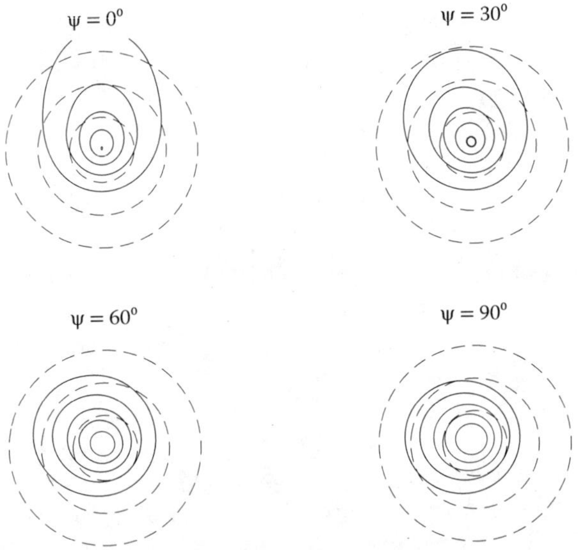

Figure 18.2 Contour plots of the velocity distribution functions for electrons in carbon dioxide in electric and magnetic fields at various ψ in the plane defined by $\phi = 0, \pi$. The value of the solid line contour heights from largest to smallest radii is 0.5, 1, 2, 4, 8 $eV^{-3/2}$, respectively. The energy scale is indicated by the dashed concentric circular plots of increasing radii referring to 0.3, 0.6, and 0.9 eV, respectively ($E/n_0 = 5$ Td, $B/n_0 = 200$ Hx, $T_0 = 293$ K). (From R. D. White et al., *Journal of Physics D: Applied Physics*, 2001 © IOP Publishing and Deutsche Physikalische Gesellschaft. CC BY-NC-SA.)

are generally required to describe f, since the fields can be oriented arbitrarily. Given the additional complexity introduced by a magnetic field, it is important to identify any symmetries at the outset in order to aid physical understanding and to minimise the numerical computational effort.

18.4 Symmetries

18.4.1 Hydrodynamic regime: Transport coefficients

In the hydrodynamic regime, the space–time properties of $f(\mathbf{r}, \mathbf{v}, t)$ are projected on to the number density $n(\mathbf{r}, t)$, typically represented by a density gradient expansion, and the same goes for the velocity moments of f. In particular, the particle flux is represented by the density gradient expansion to first order (Fick's law),

$$\Gamma(\mathbf{r}, t|\mathbf{E}, \mathbf{B}) \equiv \int f(\mathbf{r}, \mathbf{v}, t)\, \mathbf{v}\, d\mathbf{v} = \mathbf{v}_d(\mathbf{E}, \mathbf{B})\, n(\mathbf{r}, t) - \mathsf{D}(\mathbf{E}, \mathbf{B}) \cdot \nabla n(\mathbf{r}, t). \quad (18.6)$$

This effectively defines the drift velocity vector

$$\mathbf{v}_d = (v_{d,x}, v_{d,y}, v_{d,z})$$

and diffusion tensor

$$\mathsf{D} = \begin{pmatrix} D_{x,x} & D_{x,y} & D_{x,z} \\ D_{y,x} & D_{y,y} & D_{y,z} \\ D_{z,x} & D_{z,y} & D_{z,z} \end{pmatrix},$$

respectively, in the Cartesian coordinate system defined in Figure 18.1. (Strictly speaking, these are flux transport coefficients, but for present purposes, we need not make the distinction with bulk quantities.) The implications of symmetry for tensor structure are discussed in Ref. [95]. For a general orientation angle ψ, the tensors are full (no elements vanish) but there are significant simplifications in special cases:

1. *Crossed fields* ($\psi = \pi/2$)

$$v_{d,y} = 0$$
$$D_{x,y} = 0 = D_{y,x}$$
$$D_{y,z} = 0 = D_{z,y}$$

2. *Parallel fields* ($\psi = 0$)

$$v_{d,x} = 0 = v_{d,y}$$
$$D_{x,z} = 0 = D_{z,x}$$
$$D_{y,z} = 0 = D_{z,y}$$
$$D_{x,y} = -D_{y,x}$$

Full details can be found in Ref. [95]. Note that these properties emerge in a natural way from the fluid analysis presented in Section 18.5.

18.4.2 Symmetries in velocity space: A numerical example

A rigorous tensor analysis is required to examine the effect of symmetries in velocity space, and the reader is referred to the original research articles [95] for details. Here, we illustrate the nature of these symmetries by way of a numerical example for spatial homogeneity using the techniques described in Chapter 12. Thus, the results of a multi-term solution of the Boltzmann equation (Equation 18.4) are shown in Figures 18.2 through 18.4 for electrons in carbon dioxide* where we have set the electric

* The cross sections used are those attached to the MAGBOLTZ code developed by Biagi [137].

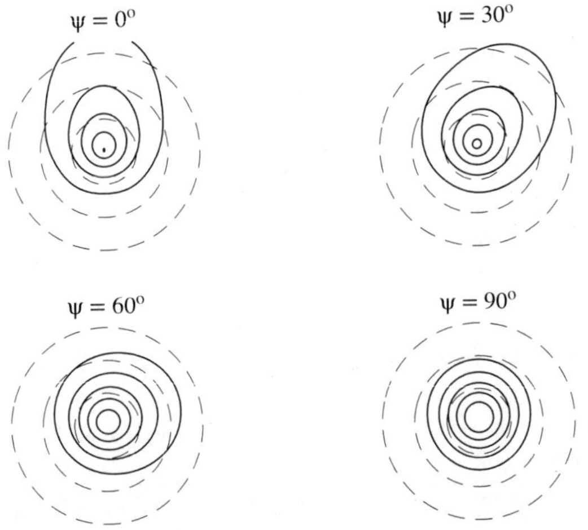

Figure 18.3 Contour plots of the velocity distribution functions for electrons in carbon dioxide in electric and magnetic fields at various ψ in the plane defined by $\phi = \pi/2, 3\pi/2$. The value of the solid line contour heights from largest to smallest radii is 0.5, 1, 2, 4, 8 $eV^{-3/2}$, respectively. The energy scale is indicated by the dashed concentric circular plots of increasing radii referring to 0.3, 0.6, and 0.9 eV, respectively ($E/n_0 = 5$ Td, $B/n_0 = 200$ Hx, $T_0 = 293$ K). (From R. D. White et al., *Journal of Physics D: Applied Physics*, 2001 © IOP Publishing and Deutsche Physikalische Gesellschaft. CC BY-NC-SA.)

field $E/n_0 = 5$ Td (1 townsend = 1 Td = 10^{-21} V m^2) and the magnetic field $B/n_0 = 200$ Hx (1 huxley = 1 Hx = 10^{-27} T m^3). In Figures 18.2, 18.3 and 18.4, we slice the distribution in the planes defined by $\phi = 0, \pi$, $\phi = \pi/2, 3\pi/2$, and $\theta = 90$ deg, respectively yielding f as a function of θ and ϕ in these planes. In each plot, the angle ψ is varied between 0 and $\pi/2$. A value of $l_{max} = 5$ was implemented in these plots, achieving an accuracy to within 1% or better. The dashed circular contours represent lines of constant energy.

The following qualitative represents various features of the velocity distribution function and its variation with ψ.

- Symmetries:
  - For parallel fields, there is an axis of rotational symmetry defined by the field directions as evidenced by the concentric circular contours centered on the origin of the $\psi = 0$ plot in Figure 18.4.
  - For perpendicular fields: (1) the electric field represents a two-fold axis of symmetry in the velocity distribution in the $\phi = \pi/2, 3\pi/2$ plane (see Figure 18.3); (2) the $\mathbf{E} \times \mathbf{B}$ direction represents a two-fold axis of symmetry in the $\theta = \pi/2$ plane (see Figure 18.4).

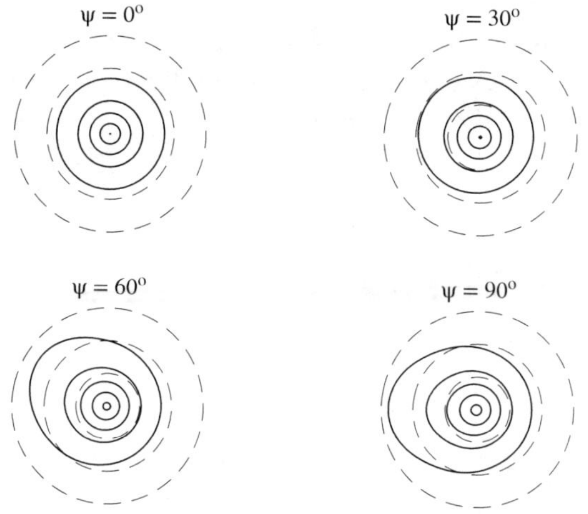

Figure 18.4 Contour plots of the velocity distribution functions for electrons in carbon dioxide in electric and magnetic fields at various ψ in the plane defined by $\theta=\pi/2$. The value of the solid line contour heights from largest to smallest radii is 0.5, 1, 2, 4, 8 $eV^{-3/2}$, respectively. The energy scale is indicated by the dashed concentric circular plots of increasing radii referring to 0.3, 0.6, and 0.9 eV, respectively ($E/n_0=5$ Td, $B/n_0=200$ Hx, $T_0=293$ K). (From R. D. White et al., *Journal of Physics D: Applied Physics*, 2001 © IOP Publishing and Deutsche Physikalische Gesellschaft. CC BY-NC-SA.)

   - The rotational symmetry about the electric field is destroyed as ψ is increased.
   - For $\psi \neq 0, \pi/2$, there is, in general, no axis of symmetry.
- Rotation of anisotropy:
  As discussed in Chapters 2 and 12, the velocity distribution in electric fields can be significantly anisotropic. When magnetic fields are applied, we observe the direction of elongation of the velocity distribution contours, as ψ is increased:
   - In the $\phi=0,\pi$ plane (Figure 18.2), the elongation direction is modified as of the contours is progressively rotated away from the **E** direction and towards the $\mathbf{E}\times\mathbf{B}$ direction.
   - In the $\phi=\pi/2, 3\pi/2$ plane (Figure 18.3), the contour elongation direction is rotated away from the electric field towards the **B** direction before rotating back to the **E** direction when the fields are orthogonal.
   - In the $\theta=\pi/2$ plane (Figure 18.4), the elongation in the contours develops and is rotated towards the $\mathbf{E}\times\mathbf{B}$ direction as ψ is increased.

It is also important to notice that for a given $\psi \neq 0$, the angle of elongation of the individual contours in the $\phi=0,\pi$ plane (see

Figure 18.2) increases with decreasing value of the contour (or equivalently the higher energy component of the velocity distribution). The direction of elongation for the high valued contours (low energy electrons) is approximately that of the drift velocity (in this plane) while the low valued contours are generally greater than the drift direction. Thus, the drift velocity direction *does not* represent an axis of rotational symmetry—an assumption used in previous Legendre expansion theories [139,140].

- Displacement:
  The displacement of the peak in the velocity distribution function is also modified through the application of a magnetic field. In particular
  - The displacement of the velocity distribution in the **E** direction monotonically decreases with increasing ψ.
  - The displacement of the velocity distribution in the $\mathbf{E}\times\mathbf{B}$ direction (see Figures 18.3 and 18.4) monotonically increases with increasing ψ.
  - The displacement of the velocity distribution in the **B** direction has a maximal property with ψ, increases from 0 for $\psi=0$ before decreasing again to 0 for $\psi=\pi/2$.

  Again, these general properties are easily verified by considering free charged particle orbits detailed above and we refer the reader to Refs. [95,138,141] for further details.
- Other properties:
  - Increasing ψ acts to cool the swarm, that is, to reduce the spacing between the contours and produce an increase in the height of the maximum contour at low energies. The mechanism for the cooling action of a perpendicular component of the magnetic field is well known.

The lack of symmetry in the velocity distribution function has important implications for the choice of basis functions used to represent f (see Chapter 12) and on our ability to visually display the velocity distribution function. Obviously, a full representation would require further slices in different planes or some other 3D visualization tools to be employed.

18.5 The Fluid Approach

Solution of the Boltzmann equation (Equation 18.4) for $f(\mathbf{r},\mathbf{v},t)$ is a complicated and computationally expensive task, and often yields more information than is actually required, since, in general, only a few low order moments are required: the drift velocity vector, diffusion tensor, mean energy, and temperature tensor. Thus, we now focus on the fluid approach, using the approximations of momentum transfer theory, as discussed in Chapter 7. The price paid for avoiding rigour is a reduction in the level of precision to around 10% or so, but this is nevertheless satisfactory for many applications, including the particle detectors discussed below.

The starting point for the discussion are the three basic equations of fluid theory, which in the hydrodynamic regime, to first order in ∇n, can be written as

1. The continuity equation

$$\frac{\partial n}{\partial t} + \nabla \cdot \Gamma = 0 \tag{18.7}$$

2. Momentum balance equation

$$-k_B \mathbf{T} \cdot \nabla n + nq\left[\mathbf{E} + \langle \mathbf{v} \rangle \times \mathbf{B}\right] = n\mu\nu_m(\langle \epsilon \rangle)\langle \mathbf{v} \rangle \tag{18.8}$$

3. Energy balance equation

$$\langle \epsilon \rangle = \frac{3}{2}k_B T_0 - \frac{1}{2}m_0\langle \mathbf{v} \rangle^2 + \Omega(\langle \epsilon \rangle) - \frac{\mathbf{Q}}{n\nu_e(\langle \epsilon \rangle)} \cdot \nabla n\,. \tag{18.9}$$

Here, ϵ denotes the energy in the centre-of-mass (CM) frame while $\langle\rangle$ denotes the average over all ion velocities. The neutral molecules of mass m_0 are in thermal equilibrium at a temperature T_0. The reduced mass of the particle–neutral system is μ. The collision frequencies for momentum and energy transfer are denoted by $\nu_m(\epsilon)$ and $\nu_e(\epsilon)$, respectively, while inelastic and superelastic processes are accounted for through the term $\Omega(\epsilon)$. Note that **B** only appears in Equation 18.8.

The ion temperature tensor **T**

$$k_B \mathbf{T} = m\langle(\mathbf{v} - \langle \mathbf{v} \rangle)(\mathbf{v} - \langle \mathbf{v} \rangle)\rangle \tag{18.10}$$

and the heat flux per particle,

$$\mathbf{Q} = \frac{1}{2}m\langle(\mathbf{v} - \langle \mathbf{v} \rangle)^2(\mathbf{v} - \langle \mathbf{v} \rangle)\rangle \tag{18.11}$$

can be calculated from spatially homogeneous higher order moment equations, as in Section 7.5.5. However, when inelastic processes are included, no closed form representation of **Q** is possible and **Q** is often treated as a parameter. It is often simply neglected, but for conditions under which negative differential conductivity are favoured, this is physically incorrect (see Chapter 8).

Equations 18.7 through 18.9 are solved by substituting

$$\langle \mathbf{v} \rangle = \mathbf{v} - \mathsf{D} \cdot \frac{1}{n} \nabla n, \tag{18.12}$$

$$\bar{\varepsilon}(\mathbf{r}, t) = \varepsilon + \gamma \cdot \frac{1}{n(\mathbf{r}, t)} \nabla n(\mathbf{r}, t), \tag{18.13}$$

(see also Equations 7.63 and 7.66) into Equations 18.7 through 18.9, linearizing in ∇n, and equating coefficients of ∇n. This yields equations for mean energy and the elements of the drift velocity, diffusion tensor, and gradient energy vector. The form and solution of these equations are discussed in the following sections.

18.5.1 Spatially homogeneous conditions: Wannier relation, extended Tonk's theorem, and equivalent field concept

The spatially homogeneous members of the hierarchy found by equating zeroth order powers in the density gradient are the following coupled non-linear equations for the drift velocity and spatial-averaged energy:

$$\mathbf{v}_d = \frac{q}{\mu \nu_m(\varepsilon)} \left[\mathbf{E} + \mathbf{v}_d \times \mathbf{B} \right] \tag{18.14}$$

$$\varepsilon = \frac{3}{2} k_B T_0 + \frac{1}{2} m_0 v_d^2 - \Omega(\varepsilon). \tag{18.15}$$

Equation 18.15 represents a relation between the swarm averaged energy in the CM frame and the swarm averaged velocity (drift velocity). An equivalent expression was derived previously for swarms in DC electric fields considering only elastic ion–molecule collisions interacting via the Maxwell model of interaction. We see that the form of the Wannier energy relation carries over to electric and magnetic fields crossed at arbitrary angles to each other. The Wannier energy relation continues to be used empirically and while its accuracy is well documented, there does exist more sophisticated formula for other interaction models. We would expect that the range of success of this relation would carry over to the present situation.

In the steady and uniform state, for arbitrary angles between **E** and **B**, the components of the drift velocity for the coordinate system defined above is

$$v_{d,x} = -\left[\frac{qE}{\mu\nu_m}\left(\frac{1}{1+(\Omega_L/\nu_m)^2}\right)\right]\left(\frac{\Omega_L}{\nu_m}\right)\sin\psi \tag{18.16}$$

$$\begin{aligned} v_{d,y} &= \left[\frac{qE}{\mu\nu_m}\left(\frac{1}{1+(\Omega_L/\nu_m)^2}\right)\right]\left(\frac{\Omega_L}{\nu_m}\right)^2 \sin\psi\cos\psi \\ &= -v_{d,x}\left(\frac{\Omega_L}{\nu_m}\right)\cos\psi \end{aligned} \tag{18.17}$$

$$v_{d,z} = \left[\frac{qE}{\mu\nu_m}\left(\frac{1}{1+(\Omega_L/\nu_m)^2}\right)\right]\left(1+\frac{\Omega_L^2}{\nu_m^2}\cos^2\psi\right), \tag{18.18}$$

where it is implied from here forth that

$$\nu_m = \nu_m(\varepsilon)\,. \tag{18.19}$$

Equivalently representing the drift velocity in terms of its spherical polar coordinates: the drift speed v_d is given by

$$v_d = \frac{qE}{\mu\nu_m}\sqrt{\frac{1+(\Omega_L/\nu_m)^2\cos^2\psi}{1+(\Omega_L/\nu_m)^2}}\,. \tag{18.20}$$

The polar angle (α_{Lorentz}) of the drift velocity or equivalently the Lorentz angle (the angle between the drift velocity and the electric field), is given by

$$\tan\left(\alpha_{\text{Lorentz}}\right) = \left(\frac{(\Omega_L/\nu_m)\sin\psi}{\sqrt{1+(\Omega_L^2/\nu_m^2)\cos^2\psi}}\right) \tag{18.21}$$

while the azimuthal angle ψ_{Lorentz} for the drift velocity is given by

$$\tan\psi_{\text{Lorentz}} = -\frac{\Omega_L}{\nu_m}\cos\psi. \tag{18.22}$$

For DC electric fields, we have shown that the drift velocity takes the form:

$$v_d = \frac{qE}{\mu\nu_m}\,. \tag{18.23}$$

Comparing Equations 18.23 and 18.20, it is evident that the drift speed associated with any combination of **E**, **B** and ψ can be generated by considering a pure equivalent electric field E_e, that is,

$$v_d = \frac{qE_e}{\mu\nu_m}, \tag{18.24}$$

whose magnitude is given by

$$E_e = E\sqrt{\frac{1+(\Omega_L/\nu_m)^2\cos^2\psi}{1+(\Omega_L/\nu_m)^2}} = E\cos\left(\alpha_{\mathrm{Lorentz}}\right). \tag{18.25}$$

Physically, this term is proportional to the power input by the electric field ($\mathbf{E}\cdot\mathbf{v}_d$). In the limit of $\psi=\pi/2$, this reduces to Tonk's theorem. Equations 18.24 and 18.25 thus constitute a generalisation of Tonk's theorem to electric and magnetic fields crossed at arbitrary angles. The explicit effect of the electric and magnetic fields in the energy balance equation is contained solely in the drift speed, and thus it follows that the mean energy is also expressible in terms of the equivalent electric field (Equation 18.25). That is,

$$\varepsilon(E,B,\psi)=\varepsilon(E_e) \tag{18.26}$$

$$v_d(E,B,\psi)=v_d(E_e)\,. \tag{18.27}$$

The utility of the equivalent electric field concept is as follows: in principle, if one has the data for (or can solve for) drift speed and energy as a function of the applied field in the electric field only case, then the drift speed and energy for any specified configuration of applied **E** and **B** can be determined. The details of the implementation of the equivalent field concept are discussed in Ref. [142]. Once ε is known, the polar angles can then be found using Equations 18.21 and 18.22 and the drift velocity is then fully specified.

18.5.2 Spatially inhomogeneous conditions: GER, gradient energy vector

The following system of coupled equations for elements of the diffusion tensor and gradient energy parameter are generated from the first order members of the hierarchy:

$$\mathbf{D}=\frac{k_B\mathbf{T}^{(0)}}{\mu\nu_m(\varepsilon)}-\frac{q}{\mu\nu_m(\varepsilon)}[\mathbf{B}\times\mathbf{D}]+\frac{\nu_m'(\varepsilon)}{\nu_m(\varepsilon)}\mathbf{v}_d\gamma \tag{18.28}$$

$$\gamma=\frac{1}{1+\Omega'(\varepsilon)}\left[-m_0\mathbf{v}_d\cdot\mathbf{D}-\frac{\mathbf{Q}^{(0)}}{\nu_e(\varepsilon)}\right], \tag{18.29}$$

where the dashed quantities refer to derivatives with respect to ε and

$$[\mathbf{B}\times\mathbf{D}]_{ij}=\epsilon_{ikl}B_kD_{lj} \tag{18.30}$$

and ϵ_{ikl} is the traditional Levi-Cevita symbol.

The physical origin of anisotropic diffusion for swarms in uniform electric and magnetic fields arises from two sources:

- Magnetic anisotropy due to the charged particle orbits;
- Electric anisotropy due to the spatial variation of the mean energy throughout the swarm, in association with a energy dependent collision frequency.

To aid in the physical understanding of the variation of the diffusion tensor elements, it is of particular interest to consider the relationship between the diffusion tensor elements and the gradient energy vector. Re-arrangement of Equation 18.28 yields

$$\mathbf{D} = \left[1 + \left(\frac{\Omega_L}{\nu_m}\right)^2\right]^{-1} \left\{ \left(\mathbf{I} - \left(\frac{\Omega_L}{\nu_m}\right) \hat{\mathbf{B}} \times + \left(\frac{\Omega_L}{\nu_m}\right)^2 \hat{\mathbf{B}}\hat{\mathbf{B}} \cdot \right) \left(\frac{k_B \mathbf{T}}{\mu \nu_m} + \frac{\nu'_m}{\nu_m} \mathbf{v}_d \gamma \right) \right\}, \quad (18.31)$$

where the hatted quantities denote unit vectors and the "·" and "×" denote operations to the right of the enclosed brackets. If one eliminates γ in Equations 18.29 through 18.31, then after some algebra we find the following expression for the generalized Einstein relation (GER):

$$\mathbf{D} = \mathbf{K} \cdot \frac{k}{q} \bar{\mathbf{T}} \quad (18.32)$$

where

$$k_B \bar{\mathbf{T}} = k_B \mathbf{T} - \left(\frac{\mu \nu'_m}{\nu_e}\right) \frac{\mathbf{v}_d \mathbf{Q}}{1 + \Omega'} \quad (18.33)$$

while the elements of the differential mobility tensor **K** are given by

$$[\mathbf{K}]_{ij} = \frac{\partial v_{d,i}}{\partial E_j}. \quad (18.34)$$

As discussed previously, the GER are a group of formulae which link the diffusion coefficients with the field derivatives of the drift velocity. They have been used extensively in DC electric field systems only and although their form for electric and magnetic field crossed at arbitrary angles to each other may not appear particularly simple, they can still provide important empirical information.

18.6 Gaseous Radiation Detectors

Ionizing particles, both charged and neutral, are detected through their interaction with matter [143,144]. Independently of whether this matter is gaseous, liquid, or solid for all these different kind of radiation detectors, the transport of charge carriers under the influence of electric or electric and magnetic fields is of fundamental interest. In the following, we focus only on so-called gaseous detectors where the sensitive detection volume is filled with gas.

For more than a century gaseous particle detectors have played a fundamental role in the detection of ionizing particles. The first single wire proportional counter to study natural radioactivity was described by Rutherford and Geiger in 1908 [145] and 1913 [146], respectively, and the first counter with single electron sensitivity was presented by Geiger and Müller in 1928 [147–150]. Applications in nuclear and particle physics experiments as well as medical imaging resulted in further new developments. Nowadays, a huge variety of different types of gaseous counters exist, in particular, the multi-wire proportional chamber (MWPC) [151], drift chamber [152], time projection chamber (TPC) [153], resistive plate chamber [154], and a large "family" of different so-called micro-pattern gaseous detectors (MPGDs) [155]. More details can be found in Refs. [16,156–159].

As common requirements, a gaseous detector operated in particle physics experiments should provide good proportionality, good spatial and timing resolution, and a stable and sufficiently large electron multiplication process at a low working voltage. Furthermore, the detector has to cope with a high rate environment and should be able to tolerate a long lifetime without radiation-induced degradation.

18.6.1 Basic processes

The operation of a gaseous detector is based on four processes:

- The creation of primary charge in the sensitive volume and the separation of the charged ion pairs
- The drift and diffusion of electrons and positive ions in the gas
- The multiplication of the primary electrons in an electron avalanche process, and finally
- the collection of the charge on detector electrodes.

The collection of the charge is mainly determined by the geometry of the detector, shape of the electrical field, and shape of the electrodes, whereas the first three processes strongly depend on the gas filling. Consequently, a detailed knowledge of gaseous electronics is necessary for the design, optimization, and operation of a gaseous detector. Depending on the size of the electron multiplication process, gaseous detectors are essentially classified into ionization chambers, proportional counters, and Geiger–Müller counters.

18.6.2 Choice of gas filling

The gas filling of a gaseous radiation detector has to fulfill various properties and the choice of the appropriate gas or gas mixture is often characterised by contradictory requirements. Basically, any gas or gas mixture may be used for a gaseous detector operated in the proportional mode. However, there are limitations. Certain physical properties are required, for example, on electron transport properties to ensure an optimized performance. There are also safety rules in research laboratories which often forbid the usage of flammable gases.

In the following, we list some of the main characteristics of a typical gas filling and mention the conflicting properties which need to be balanced. All these characteristics may be a function of the particle number density n_0, that is, T and p, the gas components and their mixing ratio, the electrical or electrical and magnetic fields, and the rate environment. Furthermore, these properties are connected to each other, often by contradictory relations—the choice and optimization of the gas filling is a difficult trade-off and delicate task.

18.6.2.1 Primary ionization n_T

Ionizing radiation which passes the sensitive volume of a gaseous detector dissipates a certain amount of energy by generating electron–ion-pairs. The energy loss dE/dx is thereby described by the Bethe–Bloch formula [16,143,144]. The total number of primary electrons which is created is $n_T = \Delta E/W_i$, where ΔE is the energy lost in the sensitive volume and W_i is the energy which is necessary to create an electron–ion pair. It is worth noting that W_i is substantially higher than the first ionization potential E_i as excitation processes of inner shells might be involved as well in the process of ionization which do not yield in free electrons. For gases normally used in gaseous detectors, the values for W_i are typically between 25 and 30 eV; actually they range from 22 eV for xenon and 23–25 eV for high molecular masses like isobutane up to 41 eV in case of helium.

The typical number n_T of electrons created by a so-called minimum ionizing particle (MIP) in a 1 cm thick gas gap is around 50–100 electrons. This amount of charge is too small to be detected. Consequently, an amplification process already in the gas of up to 10^6 might be necessary to achieve a reasonable signal to noise ratio. This amplification process usually happens in the vicinity of the thin anode wire or strip. However there are also experimental techniques to achieve electron multiplication along the drift direction towards the anode and therefore, to decouple the charge amplification and the charge detection at the anode by using a so-called gas electron multiplier (GEM) foil [160] in the gas volume.

A stable electron multiplication process, in particular, in a high rate environment, is a crucial task and therefore, it is desirable to keep it as low as possible. Consequently, a high primary ionization is preferable, in particular, for thin detectors where the track length of the charged particle and the resulting deposited energy is small.

Hydrocarbons, for example, ethane or isobutane, offer a large amount of primary ionization and are used very often for this reason. But on the other hand, for the measurement of low-momentum particles, multiple scattering should be minimised and consequently, high Z constituents—as for example, long-chained hydrocarbons—are incompatible. Furthermore, hydrocarbons might cause radiation induced degradation due to plasma polymerisation and as they are flammable they are often ruled out due to safety rules.

18.6.2.2 Drift velocity v_d

In order to cope with high rate environment or to receive a fast detector response, a high value of the drift velocity is preferable. And to avoid large variations of the drift velocity in case of small variations of the electrical field, pressure, or temperature, a constant drift velocity over a certain range of E/n_0 is advantageous.

One of the most commonly used filling gases in the 1970s and 1980s was a 90% Ar + 10% CH_4 gas mixture labelled P-10. This gas mixture offers a high value of drift velocity of more than 5 cm/μs, but this value is only limited over a very narrow E/n_0 range (see Figure 18.5). This restricts the values for the applied electrical field and limits the pressure and temperature ranges during operation. In particular, for drift chambers where we rely on a stable so-called "*x*-*t*-relation" within a drift cell, a drift velocity showing a constant plateau in a certain range of E/n_0 is necessary.

Figure 18.5 shows the electron drift velocities as a function of E/n_0 for three different gas mixtures which show a rather different behaviour. The He-C_2H_6 (50-50) gas mixture shows a constant drift velocity over a rather wide range of E/n_0 which is required for usage in a so-called drift chamber (see Section 18.6.3).

18.6.2.3 Longitudinal and transversal diffusion $D_{\parallel}$ and $D_{\perp}$

While drifting through the filling gas under the influence of external electric and magnetic fields, the electron swarm also undergoes a certain diffusion in longitudinal and transversal directions. In case of a long drift path of the electrons in the sensitive volume, for example, in case of a TPC, the diffusion has significant impact on the spatial and temporal resolutions of the detector and might even limit these properties. In general, electrons in noble gases show very large values of diffusion up to several hundreds of

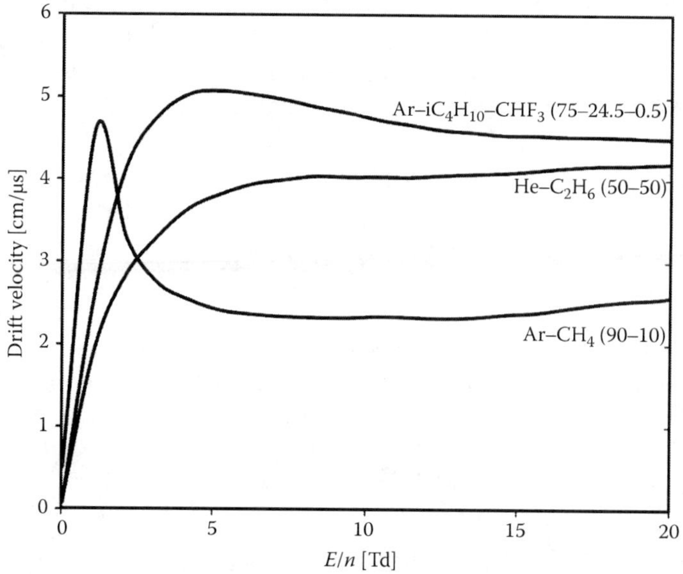

Figure 18.5 This plot shows the electron drift velocities as a function of the reduced electrical field E/n_0 in the gas mixtures Ar–iC_4H_{10}–CHF_3 (75.0–24.5–0.5), He–C_2H_6 (50–50), and Ar–CH_4 (90–10).

μm per $\sqrt{\text{cm}}$ drift path, whereas in molecular gases, the diffusion coefficient might even be close to the thermal limit. Intermediate values for the diffusion prevail in gas mixtures.

To cite helium as an example, this gas offers, in principle, the possibility of good momentum resolution for the measurement of charged particle tracks due to its low Z composition and long radiation length X_0, but shows a large diffusion in both directions and therefore sets certain limits on the achievable resolutions over long drift distances.

18.6.2.4 Magnetic deflection (Lorentz) angle α_{Lorentz}

In the presence of an electric and a magnetic field, the Lorentz force is acting on the moving charged particles and consequently, the drift velocity has not only a component in the direction of the E-field, but also in the direction of the B-field and perpendicular to the E- and B-fields. The magnetic deflection (or Lorentz) angle α_{Lorentz} is defined as the angle between the electrical field and the drift velocity of the electrons in the presence of electric and magnetic fields.

Basically, a small magnetic deflection angle is preferable, in particular, for gaseous detectors with micro-pattern readout structures as the drift under a certain angle enlarges the region where the electron swarm reaches the readout structure of the detector and therefore increases the so-called multiplicity (of readout electrodes). This can reduce the achievable spatial resolution. In general, small Lorentz angles will be achieved with so-called "cold" gases, but on the other hand, these gases offer only low values for the drift velocity and the amplification factor.

18.6.2.5 High momentum resolution, long radiation length X_0

The tracking of a charged particle, for example, for the determination of the momentum on a bended trajectory in a magnetic field, requires a rather low-mass gas composed of low Z and long radiation length gas components to minimise the multiple scattering which deteriorates the spatial or momentum resolution. Low Z gases like helium are preferably used in this case. Unfortunately, helium shows large statistical fluctuations of the primary ionization due to the high W-value of 41 eV and the electron drift suffers from a rather large diffusion of several hundreds of μm per $\sqrt{\text{cm}}$ drift path which on the other hand may deteriorate the spatial resolution.

18.6.2.6 First Townsend coefficient α_T

As the charge from the primary ionization is very small, it needs to be "amplified" already in the gas to create a reasonable signal size. In the operation mode of a proportional counter, this electron multiplication is typically in the order of 10^4–10^5 to achieve an appropriate signal to noise ratio, sometimes even factors of up to 10^6 are necessary.

This amplification takes place in an electron avalanche process, for example, close to the thin anode wire where the electrical field strength is increasing due to the $1/r$ dependence to very high values. At electrical fields around 10–15 kV/cm, the electrons gain so much energy from the field between two collisions that they can produce further electrons by impact ionization. This process continues with the newly produced electrons and an electron avalanche is built up. The number of electron–ion pairs produced per unit length by one electron in the avalanche process is called the first Townsend coefficient α_T.

In the case of noble gases, this avalanche process happens at much lower electrical fields than in complex molecules. Consequently, a noble gas is usually the main part of the gas filling. But vacuum ultraviolet (VUV) photons which are emitted in the multiplication process from the de-excitation of excited atoms might lead to instabilities as they can trigger new multiplication processes somewhere else in the counter and this might launch discharges. Additional gas components or admixtures are needed to provide a stable operation.

18.6.2.7 Stable multiplication process and high-gain operation

To prevent the above-mentioned potential triggers for discharges and to allow a stable operation, it is necessary to add gas components which absorb radiation less and "quench" these VUV photons originating from the electron avalanche process. Molecular gas components which offer a lot of rotational or vibrational excitation levels to absorb these photons, like hydrocarbons, offer appropriate properties and are usually used for this purpose. Hydrocarbon molecules dissipate their excess energy by elastic collisions or dissociation into simpler radicals. Unfortunately, these radicals may lead to polymerisation in the plasma of the avalanche process and may lead to so-called radiation induced degradation of the detector due to Malter effect [161] or coatings on anodes and cathodes.

Small admixtures of electronegative gases also help to allow for a stable high gain operation. In the 1970s, the so-called "magic gas" mixture, consisting of 70% Ar + 29.6% iC_4H_{10} + 0.4% freon-13B1 (CF_3Br), was often used and multiplication factors up to 10^8 could be reached without entering the Geiger operation regime which does not offer proportionality of deposited energy and pulse height anymore. Nowadays, freon is no longer allowed due to its potential role in ozone depletion in the atmosphere.

The voltage to trigger an electrical breakdown over a gas gap between two electrodes depends amongst other things on the charge density in the gas gap. This upper charge density is given by the so-called Raether limit [162] and is rate dependent. Consequently, this has to be taken into account in the design and operation of gaseous detectors in high rate particle physics experiments. Detectors operated in high rate environments, in particular, in heavy ionizing environment, might need a larger fraction of gas components with quenching capabilities than detectors in other applications. As already mentioned, hydrocarbons show good properties to increase the high rate stability, but on the other hand, they increase the risk of radiation induced degradation due to deposits originating from plasma polymerisation.

18.6.2.8 Sensitivity to radiation induced degradation (aging)

The degradation of the performance of a gaseous detector under exposure with ionizing radiation is called "aging." Degradation, in this context, means, for example, the decrease of the pulse height, the broadening of the pulse height distribution which leads to a decrease of the energy resolution, the appearance of dark currents, or high voltage (HV) instabilities due to discharges. The changes of these general characteristics were reported for the first time already in the 1940s for self-quenching Geiger–Müller counters after long-term operation [163–165] and later the operation of wire chambers in high energy particle physics since the 1970s launched an intensive research effort [166,167] for a fundamental understanding of these effects, including the connection to plasma chemistry [168,169].

The main reasons for the observed deterioration of the performance of gaseous detectors are, for example, damaged or etched anode or cathode surfaces, deposits on the anode or the cathode and conductive instead of insulating surfaces or vice versa.

Not all reasons and mechanisms are completely understood and some "magic" remains. There is a common understanding that the region very close to the anode wire and its "gaseous composition" plays an important role: due to the high electrical field strength the mean energy of the drifting electrons may reach values up to several tens of eV in the avalanche process. This allows impact ionization and creates a kind of plasma consisting of neutral gas atoms or molecules, electrons, and positive ions.

If the mean energy of the electrons in the avalanche process exceeds the dissociation energy of the molecules, which are typically in the order of a few eV, free radicals may be created, for example, by breaking the C-chains of hydrocarbon molecules. These radicals may form carbon polymers which build up on anodes or cathodes and this is one of the main reasons for instabilities (e.g., due to Malter effect [161]) and a decrease of the detector performance. In addition, traces of impurities contained in the gas filling or originating from outgassing of materials and glues, contamination during assembly procedure or even back diffusion from pumps or "oil bubblers" mounted downstream of the detector may enhance the plasma polymerisation and lead to deposits.

Furthermore, the positive ions will drift to the cathode with a typical drift velocity some 10^2–10^3 times slower than the electron drift velocity and, depending on their mean energy at the impact and the material of the cathode, they may cause damage or knock out electrons from the surface which may trigger unwanted avalanche processes or even discharges.

It was found out—and it is supported by plasma chemistry—that certain organic compounds which contain oxygen, like certain alcohols or methylal, help to prevent polymerisation. In their reaction with hydrocarbons, the end products are stable and volatile molecules which can be flushed out from the chamber with a steady gas flow.

It is worth mentioning that even small admixtures of these organic compounds may have a huge impact on the transport properties of the "initial" gas mixture as these compounds usually have large inelastic cross sections.

18.6.3 Working principle of a drift chamber

The drift chamber is a further development of the proportional chamber and allows a much better spatial resolution for the detection of particle tracks. In this case, the electron drift velocity v_d plays the key role and this is the reason why this detector type is called "drift chamber."

A multi-wire proportional chamber (MWPC) is basically a planar or cylindrical layer of single-wire proportional chambers without separating walls but with all wires placed adjacent to each other between two common cathode planes. The wire plane is usually mounted perpendicular to the trajectory of the particle which will be measured.

The charged particle traverses the sensitive volume of the MWPC and creates primary ionization along its path. Due to the applied electrical field—a positive voltage is applied to the wires—the electrons from the primary ionization move towards the anode wire, are multiplied in an avalanche process and this total charge is collected on the wire and creates a signal.

The spatial resolution which can be achieved with this configuration is determined by the wire distance and is typically a few millimeter at best. Usually, it is not possible to further reduce the wire pitch as this will cause electrostatic instabilities of the wires.

Beginning in the 1970s, a fundamental step in detector development was achieved by using the information of the time lag—or drift time—of the detected signal on a wire and an external trigger indicating the passage of the incident particle, as illustrated in Figure 18.6. At a time t_0, the

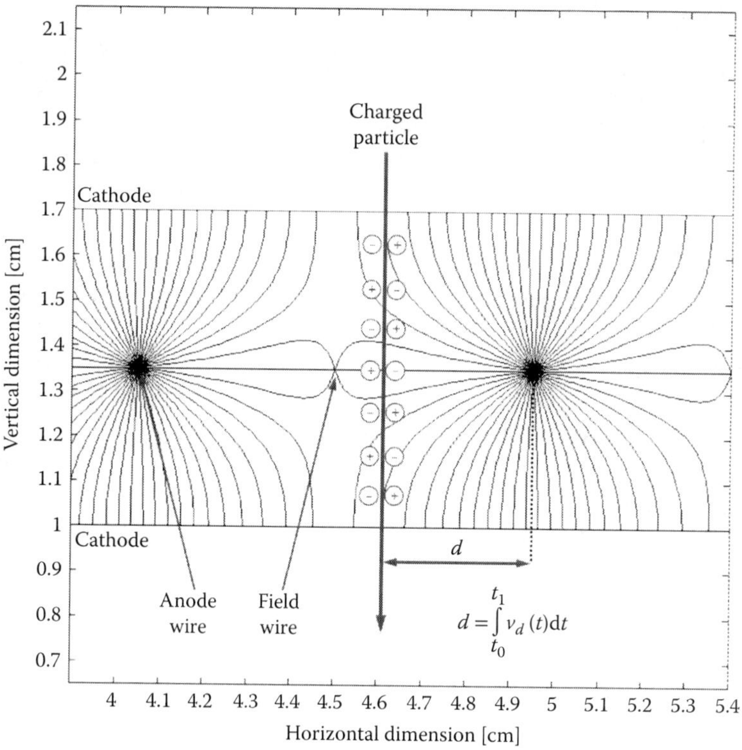

Figure 18.6 This figure shows the field configuration inside a drift chamber with alternating anode and field wires between two cathodes. The trajectory of a charged particle is indicated by the arrow. The distance of the track to the anode wire is calculated by using the drift velocity of the electrons as a function of the electrical field and the measured drift time.

electrons comprising the primary ionization are released in the filling gas along the track of the charged particle, and move to the anode where they are detected at a time t_1. Now the distance of the incident track from the anode wire can be determined from the time difference $t_1 - t_0$ using the so-called x-t relation of the drift cell which was either directly measured or computed using the electric field map and the field-dependence of the drift velocity $v_d = v_d(E)$.

In case the drift velocity changes along the drift path, the distance d of the particle trajectory to the anode wire is given by

$$d = \int_{t_0}^{t_1} v_d(t)dt. \tag{18.35}$$

In case of a constant drift velocity v_d, one obtains the linear relation

$$d = v_d(t_1 - t_0) = v_d \cdot \Delta t \tag{18.36}$$

The requirement to achieve a constant drift velocity along the drift path of the electrons is a constant electrical field strength along this path. Even if the field is not constant, a drift velocity showing a plateau over a certain range on E/n_0 may compensate for possible small variations in the field strength (see also Figure 18.5). Unfortunately, the field configuration in a conventional MWPC does not allow for such a constant drift velocity as the field strength is reduced in the middle between two anode wires. Only the implementation of a field or potential wire on ground or even negative potential between the two anode wires creates in first approximation, a linear relationship between drift time and drift path. Due to an optimisation of the shape of the electric field and a precise knowledge of the electron drift properties, a spatial resolution below 100 μm was ultimately achieved [170] even though the anode wire distance is in the order of several millimeters. It is worth noting that the time resolution of the electronics and the diffusion of the electrons on their way to the anode were the main limiting factors for further improvement.

CHAPTER 19

Muons in Gases and Condensed Matter

19.1 Muon versus Electron Transport

Muons μ^- like electrons e^- are part of the lepton family and, for present purposes, can be simply regarded as "heavy" electrons, with a mass $m_\mu \approx 207\ m_e$. Nevertheless, the muon to atom (or molecule) mass ratio m_μ/m_0 is still small enough for muons to be classified as light particles in kinetic theory, with a "two-term" spherical harmonic representation of the muon velocity distribution function often being a reasonable approximation. As for electrons, the momentum transfer collision frequency ν_m is replaced by its structure-modified counterpart $\widetilde{\nu}_m$ when dealing with a condensed matter medium.

Negatively charged, low-energy muons can undergo atomic capture and form muonic atoms by replacing a bound electron after colliding with a neutral atom. An important consequence of the muon being heavier than the electron is that the Bohr radius of this muonic atom is reduced by a factor of $m_\mu/m_e = 207$ in comparison with the normal atom. In the case of a deuterium and tritium mixture such a muonic atom, namely the muonic deuterium dμ or the muonic tritium $t\mu$, is in fact the first "leg" of the muon-catalyzed fusion cycle—see Equation 19.11. From the point of view of transport analysis, this process may be treated in the same way as any other "reactive" collision, using the fluid and kinetic theoretical framework developed previously. While we can apply much of the transport theory already developed in Chapters 7 through 9 to analyse muons, there are two additional factors which must be taken into consideration:

- A muon is unstable and decays according to

$$\mu^- \to e^- + \nu_\mu + \bar{\nu}_e, \tag{19.1}$$

  with a life-time of $\tau_\mu \approx 2.2 \times 10^{-6}$ s.
- Production of muons by pion decay,

$$\pi^- \to \mu^- + \bar{\nu}_\mu \tag{19.2}$$

  is energetically expensive and technologically challenging. The leading facilities [171,172] for experiments using high intensity pion and muon beams utilize proton drivers with proton beam energies between 0.5 and 3 GeV and beam currents of up to 2.4 mA to produce pions through nuclear interactions. We come back to this issue in Section 19.4.3.

Thus, while electrons are stable and cheap to produce, experiments and applications involving muons operate under quite severe limitations.

Just as for e^- and e^+, transport processes involving μ^- and its anti-particle μ^+ can be considered together. Kinetic theory and fluid analysis does not formally distinguish between a particle and its anti-particle, although the respective contributions to $\left(\frac{\partial f}{\partial t}\right)_{\mathrm{col}}$ may differ in detail.

19.2 Muon Beam Compression

A novel scheme for compressing μ^+ beams in He gas using a two-stage compression scheme has been recently proposed by Taqqu [173]. The second stage, namely the longitudinal compression, was already successfully demonstrated by Bao et al. [174]. We shall discuss here only the first stage of the procedure, where the beam is compressed in the transverse direction, through the combined action of a gradient ∇n_0 in the gas density and an imposed electric field **E**, both directed normal to the beam direction, and the longitudinally applied magnetic field **B** (see Figure 19.1). Since **E** and **B** are orthogonal ($\psi = 90$), we can re-express Equations 18.16 and 18.18 quite generally for the muon drift velocity,

$$\mathbf{v_d} = v_{\|}\widehat{\mathbf{E}} + v_{\perp}\widehat{\mathbf{E}} \times \widehat{\mathbf{B}}, \tag{19.3}$$

where

$$v_{\|} = \frac{KE}{1+(KB)^2}; \quad v_{\perp} = KBv_{\|} \tag{19.4}$$

and $K = e/m\nu_m$ is the muon mobility in an electric field only. Since $\nu_m \sim n_0$, then K is inversely proportional to the gas number density n_0. Details of the cross section are not needed for present purposes.

These expressions are, strictly speaking, valid only for spatially uniform conditions but, as in Refs. [173,174], we assume them to hold, at least to a first approximation, when n_0, and hence $K, v_{\|}$ and $v_{\perp}$ vary with position. The principle of the transverse compression scheme is shown schematically in Figure 19.1.

At high density, $KB < 1$, the influence of the magnetic field is weak, and the predominant contribution to drift velocity comes from the first term in Equation 19.3. Thus, in the upper region, $\mathbf{v}_d$ is approximately parallel to **E**. Conversely, at lower densities, $KB > 1$, the influence of the magnetic

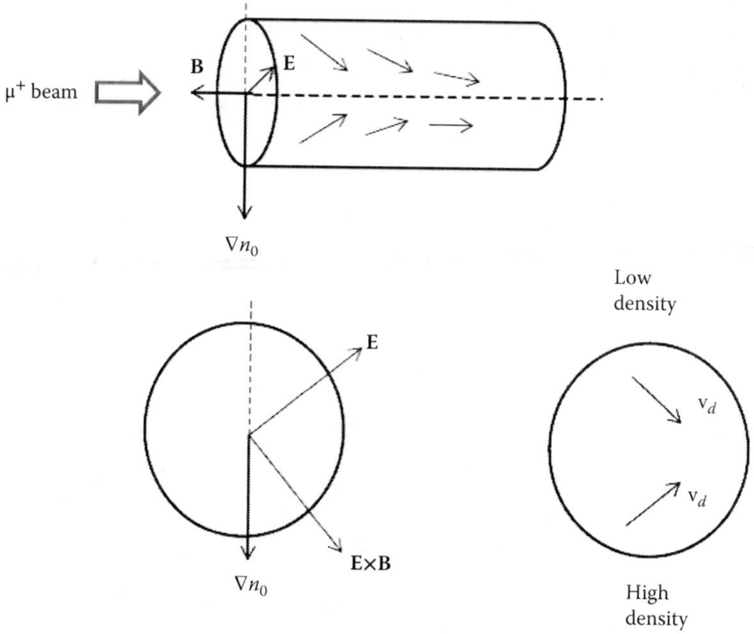

Figure 19.1 A beam of muons directed anti-parallel to **B** enters a region containing He gas, whose density n_0 increases vertically downward. An electric field **E** is applied upward at an angle of 45° to the vertical, resulting in an $\mathbf{E} \times \mathbf{B}$ drift downward at 45° to the vertical. This arrangement produces drift velocities with inwardly directed components (see text), resulting in compression of the beam.

field is significant, and the second term of Equation 19.3 dominates. Thus, in the upper region, $\mathbf{v}_d$ is approximately parallel to $\mathbf{E} \times \mathbf{B}$. The net result of the density gradient is, thus, to produce a drift velocity profile with components directed toward the axis, as shown in Figure 19.1. The beam is compressed in the transverse direction in this way.

This is only an elementary introduction to the experiment. In more detailed analysis, muon–helium cross sections are required. More details of the experimental arrangement, including the next stage of longitudinal compression, can be found in Refs. [173,174].

19.3 Aliasing of Muon Transport Data

19.3.1 Why aliasing is necessary

A number of experiments involving muons in gases, such as μ^+ in H_2 in the MuCap experiment [175,176] , and μ^+ in He, in the high quality beam proposed by Taqqu [173], require muon transport data, but none is available. In such cases, one can improvise by taking the known transport data for another, charged species in the same gas, and adapting it to the species at hand, using the approximate formulas of the fluid analysis of Chapters 7 through 9, plus one key assumption. We refer to this procedure as "aliasing" swarm data.

19.3.2 The general prescription for aliasing

In what follows, it is assumed that particles undergo predominantly elastic collisions with the neutral gas. In that case, the mobility of charged particles of arbitrary mass m in a gas of particles of mass m_0, number density n_0, and temperature T_0 is given by the fluid analysis of Chapter 7 as

$$K = \frac{v_d}{E} = \frac{e}{\mu \nu_m(\varepsilon_{CM})}, \tag{19.5}$$

where $\mu = mm_0/(m + m_0)$ is the reduced mass, and the mean energy in the centre of mass (CM) is given by the Wannier formula

$$\varepsilon_{CM} = \frac{3}{2}k_B T_0 + \frac{1}{2}m_0 v_d^2. \tag{19.6}$$

The subscript CM has been added for emphasis. The average momentum transfer collision frequency is defined in terms of the average momentum transfer cross section by

$$\nu_m(\varepsilon_{CM}) = n_0 \sqrt{\frac{2\varepsilon_{CM}}{\mu}}\sigma(\varepsilon_{CM})$$

and hence the reduced mobility coefficient is

$$\mathcal{K} \equiv n_0 K = \frac{e}{\sqrt{2\mu\varepsilon_{CM}}\mu\sigma_m(\varepsilon_{CM})}. \tag{19.7}$$

We now wish to apply these general expressions to two different charged species 1 and 2 in the same gas (we have in mind μ^+ and H^+, respectively in H_2) to find the mobility of species 1, purely on the basis of known mobility data for species 2.

If the nature of the interactions between species 1 and 2 and a given gas molecule are known from basic physical considerations to be similar, then the momentum transfer cross sections for the two different charged species in the same gas may be assumed to be approximately the same, that is,

$$\sigma_m^{(1)}(\varepsilon_{CM}) \approx \sigma_m^{(2)}(\varepsilon_{CM}) \tag{19.8}$$

for all CM energies. This assumption then implies, with Equation 19.7, that the ratio of reduced mobilities of the two species (distinguished by superscripts (1) and (2) in what follows) *at the same mean CM energy* is

$$\frac{\mathcal{K}^{(1)}}{\mathcal{K}^{(2)}} = \frac{n_0^{(1)} K^{(1)}}{n_0^{(2)} K^{(2)}} = \sqrt{\frac{\mu^{(2)}}{\mu^{(1)}}}. \tag{19.9}$$

The mass m_0 of the neutral gas molecule is the same in each case and, for simplicity, we will assume that the gas temperatures are also the same. (This assumption can be relaxed if desired.) However, the reduced fields E/n_0 will be different for 1 and 2 even though the energy is the same. Using the Wannier relation (Equation 19.6) for each species, and equating the mean energies, we find

$$\varepsilon_{CM} = \frac{3}{2}k_B T_0 + \frac{1}{2}m_0\left(K^{(1)}E^{(1)}\right)^2 = \frac{3}{2}k_B T_0 + \frac{1}{2}m_0\left(K^{(2)}E^{(2)}\right)^2$$

from which it follows that

$$\frac{(E/n_0)^{(1)}}{(E/n_0)^{(2)}} = \frac{\mathcal{K}^{(2)}}{\mathcal{K}^{(1)}} = \sqrt{\frac{\mu^{(1)}}{\mu^{(2)}}}. \tag{19.10}$$

Thus, if we know the reduced mobility for species 2 at a particular reduced field, $(E/n_0)^{(2)}$, we may use Equations 19.9 and 19.10 to find the reduced mobility of species 1, at a different value $(E/n_0)^{(1)}$, of reduced field. The procedure is illustrated in the following example.

19.3.3 Calculation of the mobility of μ^+ in H_2

No mobility data exists for μ^+ in H_2, but nevertheless, it is needed for the MuCap experiment [175]. Of the two possible candidates for aliasing, e^+ and H^+, the mass of μ^+ is closest to the latter, and hence, as far as scaling is concerned, the $\mu^+ - \mathrm{H}$ system is more like $H^+ - H_2$ than $e^+ - H_2$.

Assuming then that a muon and a proton interact with a hydrogen molecule in a similar fashion, the momentum cross sections may be assumed to be approximately the same, as in Equation 19.8. Equations 19.9 and 19.10 may be applied directly to estimate the unknown mobilities using known swarm mobility data for H^+ in H_2.

With the notation that superscripts (1) and (2) refer to muons and protons, respectively, the reduced masses are

$$\mu^{(1)} \approx m_{\mu^+} \quad \text{and} \quad \mu^{(2)} \approx \frac{2}{3}m_H$$

respectively, and hence $\mu^{(1)}/\mu^{(2)} \approx 1/6$. If we require the muon mobility at (say) $(E/n_0)^{(1)} = 0.8$ Td then Equation 19.10 tells us to look up the proton mobility at $(E/n_0)^{(2)} = 0.8 \times \sqrt{6} \approx 2$ Td. From published tables, the reduced mobility of protons in H_2 at this value of field is $\mathcal{K}^{(2)} = 2.0$ V cm^2 s^{-1} and hence by Equation 19.9,

$$\mathcal{K}^{(1)} = \sqrt{6} \times 2 \approx 4.9 \text{ V cm}^2 \text{ s}^{-1}$$

is the reduced mobility of μ^+ in H_2 at 0.8 Td. The procedure can be repeated at other reduced fields for which H^+ mobilities are available.

This method could also be used to investigate the transport properties of μ^+ in He, by aliasing (H^+, He) mobility data [177], as suggested by Taqqu [173] who, however, has used a different, purely empirical scaling procedure. This is a particularly interesting case in its own right because of the "runaway" phenomenon produced by the sharply falling cross section (see Problem 19), but a detailed discussion would take us too far afield.

19.4 Muon-Catalyzed Fusion

19.4.1 Cold versus hot fusion

The remarkable ability of a single muon μ^- to catalyze many fusion reactions in gaseous and condensed hydrogen and its isotopes at low temperatures T_0 (typically around room temperature or less) has been firmly established for some time [178,179]. The process was briefly revived in the 1980s as a serious contender for a viable energy source, following investigations by Russian theoretical physicists, and motivated by the experimental observation that fusion yield is quite sensitive to temperature, density, and isotopic composition of the medium. The first investigations were carried out with muons in liquid and gaseous hydrogen and deuterium and, although the catalysis cycle turned out to be inefficient (in the sense that energy output is much less than energy input), the observed strong temperature and density dependences provided important clues as to how greater efficiencies might be achieved. Just as for the "hot" fusion program ($T_0 \sim 10^8$ K) attention then turned to gaseous mixtures of deuterium and tritium, whose nuclei, d and t, respectively, undergo fusion to produce an α particle (the nucleus of ^{4}He), a neutron n and 17.6 MeV per fusion event. However, there are fundamental differences in the underlying physics between hot and cold fusion.

In hot fusion, the medium is a fully ionised gaseous *plasma*, and the positively-charged, bare deuterium and tritium nuclei can overcome the Coulomb barrier and fuse only by virtue of their high thermal energies. "Cold" fusion on the other hand takes place in a *swarm* of low density muons interacting with a neutral medium, in which the muon overcomes the repulsive Coulomb barrier by binding the nuclei together. The process involves only dilute negatively charged muons and their derivatives, and consequently, there are no space-charge effects, and no instabilities which make the confinement of hot plasmas so difficult. Most importantly, processes may be analysed theoretically in the same way as a swarm, using the kinetic and fluid equations introduced in Parts II and III.

19.4.2 μCF cycle

The cycle is actually quite complex and, as a matter of practicality, only the most important details are discussed in what follows. For example, while both $t\mu$ and $d\mu$ muonic atoms are formed in deuterium and tritium mixtures, it is the $t\mu$ atoms which go on to make the most significant contribution to the cycle. Thus, the main contribution to the muon-catalyzed fusion cycle in a mixture of deuterium and tritium comes from the process

$$d + t + \mu \rightarrow t\mu + d \rightarrow dt\mu \rightarrow \alpha + n + \mu + 17.6\ \text{MeV} \tag{19.11}$$

where, for the sake of brevity, only the nuclei of the participating atoms and molecules are shown explicitly.

In the first stage of the cycle, muons produced by pion decay (Equation 19.2) may be captured in collisions with deuterium and tritium, with the muon replacing an atomic electron. Since the muon binds more tightly to the nucleus than the electron (Bohr radius smaller by a factor of m_e/m_μ), the resulting muonic atoms $d\mu$ and $t\mu$ are small. Muonic atom formation takes place rapidly in a time $\sim 10^{-6}\tau_\mu$. There is also an isotopic exchange reaction

$$d\mu + t \rightarrow t\mu + d \tag{19.12}$$

which is energetically favoured and proceeds quickly in a time $\sim 10^{-3}\tau_\mu$, to form even more $t\mu$ atoms. In effect, only $t\mu$ atoms participate further in the cycle, as indicated in Equation 19.11.

The small muonic atom then penetrates the electronic shell of a deuterium or a tritium molecule in a time $\sim 10^{-2}\tau_\mu$, to form a muomolecular complex, for example,

$$t\mu + D_2 \rightarrow \left[(dt\mu)\, d2e\right]^* \tag{19.13}$$

with a nucleus consisting of a mesomolecular ion $dt\mu$ (see Figure 19.2). Fusion between the closely bound d and t nuclei then takes place rapidly in a time $\sim 10^{-6}\tau_\mu$, producing an energy 17.6 MeV. In most cases, the muon is regenerated and recycled to mediate further fusion events. In about 0.5%–0.6% of cases, however, it "sticks" to the α-particle (the nucleus of helium),

$$dt\mu \rightarrow \alpha\mu + n \tag{19.14}$$

and is lost for a further cycle.

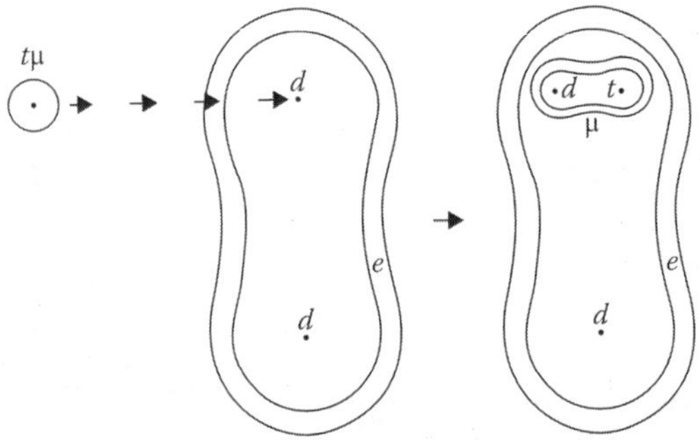

Figure 19.2 The resonant muomolecular formation process (Equation 19.13), in which d and t are bound by a muon within a deuterium molecule, overcoming the Coulomb force of repulsion, and allowing fusion to proceed rapidly.

The process (Equation 19.13), which is the dominant (and slowest) of the reaction channels, actually incorporates a resonant mechanism to the extent that the muonic atom kinetic energy $u_{t\mu} \equiv \frac{1}{2}m_{t\mu}v^2 \sim 1/20$ eV of the $t\mu$ atom plus the binding energy $E_B \approx 0.85$ eV of the $dt\mu$ ion, must match one of the excited energy levels of muomolecular complex molecule. Since E_B is fixed, the rate of the reaction (Equation 19.13), and hence the overall efficiency of the μCF cycle, depend critically on the distribution of energies of the recoiling $t\mu$ atoms.

This is only a very brief outline of the basic muon-catalysed fusion cycle. The atomic and molecular physics is well understood and free from any controversy, unlike the supposed electrochemically produced cold fusion process, reported in the late 1980s, which has been largely dismissed in the meantime. If the integrity of μCF is unchallenged, and the physics largely understood, why then is it not considered a viable alternative to the more expensive and sometimes problematic hot fusion program?

19.4.3 Factors limiting the efficiency of μCF

In the quest towards scientific break even (i.e., energy out = energy in) the following fundamental limitations must be addressed:

- The finite life-time ($\tau_\mu = 2.2$ μs) of the muon;
- The sticking loss of the muon to the alpha particle (Equation 19.14) which prevents it from playing any further role in the cycle; and
- Cost of muon production through pion decay $\pi^- \rightarrow \mu^- + \bar{\nu}_\mu$.

If one assumes the latter to correspond to an equivalent of 5 GeV (estimated from the initial beam energy, pion production probability, and muon yield, [180]) it can be seen that each muon has to produce at least 300 reactions of the type (Equation 19.11) before it is lost, by either decay or sticking, in order to achieve scientific break-even. At the time of writing, the best that has been achieved is around half [179,181].

Further progress depends on understanding how the macroscopic reaction rates corresponding to the microscopic, atomic scale processes can be optimized through adjustment of experimental parameters. This link is provided by the kinetic theory and fluid analysis introduced in this book.

19.4.4 Kinetic and fluid analysis

19.4.4.1 Factors influencing the muomolecular formation rate

So far in this book, we have analyzed reactive collisions of the type *charged particle + neutral → products* from the point of view of the their effect on the original charged particles only, without being concerned about the products (see Chapter 9). However, for a chain of reactions such as Equation 19.11, it is necessary to consider the products of each process, and their effect on the next segment of the process, that is,

$$\text{Charged particle + neutral} \rightarrow \text{products 1} \rightarrow \text{products 2} \rightarrow \ldots$$

In other words, it is essential to consider the cycle, as a whole, by solving coupled kinetic or fluid equations for all the species in the cycle, as outlined in Ref. [182]. This is a major task, beyond the scope of this book. The discussion which follows therefore focuses on the first two parts of the cycle only.

The "bottleneck" comes at the second stage (muomolecular formation—see Figure 19.2) of the cycle which, because of its resonant nature depends sensitively upon the kinetic energy $u_{t\mu}$ of the incoming $t\mu$ atom produced in the first stage. This, in turn, is controlled by the mole fractions x_t and $x_d = 1 - x_t$ of tritium and deuterium, respectively, as well as the post-collision recoil energy $u^+_{t\mu}$ (rec) of the $t\mu$ atom. Ultimately, it is the mean energy

$$\langle\epsilon\rangle = \left(m_0 \left\langle u_{t\mu}\right\rangle + m_{t\mu}\frac{3}{2}k_B T_0\right) / \left(m_{t\mu} + m_0\right) \tag{19.15}$$

in the CM of the colliding $t\mu$ atom and deuterium or tritium molecule of mass m_0 which determines the muomolecular formation rate. Typically, $\langle\epsilon\rangle$ is a few tenths of an eV. This may be controlled experimentally through the medium temperature T_0 and tritium fraction x_t both explicitly and implicitly through the mean energy $\left\langle u_{t\mu}\right\rangle$ of the $t\mu$ atom.

Using energy balance equations for each of the species, in the fluid model of Ref. [182], it can be shown that in the steady state and the average energy of a muonic atom is given by

$$\left\langle u_{t\mu}\right\rangle \approx \left\{\frac{3}{2}k_B T_0 + r\left[x_t\, u^{+}_{t\mu}(\text{rec}) + x_d\, u^{+}_{t\mu}(\text{exch})\right]\right\} / (1+r) \tag{19.16}$$

where $u^{+}_{t\mu}(\text{exch})$ is the mean energy of $t\mu$ atom produced in the isotopic exchange reaction (Equation 19.12), and r is the ratio of the muomolecular formation rate to the non-reactive (i.e., elastic and inelastic) collision rates. Note that since r depends on $\left\langle u_{t\mu}\right\rangle$, by virtue of the energy dependence of the cross sections, the solution of Equation 19.16 is not straightforward. Nevertheless, it provides a means of predicting the way in which $\left\langle u_{t\mu}\right\rangle$, $\langle\epsilon\rangle$, and hence the muomolecular formation rate depend on T_0 and x_t. By assuming simplified model cross sections and $u^{\dagger}_{t\mu}(\text{rec}) = 0.1$ eV and $u^{\dagger}_{t\mu}(\text{exch}) = 1.4$ eV, the theoretical prediction [182] is that the muomolecular rate reaches a maximum with $x_t = 0.6$ and $T_0 \approx 1200$ K. Temperatures of this magnitude may not be practicable, so other methods of controlling $\left\langle u_{t\mu}\right\rangle$ may need to be tried. Application of an electric field offers an alternate way of increasing $\left\langle u_{t\mu}\right\rangle$ and hence $\left\langle \epsilon_{CM}\right\rangle$, without the need to go to such high temperatures.

19.4.4.2 Influence of an electric field

In the experiments discussed in Chapter 1, a very low density, low energy ($\lesssim 1$ eV) swarm of ions or electrons is confined in a vessel containing a neutral gas of number density n_0, and is subject to an external electric field E. The experiment yields transport quantities as functions of the gas temperature T_0 and/or the reduced field E/n_0, data which may be unfolded via Boltzmann's equation to give scattering cross sections as a function of the CM energy or, if desired, interaction potentials as a function of interparticle separation. By varying E/n_0, the experiment effectively scans charged particle–neutral interactions over a range of CM energies, without the need to vary the gas temperature T_0.

Similarly, μCF involves a swarm of low density particles, whose mutual interaction is negligible in comparison with their interaction with the constituent molecules of the medium. Hitherto, the temperature T_0 has been varied to optimise the fusion yield, but it is reasonable to ask whether application of an electric field might have a similar effect, while keeping T_0 fixed. This question was addressed in Ref. [183] by solving Boltzmann's equation for muons in deuterium gas, with aliasing of the (μ, D_2) interaction based (in part) on (e, D_2) cross sections. The muon capture cross section was modelled as a constant $\sigma^* = 3.5 \times 10^{-20} m^2$ up to 25 eV, and zero beyond that.

The production of muonic atoms is found to vary with E/n_0 for fixed gas temperature $T_0 = 293$ K according to Figure 19.3 This shows that the muonic atom formation rate is considerably enhanced by the field, peaking at $E/n_0 \approx 850$ Td, with a value some 20 times larger than in the absence of a

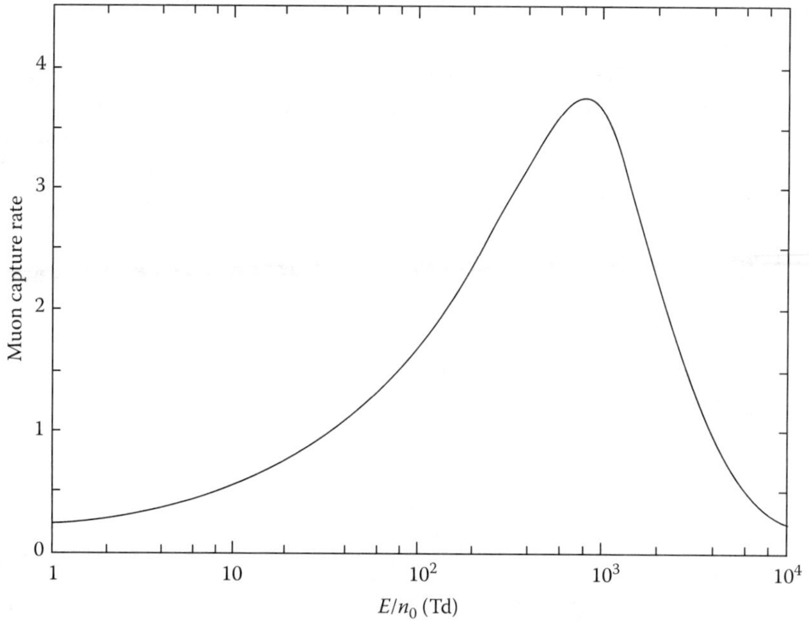

Figure 19.3 Muon capture rate (vertical axis, arbitrary units) in deuterium at room temperature as a function of reduced electric field [183].

field [183]. Application of a field can also influence the meso-molecular formation rate, by controlling the CM energy (Equation 19.15) via the mean energy and recoil energy of the muonic atom after capture. Whether such "field-tuning" of the "bottleneck" process shown in Figure 19.2 is a practicable proposition remains to be seen.

19.4.5 Observations and challenges for μCF

We have given an overview of the muon-catalyzed fusion cycle and compared it with a swarm experiment, using the kinetic theory and fluid modelling discussed in this book. The focus has been on a gaseous medium, but the calculations for the light muon component, could also be extended to liquids using structure-modified cross sections, as explained in Chapter 6. However, given the observed density dependence for μCF in liquids, it is clearly important to allow for structure effects for the heavier components in the cycle as well, particularly in the muomolecular formation process of Figure 19.2. There is clearly incentive to develop a Boltzmann collision operator for particles of arbitrary mass in a liquid, by using the formalism of Chapter 6.

Given the present impasse in that energy break-even seems as far off as ever, radical new ideas are required to improve the efficiency of μCF.

We have pointed out that an external electric field can influence the properties of a muon swarm and their reactive derivatives, and thus, enhance the efficiency of the cycle. However, it remains to be seen if it is a practicable proposition to field-tune the μCF cycle at the high fields required. Since n_0 is typically around liquid hydrogen densities, very large fields, $E \sim 10^9$ V/m, would be required to achieve the desired outcome. This is readily achievable in small regions of intense laser focus.

Finally, we leave the question open as to whether a field might also help solve the problem of muon sticking, by somehow enhancing the dissociative reaction $\alpha\mu \rightarrow \alpha + \mu$.

CHAPTER 20

Concluding Remarks

20.1 Summary

This book deals with the fundamentals of transport theory for classical, non-relativistic dilute particles in neutral gases and condensed matter, using:

- Kinetic theory, in which properties of the particles are represented by their phase-space distribution function $f(\mathbf{r}, \mathbf{v}, t)$, which is the solution of a kinetic equation; and
- Fluid modelling, in which the quantities of physical interest are found by solving moment equations obtained by integrating the kinetic equation over $\mathbf{v}$.

These two approaches are complementary, providing, respectively, rigorous, accurate values of transport properties on the one hand and a semi-quantitative, physical picture on the other.

The starting point is the classical 1872 Boltzmann equation, generalized to include inelastic processes through the Wang Chang–Uhlenbeck–de Boer (WUD) collision operator. Nonconservative loss processes, such as ion–atom reactions, positron annihilation, and electron attachment, are subsequently included. These kinetic equations and their fluid counterparts provide the platform for a unified treatment of all types of particles, that is ions, electrons, positrons, muons, and so on, regardless of mass and charge.

Other kinetic equations follow as special cases:

- The Fokker–Planck equation for particles undergoing small angle (e.g., Coulomb) scattering;
- The hierarchy of kinetic equations for the spherical components of f for light particles in both gases and soft-condensed matter;
- The Rayleigh model for heavy particles in gases;
- The Bhatnagar–Gross–Krook (BGK) relaxation time model for ions in their parent gas undergoing charge exchange collisions, and its counterpart for charge carriers in amorphous semiconductors; and
- Electron-impact ionizing collisions.

Numerical and analytic procedures for solving the respective kinetic equations have been outlined. Cross sections $\sigma(g, \chi)$ are input data, along with the structure factor $S(K)$ or the relaxation function $\phi(\tau)$ of the material, to account for coherent scattering and trapping in localized states, respectively. Details of numerical calculation of classical ion–atom cross sections have been given, but otherwise $\sigma(g, \chi)$, $S(K)$, and $\phi(\tau)$ are required from other sources.

We have considered a diverse range of examples, including

- Hot atom chemistry,
- Negative differential conductivity,
- Diffusion cooling and heating of electrons in a gas in finite geometry,
- Analysis of swarm experiments in gases,
- Electrical conduction in organic semiconductors,
- Periodic structures in the Franck–Hertz experiment,
- Positron emission tomography (PET), and
- Muon-catalyzed fusion.

However, this list is not meant to be regarded in any way as exhaustive. There are many other situations where the kinetic theory outlined in this book can be applied, either directly or with minimal modifications, or, with some lateral thinking, to problems which at first glance are unrelated, such as dispersion of a "passive additive" in a turbulent boundary layer [184]. The common picture is one of dilute, non-interacting particles (or in the latter example, macroscopic "fluid particles") in a medium with specified properties.

A number of important questions have emerged which been left unanswered. Moreover, there are several long-standing issues which we have not addressed, and which should be mentioned before drawing to a close. These points are discussed briefly below.

20.2 Further Challenges

20.2.1 Heavy particles in soft matter

In the most important "leg" of the muon-catalyzed fusion (μCF) cycle discussed in Chapter 19, a muonic atom $t\mu$ is incident on a deuterium molecule, and results in the formation of a complex muonium molecule (see Figure 19.2), if the energy in the centre-of-mass of the colliding partners falls within a certain range. The formation rate can be calculated from the distribution function $f_{t\mu}$ of the $t\mu$ atoms, which, in turn, is found as the solution of an appropriate kinetic equation. For a gaseous medium, treating $(t\mu)$ as a point particle with no internal structure (see, however, the remarks below), the collision term $\left(\frac{\partial f}{\partial t}\right)_{\text{col}}$ is prescribed as in Chapter 5.

However, for μCF experiments conducted in a liquid medium at low temperatures [178], the structure of the medium comes into play, and the $t\mu - D_2$ reaction rate may be affected [181]. Since the expression for $\left(\frac{\partial f}{\partial t}\right)_{\text{col}}$, obtained in Chapter 6, is valid only for light particles ($m \ll m_0$) in soft matter, there is strong incentive to find a more general form of $\left(\frac{\partial f}{\partial t}\right)_{\text{col}}$ for heavier particles ($m \sim m_0$).

20.2.2 Beyond point particles

This book has dealt with transport processes associated with structureless point particles in

- Gases, using the Boltzmann and Wang-Chang et al. kinetic equations (Chapters 3 and 5) and
- Soft-condensed matter and amorphous materials, using the kinetic equations developed in Chapter 6.

The extension to particles with internal states (e.g., molecular ions) is the logical next step:

- For a *gaseous medium*, we could use a generalization of the semi-classical Wang-Chang et al. equation or adopt the quantum Waldmann–Snider kinetic equation, or alternatively, follow Viehland et al. [32,185] and employ a purely classical kinetic equation which treats closely spaced rotational states as continuous, rather than quantized; but
- For a *condensed matter*, a rigorous kinetic theory has yet to be developed.

Thus, for example, we can analyze positrons in the soft matter medium of a PET environment up to the point of positronium *Ps* formation (Chapter 17), but not the subsequent motion of the *Ps* atom, given that scattering may excite its internal states. Until an appropriate kinetic equation is available, the energy and momentum of the *Ps* atom cannot be accurately estimated at the point of (e^+, e^-) annihilation, and hence the all-important question of non-collinearity of the emitted gamma rays cannot be addressed.

20.2.3 Relativistic kinetic theory

Another aspect of the PET analysis in Chapter 17 requiring attention is that the positrons have been treated non-relativistically, even though near the source the positron kinetic energy may be comparable with its rest mass mc^2. In relativistic kinetic theory, the particle distribution function

$f = f(\mathbf{r}, \mathbf{p}, t)$ in phase space $(\mathbf{r}, \mathbf{p})$ satisfies the relativistic kinetic equation,

$$(\partial_t + \mathbf{v} \cdot \nabla + \partial_{\mathbf{p}} \cdot \mathbf{F}) f = \left(\frac{\partial f}{\partial t} \right)_{\text{col}}, \tag{20.1}$$

where $\mathbf{v} = \mathbf{p}/(\gamma m)$, m is the particle rest mass, $\gamma = \sqrt{1 + (p/mc)^2}$, c is the speed of light, and $\mathbf{F} = e\,[\mathbf{E} + \mathbf{v} \times \mathbf{B}]$ is the force due to electric and magnetic fields $\mathbf{E}$ and $\mathbf{B}$. Relativistic kinetic theory has been extensively developed in gases over the years, even using a BGK representation of the collision term [186], but has not, as far as we are aware, been applied to positrons in a soft matter environment. The challenge is to develop a suitably modified form of $\left(\frac{\partial f}{\partial t} \right)_{\text{col}}$ to replace the expression given in Chapter 17, and solve Equation 20.1 accordingly.

20.2.4 Partially ionized plasmas

This book has also focused upon calculating the transport properties of low density ions and electrons in gases subject to externally prescribed fields. The analysis applies to either single species swarms, or to both the charged species of a very weakly ionized gas. In the cases considered, the Debye length λ_D is large compared with any relevant macroscopic dimension, and consequently, particles move essentially freely, unrestrained by any internally generated space-charge fields, or interaction with other charged particles. The distribution function f_0 of the neutral gas is assumed to remain a Maxwellian, and the kinetic equations for the respective charged species are both linear. They can be solved separately using the methods outlined in Chapter 12. This is represented schematically by the upper and lower branches of Figure 20.1.

As the degree of ionization, and hence particle density n increase, $\lambda_D \sim n^{-\frac{1}{2}}$ decreases, and the internal fields generated by the charged particles begin to influence particle motion in a self-consistent way and the collective behaviour (wave motion, etc.) characteristic of a plasma then becomes evident. The kinetic equations must then be solved for the respective particle distribution functions f simultaneously with Maxwell's equations, that is, the upper and lower branches of Figure 20.1 are now connected as shown. Since Maxwell's equations involve charge densities and currents given by integrals over f, the problem as a whole is non-linear.

As the degree of ionization is increased further, Coulomb collisions between charged particles become significant, and consequently, non-linear Fokker–Planck collision operators (see Chapters 3 and 5) must be included, as represented by the connecting box in the diagram. In addition, the equilibrium state of the neutral gas becomes perturbed, and f_0 must be calculated from yet another kinetic equation, and not simply assumed to be a Maxwellian. The problem, as a whole, is now strongly non-linear,

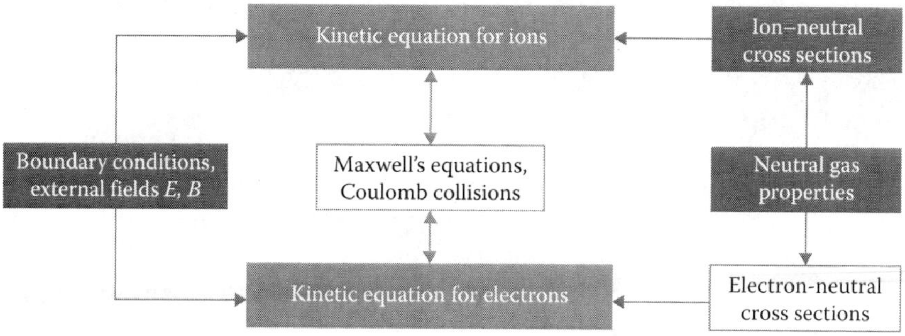

Figure 20.1 Schematic representation of the procedure for finding the transport properties of a partially ionized gas. The upper and lower branches correspond to solutions of the Boltzmann kinetic equation for equations for ions and electrons, respectively, which are independent in the low density swarm limit. At higher densities, they become coupled by Maxwell's equations and the influence of mutual Coulomb interaction between the charged species must also be accounted for.

making a rigorous kinetic treatment of low temperature, partially ionized plasma a formidable task.

Rigorous, accurate methods of *separate* solution of both the ion and electron Boltzmann kinetic equations have long been available (see Chapter 12) and can readily be assimilated into any plasma model. Further progress should then be primarily dependent on finding an efficient numerical procedure for dealing with the complexity arising from the coupling shown in Figure 20.1.

Fluid modelling (Chapters 7 through 10) offers a computationally economic alternative to kinetic analysis, for either one or both of the branches shown in the figure. However, since plasmas are generally in a highly non-equilibrium, non-hydrodynamic state, correspondingly more sophisticated, physically tenable fluid models are required. An accurate fluid model for electrons is now available [85] but not so for ions, which are often still modelled simply through a diffusion equation.

20.3 Unresolved Issues

20.3.1 The (e, H_2) controversy

The controversy surrounding the (e, H_2) vibrational (including ro-vibrational) cross section, which has prevailed for some 40 years, and defied every attempt to resolve it, is something which demands attention in the future. Thus, there is a large discrepancy, as much as 60%, between values of the cross section obtained from swarm experiments [187,188], quantum mechanical theory [62,189] and beam experiments [190,191]. In contrast, the situation for rotational excitations is far more satisfactory, with good agreement between theory and swarm-derived cross sections. The (e, H_2) controversy has been reviewed over the years [192,193] and the discrepancy continually highlighted, but to no avail.

Beam and swarm experiments involve single and multiple scattering processes, respectively, and offer independent means of investigating electron–molecule interaction cross sections. Cross sections are determined directly from a beam experiment, but only indirectly from swarm experiments, after inversion of the measured transport data using kinetic theory. In the case of H_2, information has been obtained from beam experiments down to the threshold of the first vibrational excitation, near 0.5 eV, while swarm experiments probe to even lower energies below the threshold energy for rotational excitation.

Quite independently, *ab initio* quantum mechanical theoretical calculations have furnished (e, H_2) impact cross sections which tend to support the beam experimental results, at least over a range of energies. For other diatomic gases like N_2 [194], however, the situation is not at all clear cut. This discrepancy indicates that there is something wrong with either:

- Swarm experiments;
- Beam experiments;
- Quantum mechanical scattering theory for diatomic gases; or
- The kinetic equation which has been used to "unfold" the swarm experiments.

The puzzle still remains.

20.3.2 Striations

A significant motivation for the late nineteenth and early twentieth century landmark investigations of electrical discharges [8], which laid the foundations of modern physics is due to Abria's observation, in 1843 [5], of the enigmatic phenomenon of striations, alternating bright and dark regions in the positive column of a gaseous discharge. However, in spite of much effort in the intervening period [6], questions still linger about the basic physics of striations. Indeed, the observation made over 60 years ago that "the mechanism of their (the striations) appearance or absence is still obscure" [195] still carries some weight.

Unlike the periodic structures associated with an electron swarm in the Franck–Hertz experiment, which have only recently become better understood, striations are a more complex plasma phenomenon, and are intimately associated with an internally generated, self-consistent space-charge field. There is a three-way synergism between non-locality, resonances with Franck–Hertz oscillations (see Chapter 16), and plasma screening effects. An understanding of the phenomenon can only come from the solution of both the ion and electron kinetic equations, that is, both branches on Figure 20.1 in conjunction with Poisson's equation. This is the challenge for plasma modellers.

To conclude, it seems something of an irony that the outcomes of the "golden era" of drift tube experiments of over 100 years ago continue to resonate well into the twenty-first century, while progress in understanding the phenomena associated with the original experiments themselves has been much slower.

Part V

Exercises and Appendices

Exercises

PROBLEM 1
Relaxation time model

Suppose that scattering of particles of mass m due to collisions with a medium of temperature T_0 takes place as shown in Figure 2.6, and according to the following model:

(a) The rate of scattering *out* of the phase-space element at time t is proportional to the distribution particle function $f(\mathbf{v}, t)$.

(b) The rate of scattering back *into* the element after collisions with the medium is proportional to the instantaneous equilibrium distribution function, which we write as $f_0(\mathbf{v}, t) = n(t)w(\alpha, v)$, where $w(\alpha, v) = \left(\frac{\alpha^2}{2\pi}\right)^{3/2} \exp\left(-\frac{1}{2}\,\alpha^2 v^2\right)$ and $\alpha \equiv \sqrt{m/kT_0}$.

If collisions conserve particle number, show that the collision term must be of the form

$$\left(\frac{\partial f}{\partial t}\right)_{\text{col}} = -\nu\left[f(\mathbf{v}, t) - n(t)w(\alpha, v)\right],$$

where ν is a constant with dimensions of frequency. This collision term is well known in the gas, plasma, and semiconductor literature and is sometimes called a "relaxation time" collision model (see also Chapter 14).

PROBLEM 2
Solution of the model kinetic equation

Suppose that the velocity distribution function $f = f(\mathbf{v}, t)$ of a system of uniformly distributed particles is governed by the model kinetic equation:

$$\frac{\partial f}{\partial t} = -\nu\left[f - n(t)\;w(\alpha, v)\right]$$

(a) Show that n is constant.

(b) Solve the kinetic equation for $f(\mathbf{v}, t)$, for any given initial distribution function $f(\mathbf{v}, 0)$.

(c) If $f(\mathbf{v}, 0) = n\ w\left(\alpha', |\mathbf{v} - \mathbf{v}_d|\right)$, where $\alpha' \equiv \sqrt{m/k_B T'}$, and T' and $\mathbf{v}_d$ are the initial temperature and average velocity of the particles, respectively, find the average velocity $\langle \mathbf{v} \rangle$ and average energy $\varepsilon = \frac{1}{2} m \langle v^2 \rangle$ at any subsequent time t.

PROBLEM 3
Model kinetic equation with trapping

Returning to Figure 2.6, suppose now that collisions are not instantaneous, and that particles are effectively trapped for a finite time τ before being released.

(a) Generalize the result of Problem 1 to formulate a collision term $\left(\frac{\partial f}{\partial t}\right)_{\text{col}}$ for this situation.

(b) Hence write down the kinetic equation in the spatially uniform, field free case. Is particle number conserved?

PROBLEM 4
External versus space-charge fields

Consider a partially ionized gas comprised of ions and electrons, each having the same temperature T, in a neutral gas of temperature T_0. The system is confined by two plane boundaries separated by a distance L. Initially, the system is electrically neutral, the ions and charged electrons being uniformly distributed throughout the gap, with equal number densities everywhere, that is, $n_i = n_e \equiv n$. Under these conditions, there is no net charge and no electric field at any point.

Suppose for some reason all the lighter, more mobile electrons suddenly move to the boundary at the right, leaving behind only positive charges in the gap, that is., $n_i = n$, $n_e \approx 0$. Since electrical neutrality no longer prevails, a restorative, "space-charge" electric field $\mathbf{E}$ develops, which restrains any further motion of the electrons, and simultaneously exerts a force on the ions to follow the electrons, as shown in Figure P.4.1. In the following, it is assumed that the properties of ions and electrons may vary only in a direction normal to the boundaries (the x-direction).

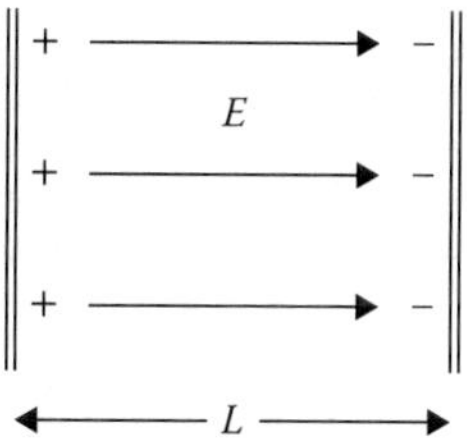

Figure P.4.1 An ionized gas confined between two conducting electrodes separated by a distance L suddenly loses all of its electron component to the right boundary, resulting in a space-charge electric field $\mathbf{E}_1$ which attempts to restore electrical neutrality.

(a) By solving Poisson's equation or using Gauss' theorem show that the restoring space-charge field at position x has magnitude

$$E = \frac{nex}{\epsilon_0} \tag{P.4.1}$$

where e is the magnitude of the electric charge and $\varepsilon_0 = 8.854 \times 10^{-12}$ F/m is the permittivity of free space, and hence estimate the potential energy of a typical ion remaining in the gap.

(b) Debye lengths:

(i) Using the fact that total energy (potential plus kinetic) must be conserved, and that the average thermal energy per particle is $\sim k_B T$, show that charge separation as represented in Figure P.4.1 is energetically possible only if the *Debye length*

$$\lambda_D \equiv \sqrt{\frac{\epsilon_0 k_B T}{ne^2}} \tag{P.4.2}$$

is larger than the dimension of the system, that is,

$$\lambda_D > L. \tag{P.4.3}$$

In this regime, ions and electrons diffuse freely and the restorative space charge field is negligible.

(ii) Show that the Debye length can be expressed as

$$\lambda_D = 69\sqrt{\frac{T}{n}} \quad \text{m}, \tag{P.4.4}$$

where T is in degrees Kelvin, and n is in m^{-3}.

(iii) Is the condition (Equation P.4.3) satisfied for charged particles in a drift tube a few cm in length, where $T \sim 10^3$K and $n \sim 10^6$ m^{-3}?

(c) Estimate the magnitude of the space-charge field in

(i) A drift tube experiment (see previous example); and

(ii) The wake of a laser wavefront moving through a plasma of density $n \sim 10^{24}$ m^{-3}, causing charge separation ~30 μm.

PROBLEM 5
Cross sections and interaction potentials

The procedure for calculating the differential cross section for a given interaction potential $V(r)$ is

- Calculate $\chi = \chi(b, g)$, or equivalently $b = b(g, \chi)$, from Equation 4.1; and
- Substitute in Equation 4.3 and calculate $\sigma(g, \chi)$.

(a) Using *dimensional analysis* for an inverse nth power law interaction potential $V(r) = K\,r^{-n}$, show that χ depends on b and g solely through the combination of parameters $b\left(\frac{\frac{1}{2}\mu g^2}{K}\right)^{\frac{1}{n}}$, and that consequently, b must be of the form

$$b = \left(\frac{\frac{1}{2}\mu g^2}{K}\right)^{-\frac{1}{n}} f(\chi),$$

where $f(\chi)$ is some function of scattering angle. (N.B. Do *not* try to evaluate any integrals!)

(b) Hence show that the differential cross section must be of the functional form

$$\sigma(g, \chi) = g^{-\frac{4}{n}} F(\chi),$$

where $F(\chi)$ is some other function of scattering angle.

(c) Hence show that all partial cross sections (Equation 3.14) must be of the form

$$\sigma^{(l)}(g) = C^{(l)} g^{-\frac{4}{n}}, \tag{P.5.1}$$

where $C^{(l)}$ is a constant.

(d) Hence find the g-dependence of the momentum transfer collision frequency for the following cases:

(i) Coulomb potential $n = 1$;
(ii) Point charge–atom interaction $n = 4$; and
(iii) Rigid sphere interaction $n \to \infty$.

PROBLEM 6
Hard sphere collisions

Consider a collision between rigid spheres, as shown in Figure 4.1. Show that the differential cross section is given by

$$\sigma(g, \chi) = \frac{1}{4}a^2$$

and hence that the total cross section is

$$\sigma^{(0)}(g) = \pi a^2.$$

PROBLEM 7
Coulomb collisions

(a) *Given* that for two charges q_1, q_2 interacting through the Coulomb potential $V(r) = \xi / r$, where $\xi = \frac{q_1 q_2}{4\pi\varepsilon_0}$, Equation 4.1 gives

$$b = \frac{\xi}{\mu g^2} \cot(\chi/2)$$

obtain the Rutherford differential section

$$\sigma_R(g, \chi) = \left(\frac{\xi}{2\mu g^2}\right)^2 \frac{1}{\sin^4(\chi/2)}.$$

(b) Show that the integral defining the corresponding momentum transfer cross section,

$$\sigma_m = 2\pi \int_0^{\pi} (1 - \cos\chi)\ \sigma_R(g, \chi)\ \sin\chi\ d\chi$$

diverges at the lower limit. Physically speaking, why does this problem arise?

(c) The standard procedure in plasma physics for removing the singularity is to "cut off" the integral at the lower limit, by replacing zero with a small angle $\chi_{\min}$, defined by setting the impact parameter equal to the Debye screening distance:

$$\lambda_D = \frac{\xi}{\mu g^2} \cot\left(\chi_{\min}/2\right).$$

Comment on this from a physical perspective.

(d) Would you expect similar divergence problems for charged particle–neutral interactions?

(e) What are the implications at high energy of the fact that the Rutherford cross section $\sigma_R(g,\chi) \sim g^{-4}$?

PROBLEM 8
Maxwellian velocity distribution

In the equilibrium state, $f = f^{(\text{equil})}$, $\ln f^{(\text{equil})}$ must be a linear combination of the summational invariants Equation 3.31. Fill in the algebraic steps to show that the equilibrium distribution is the Maxwellian velocity distribution (Equation 3.32).

PROBLEM 9
Some integrals

(a) Prove the following general identities, using the spherical polars coordinates of Figure 2.3:

$$\int_0^{2\pi} d\varphi \int_0^{\pi} d\theta \sin\theta = 4\pi \qquad \int_0^{2\pi} d\varphi \int_0^{\pi} d\theta \sin\theta\, v_i = 0$$

$$\int_0^{2\pi} d\varphi \int_0^{\pi} d\theta\, \sin\theta\, v_i v_j = \frac{4\pi}{3} v^2 \delta_{ij} \qquad \int_0^{2\pi} d\varphi \int_0^{\pi} d\theta \sin\theta\, v_i v_j v_k = 0$$

$$(i, j, k) = (x, y, z)$$

(b) Show that the energy flux vector is given by

$$\mathbf{J}_E = n\left\langle \frac{1}{2} nmv^2 \mathbf{v} \right\rangle = \frac{4\pi}{3} \int_0^{\infty} dv\, \frac{1}{2} mv^5 \mathbf{f}^{(1)}(v).$$

PROBLEM 10
Lorentz gas calculations

This and the following problems deal with the two-term representation of f in a steady state, spatially uniform electron swarm in a gas subject to an electric field only. We first consider the case of arbitrary $\nu_m(v)$, *before moving on to special cases.*

(a) Solve the coupled differential equations of the two-term approximation, as described by Equations 5.47 and 5.48, with $B=0$ to obtain the *Davydov energy distribution function,*

$$f^{(0)}(v)=A\exp\left\{-\frac{3m}{m_0}\int_0^v \frac{v\,dv}{\left(\frac{eE}{m\nu_m(v)}\right)^2+\frac{3k_BT_0}{m_0}}\right\} \tag{P.10.1}$$

where A is a constant of integration, determined from normalization.

(b) Hence find an expression for the vector part of the distribution function $\mathbf{f}^{(1)}(v)$.

(c) Evaluate Equation P.10.1 in the limits $E\to 0$, and $T_0\to 0$, respectively.

(d) Prove that for *weak fields*, the average electron velocity, or *drift velocity* $\mathbf{v}_d=\langle\mathbf{v}\rangle$, is simply proportional to E.

PROBLEM 11
Inverse fourth power law interaction potential (ν_m = constant)

(a) Find expressions for

(i) $f^{(0)}(v)$ and $\mathbf{f}^{(1)}(v)$; and

(ii) the drift velocity $\mathbf{v}_d$ and mean energy $\varepsilon=\frac{1}{2}m\langle v^2\rangle$.

(b) Prove that the mean electron energy is related to the average velocity by

$$\varepsilon=\frac{3}{2}k_BT_0+\frac{1}{2}m_0v_d^2 \tag{P.11.1}$$

and that for a cold gas, this implies $\frac{v_{\text{RMS}}}{v_d}=\left(\frac{m_0}{m}\right)^{\frac{1}{2}}\gg 1$ where $v_{\text{RMS}}=\sqrt{\langle v^2\rangle}$. Explain the physical significance of this result. Is it consistent with the assumptions underlying the two-term approximation?

PROBLEM 12

Constant cross section (hard sphere) interaction

In the cold gas limit $T_0 \to 0$:

(a) Find $f^{(0)}(v)$ and $\mathbf{f}^{(1)}(v)$; and

(b) Show that $v_d \sim E^{\frac{1}{2}}$.

PROBLEM 13

Inverse *n*th power law interaction potential

In the cold gas limit $T_0 \to 0$:

(a) Show that $v_d \sim E^{p}$, and find p in terms of n.

(b) Investigate the range of values of n for which a steady state is possible.

PROBLEM 14

Li^+ and K^+ ions in helium

(a) In Table P.14.1, the *reduced* mobility (units $cm^2V^{-1}s^{-1}$) versus reduced field (units Td) data were obtained experimentally for lithium ions in helium gas at 294 K.
Using the fluid theory of Chapter 7, obtain the momentum transfer cross section for (Li^+, He) over as wide range of energies as possible.

(b) Repeat for (K^+, He) for which the corresponding data in shown in Table P.14.2.

Table P.14.1 Reduced Mobility of Li^+ in He as a Function of Reduced Field at 294 K

E/n_0	3.0	4.0	5.0	8.0	8.0	12.0	15.0	18.0
$\mathcal{K}$	22.8	22.8	22.9	23.4	23.4	24.3	25.3	26.7
E/n_0	20.0	25.0	30.0	35.0	40.0	50.0	60.0	70.0
$\mathcal{K}$	27.6	29.6	31.2	32.1	32.5	32.3	31.7	31.4

Table P.14.2 Reduced Mobility of K^+ in He as a Function of Reduced Field at 294 K

E/n_0	3.0	7.0	10.0	15.0	20.0	25.0	30.0
$\mathcal{K}$	21.17	21.22	21.27	21.33	21.30	21.14	20.86
E/n_0	40.0	50.0	70.0	100.0	200.0	300.0	
$\mathcal{K}$	20.07	19.20	17.62	15.83	12.69	11.19	

(c) How could the range of energies probed be extended experimentally?

(d) Can you obtain *any* information at all about the nature of the interaction potential itself *directly* from these data?

PROBLEM 15
Application of generalized Einstein relations

(a) Table P.15.1 shows the transport data for electrons in He gas at 293 K. Use the generalized Einstein relations to estimate $D_\parallel/D_\perp$ from the drift velocity data alone, and compare with the last row of the above data. Comment on your results.

(b) Repeat the above exercise for electrons in water vapor at 300 K shown in Table P.15.2.

Table P.15.1 Electron Transport Coefficients in He as a Function of Reduced Field at 293 K

E/n_0 (Td)	0.003	0.006	0.012	0.024	0.046	0.121
v_d (10^3 m s^{-1})	0.0764	0.150	0.291	0.535	0.865	1.62
$D_\parallel/D_\perp$	0.995	0.981	0.938	0.835	0.723	0.580
E/n_0 (Td)	0.607	1.214	2.43	3.64	6.07	
v_d (10^3 m s^{-1})	3.79	5.34	7.59	9.49	13.0	
$D_\parallel/D_\perp$	0.479	0.471	0.493	0.529	0.612	

Table P.15.2 Electron Transport Coefficients in Water Vapour as a Function of Reduced Field at 300 K

E/n_0 (Td)	10	20	30	40	45	47.5	50
v_d (10^3 m s^{-1})	2.22	4.52	7.14	10.4	13.1	15.6	18.8
$D_{\|}/D_{\perp}$	1.01	1.06	1.2	3.24	5.62	6.18	5.24
E/n_0 (Td)	55	60	70	80	100	120	140
v_d (10^3 m s^{-1})	23.4	34.5	59.9	87.2	130	159	187
$D_{\|}/D_{\perp}$	6.36	4.5	3.08	1.81	1.07	0.979	0.904

PROBLEM 16
Electrons in a model gas

Consider electrons in two model gases defined as follows:

(a) $T_0 = 0$, $m_0 = 28$ amu, $\sigma_m(\epsilon) = 5.0 + 0.15\epsilon$ [Å^2], $\sigma_I = 0$

(b) As above, but with a "hat function" inelastic cross section:

$$\sigma_I = \begin{cases} 0 & \epsilon < 0.1 \text{ eV}, \\ 0.03 \text{ Å}^2 & 0.1 \text{ eV} < \epsilon < 3.0 \text{ eV}. \end{cases} \tag{P.16.1}$$

In each case, calculate and plot drift velocity v_d and mean energy ε of the electrons for 0.1 Td $\leq E/n_0 \leq 40$ Td. For case (b), also calculate and plot the inelastic collision term $\Omega(\varepsilon)$, and hence explain the nature of the v_d versus E/n_0 curve. (Hint: The calculations are greatly simplified if the mean energy ε is taken as the independent variable, not E/n_0.)

PROBLEM 17
Transport in crossed electric and magnetic fields, Tonks' theorem

The presence of a magnetic field $\mathbf{B}$ can be readily accounted for in the moment equations describing the steady, spatially uniform state by making the substitution $\mathbf{E} \to \mathbf{E} + \langle \mathbf{v} \rangle \times \mathbf{B}$ in the momentum balance equation, while the energy balance equation remains unaltered in form. If the electric and magnetic fields are crossed at right angles, the drift velocity is given by

$$\mathbf{v}_d = v_{\|}\hat{\mathbf{E}} + v_{\perp}\hat{\mathbf{E}} \times \hat{\mathbf{B}},$$

where

$$v_{\|} = \frac{KE}{1 + (KB)^2}, \quad v_{\perp} = KBv_{\|},$$

and $K = K(\varepsilon) \equiv q/\mu\nu_m(\varepsilon)$.

(a) Show that the magnitude of the drift velocity can be written as

$$v_d = KE_e,$$

where

$$E_e = E/\left[1+(KB)^2\right]^{\frac{1}{2}} = E\cos\varphi$$

is the *equivalent electric field*, and φ (the so-called Lorentz angle) is defined by

$$\tan\varphi = \frac{v_\perp}{v_\parallel} = KB.$$

(b) Hence prove *Tonks' theorem*, which states that the crossed field situation described above can be discussed in terms of the *equivalent electric field* situation only, that is,

$$\varepsilon(E,B) = \varepsilon(E_e,0)$$
$$v_d(E,B) = v_d(E_e,0)$$

Ensure the generality of your results by including inelastic collisions.

PROBLEM 18
Blanc's law for mixtures

Following Equation 2.12, the steady state, spatially uniform Boltzmann equation for charged particles in a mixture of two neutral gases may be written

$$\frac{q\mathbf{E}}{m}\cdot\frac{\partial f}{\partial \mathbf{v}} = \left(\frac{\partial f}{\partial t}\right)_{\text{col}}^{(1)} + \left(\frac{\partial f}{\partial t}\right)_{\text{col}}^{(2)},$$

where superscripts denote collision terms for charged particles with the respective neutral species. Only an electric field E is considered in this problem. The momentum balance Equation 7.29 can similarly be generalized to a gas mixture,

$$q\mathbf{E} = \left(\mu^{(1)}\nu_m^{(1)} + \mu^{(2)}\nu_m^{(2)}\right)\mathbf{v}_d,$$

where

$$\mu^{(s)} = \frac{mm_0^{(s)}}{m+m_0^{(s)}}$$

$$\nu_m^{(s)} = n_0^{(s)}\sqrt{\frac{2\varepsilon^{(s)}}{\mu^{(s)}}}\,\sigma_m^{(s)}$$

denote reduced masses and momentum transfer collision frequencies for $s = 1, 2$.

(a) Hence show that the mobility K of particles in a mixture of two neutral gases is given by *Blanc's law*

$$\frac{1}{K} = \frac{x^{(1)}}{K^{(1)}} + \frac{x^{(2)}}{K^{(2)}},$$

where $x^{(s)} = n_0^{(s)}/n_0$ is the mole fraction of neutral species s, $n_0 = n_0^{(1)} + n_0^{(2)}$ is the total number density of the gas mixture, and $K^{(s)}$ is the mobility coefficient in the pure gas of number density n_0.

(b) The collision frequencies of electrons in pure Ar (species 1) and pure Ne (species 2) of number density 3.5×10^{22} m^{-3} at room temperature are approximately 10^8 and 10^9 s^{-1}, respectively. Calculate the mobility coefficients for

(i) Pure Ar,
(ii) Pure Ne, and
(iii) A 60:40 mixture of Ar:Ne and Ne:Ar, respectively.

PROBLEM 19
The phenomenon of "runaway"

For "soft" interaction potentials, collision effects may not be sufficient to balance the momentum and energy imparted by the applied field, and the charged particles may "runaway" beyond a certain critical field $(E/n_0)_c$. A steady state is not then possible. The phenomenon is well known in plasma physics (the Coulomb potential is certainly soft!), but it also may occur in other cases.

Consider, for example, electrons in a gas of neutral particles with *permanent* dipole moments, for which the interaction potential is of the form $V(r) \sim \frac{1}{r^2}$, and the corresponding momentum transfer cross section can be found to be $\sigma_m(\epsilon) = p\epsilon^{-1}$, where p is a constant.

(a) Show that the mean energy is given by

$$\varepsilon = \frac{\frac{3}{2}k_B T_0}{1 - A\left(\frac{E}{n_0}\right)^2},$$

where

$$A = \frac{m_0}{4m}\left(\frac{e}{p}\right)^2$$

(b) If the gas is *water vapor*, for which $p = 30$ eV Å^2 (note the units), estimate the value of $(E/n_0)_c$. (N.B. In actual fact, inelastic processes take over and prevent runaway, but the transport properties of electrons in water vapor nevertheless show a very pronounced increase in the vicinity of $(E/n_0)_c$.

PROBLEM 20

Diffusion equation—solution for spherical geometry

N particles are created at $t = 0$ at the center of a spherical, gas-filled container of radius R, and diffuse free from any space-charge or external field, to perfectly absorbing walls. If the diffusion coefficient is D, find an expression for the density distribution $n(r, t)$, applying appropriate initial and boundary conditions, and assuming spherical symmetry. (Hint: Be sure to enforce the correct physical behavior at $r = 0$.)

PROBLEM 21

Diffusion equation—a cautionary example

Consider the case of an infinite plane source emitting charged particles at a steady rate into an infinite half-space filled by a uniform gas, with a uniform electric field directed normal to the plane. Properties vary only in this direction, taken to be the z-axis of a system of coordinates. If a steady state has been reached, then one might anticipate that the following one-dimensional diffusion equation might be sufficient to describe the system:

$$v_d \frac{\partial n}{\partial z} - D_{\parallel} \frac{\partial^2 n}{\partial z^2} = 0$$

Is the solution physically tenable? What do you conclude from this analysis?

Remember: The diffusion equation is valid if and only if:

- The hydrodynamic regime prevails, so that the space-time dependence of all quantities is carried by the number density $n(\mathbf{r}, t)$; and
- Gradients of n are weak.

Both these conditions must apply before the diffusion equation can be applied to a particular problem.

PROBLEM 22

Pulsed radiolysis drift tube experiment

(a) Consider first an idealized drift tube experiment in which N_0 electrons are created by a sharp pulse of x-rays at time t_0 in the plane

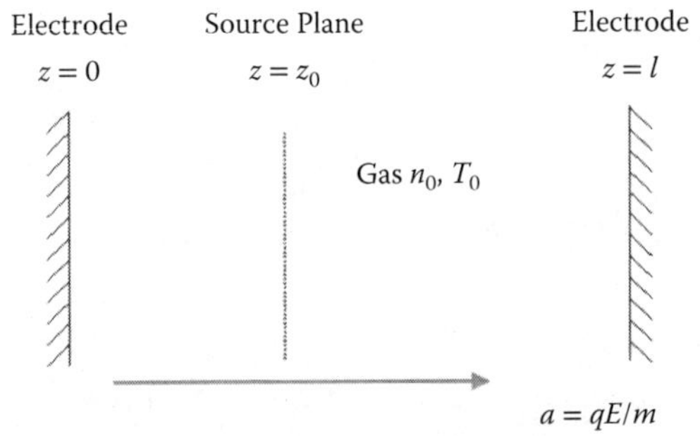

Figure P.22.1 An idealized drift tube experiment in plane-parallel geometry, in which electrons are created in the plane $z = z_0$ at time t_0.

$z = z_0$ as shown in Figure P.22.1. For finite geometry, the diffusion equation can be readily solved by the method of separation of variables. If the electrodes are assumed to be perfectly absorbing, the boundary conditions on $n(z, t)$ may be approximated by

$$n(0, t) = 0 = n(l, t) \tag{P.22.1}$$

while the initial condition at $t = t_0$ can be represented as

$$n(z, t_0) = N_0\, \delta(z - z_0) \tag{P.22.2}$$

(i) Show that the solution of the diffusion equation is

$$\begin{aligned} n(z, t) &= \frac{2N_0}{l} \sum_{j=1}^{\infty} \exp\left[\lambda(z - z_0) - \omega_j(t - t_0)\right] \sin(k_j z) \sin(k_j z_0) \\ &\equiv \frac{N_0}{l} \sum_{j=1}^{\infty} \exp\left[\lambda(z - z_0) - \omega_j(t - t_0)\right] \\ &\quad \times \left\{ \cos\left[k_j(z - z_0)\right] - \cos\left[k_j(z + z_0)\right] \right\} \end{aligned} \tag{P.22.3}$$

where

$$\omega_j = D_{\|}\left(\lambda^2 + k_j^2\right)$$
$$\lambda = \frac{v_d}{2D_{\|}}$$
$$k_j = \frac{j\pi}{l}.$$

(ii) In the limit as $l \to \infty$, show that the summation on the right hand side of Equation P.22.3 can be replaced by an integral, which can be evaluated to give

$$n(z,t) = \frac{N_0}{\sqrt{4\pi D_{\|}(t-t_0)}} \exp\left\{-\frac{[z-z_0-v_d(t-t_0)]^2}{4D_{\|}(t-t_0)}\right\}$$

the familiar travelling gaussian pulse.

(b) In the pulsed radiolysis experiment [196], the electrons are produced *uniformly* across the gap, and this can be accounted for by integrating (P.22.3) over all z_0. The experiment also actually measures the *total* number of electrons within the gap at any time t, which can be found by a further integration over z.

(i) Hence show that the total number of electrons in the gap at any time t, for a delta function pulse of ionizing radiation at some earlier time t_0, is

$$N(t-t_0) = \int_0^l \mathrm{d}z \int_0^l \mathrm{d}z_0\, n(z,t)$$

$$= \frac{4N_0}{l} \sum_{j=1}^{\infty} e^{-\omega_j(t-t_0)} \frac{k_j^2}{\left(\lambda^2+k_j^2\right)^2} \left[1+(-1)^{j+1}\cosh\lambda l\right].$$

(ii) In an actual experiment, the ionizing radiation begins at $t=0$ and continues for a *finite* length of time τ. Find, by appropriate integration over t_0, expressions for the total number of electrons $N_\tau(t)$ in the gap at any time t, both for $0 \le t \le \tau$ and $t > \tau$.

(N.B. For reasonably large gaps, $N_\tau(t)$ is *observed* experimentally to have a *linear* dependence on time, something which is not immediately obvious from the theory, but which can nevertheless be reproduced from the expressions obtained by using the *Poisson summation theorem*.)

PROBLEM 23
Reactive correction terms

Verify the right hand side of Equations 9.15 and 9.16 by carrying out the averaging processes assuming a Maxwellian distribution function for the swarm particles, that is,

$$f(\mathbf{v}) = n\, w(\alpha, |\mathbf{v} - \langle \mathbf{v} \rangle|).$$

PROBLEM 24
Bulk and flux drift velocities

Consider a uniform electron swarm in an attaching gas ($m_0 = 16$ amu) at temperature $T_0 = 293$ K, in a reduced field $E/n_0 = 0.4$ Td. Momentum transfer and attachment cross sections are given by

$$\sigma_m(\epsilon) = 5.0\,\epsilon^{-\frac{1}{2}}\,\text{Å}^2$$

$$\sigma_*(\epsilon) = b\,\epsilon^{\frac{1}{2}}\,\text{Å}^2,$$

respectively, where ϵ is in units of eV, and b is a dimensionless parameter. *Calculate* the flux and bulk drift velocities, v_d^* and v_d, respectively, and mean electron energy ε, for $b = 0, 10^{-4}, 10^{-3}$, and 10^{-2}.

PROBLEM 25
Physical interpretation of reactive corrections

Consider the gaussian pulse of electrons in the time-of-flight experiment shown in Figure 1.3, with an attaching gas in the drift tube. Recall that the center of mass (CM) of the pulse travels with velocity $\mathbf{v}_d$. For parts (a) and (b) an *intuitive* argument is sought.

(a) How does the mean electron energy $\langle\epsilon\rangle(\mathbf{r}, t)$ vary through the pulse, from the back to the front? Does it increase or decrease?

(b) If the attachment cross section and frequency are such that more low energy electrons are lost by attachment than high energy electrons, does this enhance or retard the motion of the pulse?

(c) Reconcile your physical arguments with Equation 9.55.

(d) Repeat the above for the opposite case, that is, where higher electrons are selectively lost by attachment.

PROBLEM 26
Structure-modified partial cross sections

The effective differential cross section for scattering in a structured material is given by

$$\Sigma(v, \chi) = \sigma(v, \chi)\, S\left(\frac{2mv}{\hbar}\sin\frac{\chi}{2}\right) \tag{P.26.1}$$

If we represent $\Sigma(v,\chi)$ through an expansion in terms of Legendre polynomials:

$$\Sigma(v,\chi) = \sum_{\lambda=0}^{\infty} \frac{2\lambda+1}{2} \Sigma_\lambda(v) P_\lambda(\cos\chi) \tag{P.26.2}$$

then one can make connection with the previous calculations of the collision matrix elements for dilute gaseous systems. Show that the effective partial cross sections $\Sigma_l(v)$ are defined by

$$\begin{aligned}\Sigma_l(v) &= 2\pi \int_{-1}^{1} \Sigma(v,\chi) P_l(\cos\chi) \mathrm{d}(\cos\chi)\\ &= \frac{1}{4\pi} \sum_{\lambda'\lambda''} \frac{(2\lambda'+1)(2\lambda''+1)}{2l+1} (\lambda' 0 \lambda'' 0|l0)^2 \sigma_{\lambda'}(v) s_{\lambda''}(v),\end{aligned}$$

where

$$\sigma_l(v) = 2\pi \int_{-1}^{1} \sigma(v,\chi) P_l(\cos\chi) \mathrm{d}(\cos\chi) \tag{P.26.3}$$

and

$$s_l(v) = \frac{1}{2} \int_{-1}^{1} S\left(\frac{2mv}{\hbar} \sin\left(\frac{\chi}{2}\right)\right) P_l(\cos\chi) \mathrm{d}(\cos\chi). \tag{P.26.4}$$

PROBLEM 27
Diffusion cooling in Neon

This problem and the next relate to a Cavalleri experiment (see Chapter 13) carried out in a cylindrical chamber of diffusion length Λ, given by

$$\Lambda^{-2} = 1.509 \times 10^4 \text{ m}^{-2}.$$

You will need to have access tables of electron-noble gas atom scattering cross sections, as for example in Refs. [10,197], and to carry out the required numerical calculations.

Numerical solution of the eigenvalue equation (Equation 13.22) for electrons in *neon* at 293 K using an accurate pseudo-spectral method (see Ref. [197]) furnishes $n_0 D$ and electron temperature T as shown in the following table. Calculate the same quantities using the variational method described in Section 13.4.2, and compare with the tabulated results.

Gas Pressure (kPa)	$n_0D\,(10^{24}\ \mathrm{m^{-1}s^{-1}})$	$T(K)$
6.67	6.87	235.7
8.0	7.0	253.7
10.67	7.11	271.3
13.33	7.16	279.1
16.0	7.19	283.3
20.0	7.21	286.7
26.66	7.22	289.3
100.0	7.24	292.4

PROBLEM 28
Diffusion cooling in other rare gases

(a) Calculate n_0D and T for electrons in Xe, Kr, and Ar at $T_0 = 293$ K using the variational method of Chapter 13, with pressures $p_0 = n_0k_BT_0$ as specified below, and using values of momentum transfer cross section given in the Refs. [10,197].

Gas ↓	$p_0(k\,Pa) \rightarrow$												
Xe	0.5	1.0	1.5	2.0	2.5	3.0	5.0	10.0	20.0	50.0			
Kr	0.5	1.0	1.5	2.0	2.5	3.0	5.0	10.0	20.0	100			
Ar	1.3	2.7	4.0	5.3	6.7	8.0	10.7	13.3	16.0	20.0	20.7	100	200	1000

(b) Given that the *aim* of the Cavalleri experiment is to determine the *thermal equilibrium value* of the diffusion coefficient to *better than* 1%, and that there is a *practical upper limit on the gas pressure* of (say) 100 kPa, in which of the above cases can the experiment be regarded as being successful? What *practical measures* can you suggest to improve things?

PROBLEM 29
Energy-independent reaction rate

Consider charged particles interacting with a gas in which reactive *loss* collisions are governed by a constant, energy-independent reaction rate ν_R. In this case, the Boltzmann equation takes the form

$$\left(\partial_t + \mathbf{v}\cdot\nabla + \mathbf{a}\cdot\partial_{\mathbf{v}}\right) f = \left(\frac{\partial f}{\partial t}\right)^{(C)}_{coll} - \nu_R f \tag{P.29.1}$$

where $\left(\frac{\partial f}{\partial t}\right)^{(C)}_{coll}$ denotes the non-reactive, particle-conserving collision term.

(a) *Prove* that the solution of Equation P.29.1 is given by

$$f(\mathbf{r},\mathbf{v},t)=f^{(C)}(\mathbf{r},\mathbf{v},t)\,e^{-\nu_R t},$$

where $f^{(C)}(\mathbf{r},\mathbf{v},t)$ denotes the solution of Equation P.29.1 in the absence of reactions ($\nu_R=0$), that is, for conservative collisions only.

(b) *Verify* that all velocity moments are independent of ν_R and are in fact the same as for the conservative collision case, that is,

$$\langle\phi(\mathbf{v})\rangle=\frac{\int d^3v\,\phi(\mathbf{v})\,f^{(C)}(\mathbf{r},\mathbf{v},t)}{\int d^3v\,f^{(C)}(\mathbf{r},\mathbf{v},t)}$$

and *explain* this in physical terms.

PROBLEM 30
Talmi coefficients

Using the table of low order Burnett functions given in Chapter 12, and making transformations from the lab to the CM frame, that is, $(\mathbf{v},\mathbf{v}_0)\Longrightarrow(\mathbf{g},\mathbf{G})$, in the usual way, investigate the *Talmi transformation* for the Burnett function $\phi_0^{[10]}(\alpha\mathbf{v})$, that is, the expression for this quantity in terms of a sum over products of other Burnett functions evaluated at the CM and relative velocities:

$$\phi_0^{[10]}(\alpha\mathbf{v})=\sum_N\sum_\nu \mathcal{T}\,(\mathrm{NLM},\nu\mathrm{lm})\;\phi_M^{[NL]}(\Gamma\mathbf{G})\,\phi_m^{[\nu l]}(\gamma\mathbf{g}),\qquad\text{(P.30.1)}$$

where

$$\Gamma^2=\alpha^2+\alpha_0^2\qquad\gamma^2=\frac{\alpha^2\alpha_0^2}{\Gamma^2}$$

$$\alpha^2=\frac{m}{k_BT_b}\qquad\alpha_0^2=\frac{m_0}{k_BT_b}$$

and T_b is some arbitrary basis temperature.

(N.B. All that is required is to ascertain which values of (NLM, νlm) contribute to the summation (Equation P.30.1). Do *not* attempt to find explicit expressions for the Talmi coefficients $\mathcal{T}$ (NLM, νlm).

PROBLEM 31
Decomposition of the kinetic equation

For a spatially uniform system of dilute particles of charge q and mass m in a gaseous or condensed matter medium subject to an applied electric field $\mathbf{E}$, the kinetic equation can be written as

$$\frac{\partial f}{\partial t} + \mathbf{a} \cdot \frac{\partial f}{\partial \mathbf{v}} = -J(f),$$

where the collision term $J(f)$ accounts for all types of interactions and any structure of the medium, and the force per unit mass $\mathbf{a} = q\mathbf{E}/m$ defines an axis of symmetry. For the purposes of this exercise, all we need to know about J is that it is a *linear* and *isotropic* operator acting in velocity space. The latter simply means that J preserves the tensorial rank of any quantity on which it operates, for example,

$$J\,[f^{(l)}(v) Y_m^{[l]}(\hat{\mathbf{v}})] = J_l\,[f^{(l)}(v)]\, Y_m^{[l]}(\hat{\mathbf{v}}),$$

where J_l is another operator acting in speed space only. Again, we do not need to know the explicit form of J_l for present purposes.

The *addition theorem for spherical harmonics*,

$$P_l(\hat{\mathbf{a}} \cdot \hat{\mathbf{v}}) = \frac{4\pi}{2l+1} \sum_{m=-l}^{l} Y_m^{(l)}(\hat{\mathbf{a}}) Y_m^{[l]}(\hat{\mathbf{v}}) \tag{P.31.1}$$

is needed for this and the following problem. In these equations, P_l and $Y_m^{(l)}$ denote a Legendre polynomial and spherical harmonic, respectively, and ^ denotes a unit vector.

(a) Show that in this axially symmetric case, the distribution function and the collision operator can be decomposed as follows:

$$f = f(v, \hat{\mathbf{a}} \cdot \hat{\mathbf{v}}, t) = \sum_{l=0}^{\infty} f^{(l)}(v, t) P_l(\hat{\mathbf{a}} \cdot \hat{\mathbf{v}})$$

and

$$J(f) = \sum_{l=0}^{\infty} J_l[f^{(l)}(v, t)]\, P_l(\hat{\mathbf{a}} \cdot \hat{\mathbf{v}}), \tag{P.31.2}$$

respectively.

(b) Show that the left hand side of the kinetic equation can be decomposed as follows:

$$\mathbf{a}\cdot\frac{\partial f}{\partial \mathbf{v}} = a\sum_{l=0}^{\infty}\left\{\frac{l}{2l-1}\left[\frac{\partial f^{(l-1)}}{\partial v} - \frac{(l-1)}{v}f^{(l-1)}\right]\right.$$
$$\left. + \frac{l+1}{2l+3}\left[\frac{\partial f^{(l+1)}}{\partial v} + \frac{(l+2)}{v}f^{(l+1)}\right]\right\} P_l(\hat{\mathbf{a}}\cdot\hat{\mathbf{v}}). \qquad \text{(P.31.3)}$$

The following identities for Legendre polynomials may be required:

$$\left(1-x^2\right)\frac{\mathrm{d}P_l(x)}{\mathrm{d}x} = -lxP_l(x) + lP_{l-1}(x)$$
$$xP_l(x) = \frac{1}{2l+1}\left[(l+1)P_{l+1}(x) + lP_{l-1}(x)\right].$$

(c) By combining the results Equations P.31.2 and P.31.3 show that the one kinetic equation for $f(\mathbf{v}, t)$ can be represented in the Legendre polynomial basis by the following equivalent infinite set of coupled equations for $f^{(l)}(v, t)$:

$$\frac{\partial f^{(l)}}{\partial t} + a\left\{\frac{l}{2l-1}\left[\frac{\partial f^{(l-1)}}{\partial v} - \frac{(l-1)}{v}f^{(l-1)}\right]\right.$$
$$\left. + \frac{l+1}{2l+3}\left[\frac{\partial f^{(l+1)}}{\partial v} + \frac{(l+2)}{v}f^{(l+1)}\right]\right\}$$
$$= -J_l[f^{(l)}] \qquad (l = 0, 1, 2, \ldots\infty)$$

(d) What approximations are necessary to close this set of equations in the two-term approximation?

PROBLEM 32

Spherical components of collision operator in a special case

For particles in a gas undergoing elastic collisions with neutral atoms, the classical Boltzmann collision operator of Chapter 3 applies.

(a) Show that this has spherical components

$$J_l\,[f^{(l)}] = \int\left[f^{(l)}(v)\,f_0(v_0) - f^{(l)}(v')\,f_0(v_0')\,P_l(\hat{\mathbf{v}}\cdot\hat{\mathbf{v}}')\right] g\sigma\,\mathrm{d}^2\Omega_{g'}\,\mathrm{d}\mathbf{v}_0,$$

where notation is the same as in Chapter 3.

(b) For very light particles in a heavy gas, $(m/m_0 \ll 1)$, the energy exchange is very small. Show that to zero order in m/m_0, $J_0[f^{(0)}]=0$ and

$$J_1[f^{(1)}]=\nu_m(v)f^{(1)}(v)$$

where $\nu_m(v)$ is the collision frequency for momentum transfer.

PROBLEM 33
Estimation of mean electron energy and speed

(a) A swarm of electrons is initially in equilibrium with He gas at temperature 20°C. What is the average electron energy in eV and the average electron speed in m s^{-1}?

(b) An electric field is then switched on causing the electrons to drift with an average velocity $v_d = 5.0 \times 10^3$ m s^{-1}. Using Wannier's formula, estimate the average electron energy and hence calculate the average electron speed.

PROBLEM 34
Electron multiplication in a spark chamber

The electron multiplication in an avalanche process is characterised by the first Townsend coefficient α which describes the increase of the electron number dn/dx per primary electron over the distance dx and which is a function of E/n. In the constant electrical field of a parallel-plate electrode configuration, for example, of a spark chamber, α is constant and the increase of the number of electrons is given by $dn = \alpha n dx$.

(a) Calculate the number of electrons in a multiplication distance $d=1$ cm for a neon–helium (70/30) gas mixture which is typically used to operate a spark chamber. In this mixture, the primary ionisation is $n_T \approx 50\ e^-$/cm and for an electrical field strength of 7 kV/cm, the ionisation coefficient is $\alpha \approx 35$/cm. Compare the number of electrons $N(d=1$ cm) with the Raether–Meek criterion which describes the transition from a stable avalanche process to a streamer discharge above $N_{\text{Raether}} \approx 5 \times 10^8$ electrons per centimeter.

(b) Now assume a 5% oxygen contamination in the neon–helium gas mixture. The primary ionisation increases slightly to $n_T \approx 55$ e^-/cm, but some of the electrons will be attached to the electronegative oxygen and these electrons are lost for the avalanche process.

This process is described by the attachment coefficient β which describes the decrease of the electron number over the distance dx. Calculate the number of electrons and the total number of negative charges (electrons and negative ions due to attachment) in the multiplication distance $d = 1$ cm, assuming an attachment coefficient of $\beta \approx 5/\text{cm}$ and—due to the new composition of the gas mixture—an ionisation coefficient of $\alpha \approx 32/\text{cm}$.

PROBLEM 35

Ion drift velocity

The electron avalanche process in a strong electrical field creates electrons as well as positive ions. These positive ions must ultimately be removed to avoid shielding effects of the electrodes and a distortion of the electrical field. Unfortunately, positive ions move slowly and consequently, this is a crucial issue, in particular, in drift chambers or time projection chambers.

Using the fact that the mobility of Ar^+ ions in argon environment is $K_{Ar} \approx 1.7\ \text{cm}^2/(\text{Vs})$, calculate the ion drift velocity of Ar^+ ions in an electrical field of 2 kV/cm and compare this velocity with the electron drift velocities in Figure 18.5.

PROBLEM 36

Energies of fragments after neutron capture in BF_3 and ^3He gas

Gaseous detectors are also used for neutron detection using the following neutron capture reactions in BF_3 or ^3He gas:

$$n + {}^{10}B \rightarrow {}^{7}Li + \alpha + 2.789\,\text{MeV} \tag{P.36.1}$$

$$n + {}^{3}He \rightarrow t + p + 0.764\,\text{MeV} \tag{P.36.2}$$

The charged fragments of the neutron capture reactions are then detected in the gas due to their primary ionization which they create along their path in the gas. The ranges of the fragments in the gas depend on their energies. Calculate the fractions of energy which go to each of the two fragments in each of the reactions. In both cases, the fragments are emitted back to back.

PROBLEM 37

Positron emission tomography (PET) and non-collinearity of emitted gamma rays

PET was discussed in Chapter 17 up to the point of positronium (*Ps*) formation. The subsequent decay of *Ps* into two gamma rays is, however, the key element in the imaging process. The processes may be represented as

$$e^+ + A \rightarrow A^+ + Ps; \quad Ps \rightarrow 2\gamma.$$

(a) Using conservation of momentum and energy, estimate the average energy and momentum of the *Ps* atom in the first of these equations.

(b) Using conservation of momentum in the second process, obtain an expression for the non-collinearity, that is, deviation from "back-to-back" directions, of the two emitted gamma rays. What information is required to make a numerical estimate of this important quantity?

PROBLEM 38

Production and decay of atoms in excited states

Suppose that electrons of number density n collide with the atoms A of a medium of number density n_0 and temperature T_0 to produce atoms in excited states A^*. These excited atoms then undergo elastic collisions with the medium, and subsequently decay back to the ground state:

$$e^- + A \rightarrow e^- + A^*; \quad A^* \rightarrow A + \gamma .$$

(a) If ν^* denotes the rate of production of excited atoms, which recoil from the collision with energy $u^{\dagger}_{\text{rec}}$, show that the equations of continuity and energy balance for the excited atom number density n^* and mean energy u^* in the spatially uniform case are

$$\frac{\partial n^*}{\partial t} = n\,\nu^* - \lambda_0\, n^* \tag{P.38.1}$$

$$\frac{\partial (n^* u^*)}{\partial t} = n\,\nu^* u^{\dagger}_{\text{rec}} - n^* \nu_e \left(u^* - 3k_B T_0/2\right) - \lambda_0\, n^* u^*, \tag{P.38.2}$$

respectively, where ν_e is the collision frequency for energy exchange between A^* and A, and λ_0 is the natural rate of decay of A^* back to the ground state.

(b) By eliminating $\partial n^*/\partial t$ between these equations, show that

$$n^* \frac{\partial u^*}{\partial t} = n\,\nu^* \left(u^{\dagger}_{\text{rec}} - u^*\right) - n^* \nu_e \left(u^* - 3k_B T_0/2\right).$$

(c) Hence verify that for a *steady energy state* ($\partial u^*/\partial t = 0$), the mean energy of excited atoms is given by

$$u^* = \frac{n^* \nu_e \, 3k_B T_0/2 + n\, \nu^* \, u^{\dagger}_{\text{rec}}}{n^* \nu_e + n\, \nu^*}. \qquad \text{(P.38.3)}$$

(d) If an overall steady state pertains, such that the rate of production of A^* is balanced by the rate of decay in Equation P.38.1, show that Equation P.38.3 becomes

$$u^* = \frac{3k_B T_0/2 + r\, u^{\dagger}_{\text{rec}}}{1 + r}, \qquad \text{(P.38.4)}$$

where $r \equiv \lambda_0/\nu_e$.

(e) While ν_e is proportional to n_0, the natural decay rate λ_0 is an intrinsic property of the atom, and therefore independent of gas density. What then is the mean energy of the excited atoms in the limits of high and low gas pressures, respectively?

Note: In general, ν_e is energy dependent, and Equations P.38.3 and P.38.4 are implicit equations for u^*.

PROBLEM 39
Muon-catalyzed fusion

In the muon-catalyzed fusion cycle discussed in Chapter 19, muonic atom production $\mu + te \rightarrow t\mu + e$ is followed by muomolecular formation:

$$t\mu + d \rightarrow td\mu$$

By adapting the equations and notation (e.g., $u^* \rightarrow \langle u \rangle_{t\mu}$, $\lambda_0 \rightarrow$ muomolecular formation rate) of the previous problem, obtain an expression for the mean energy of the muonic atoms and compare with Equation 19.16.

PROBLEM 40
Beam acceleration in a plasma

(a) An axially symmetric energetic beam of N electrons of mass m is driven through a plasma medium by a constant longitudinal force F_z and compressed by a transverse force $F_x = -sx$, where x is the distance from the axis of symmetry (the z-axis) and s is a

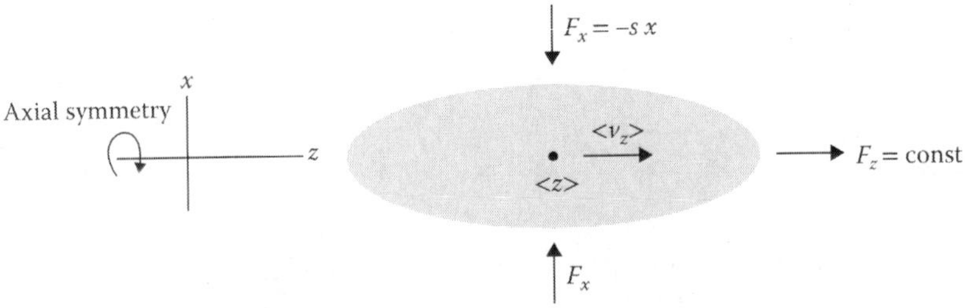

Figure P.40.1 An axially symmetric beam of particles subject to a longitudinal accelerating and transverse compression forces, respectively.

constant (see Figure P.40.1). At sufficiently high energies, collisions between charged particles may be neglected (see Problem 7), that is, $\left(\frac{\partial f}{\partial t}\right)_{\text{col}} \approx 0$ and the kinetic equation (Equation 2.10) then reduces to the *collisionless Boltzmann* or *Vlasov kinetic equation*,

$$\frac{\partial f}{\partial t} + v_x \frac{\partial f}{\partial x} + v_z \frac{\partial f}{\partial z} + \frac{F_x}{m}\frac{\partial f}{\partial v_x} + \frac{F_z}{m}\frac{\partial f}{\partial v_z} = 0,$$

where only the x coordinate need be considered in this axially symmetric problem.

(b) For an unbounded medium, the *phase-space average* (Equation 2.8) of a physical property Φ depending on phase-space coordinates $(x, z; v_x, v_z)$ is

$$\langle \Phi(x, z; v_x, v_z) \rangle = \frac{1}{N} \int_{-\infty}^{\infty} dx \int_{-\infty}^{\infty} dz \int_{-\infty}^{\infty} dv_x \int_{-\infty}^{\infty} dv_z \; \Phi(x, z; v_x, v_z) f(x, z; v_x, v_z; t)$$

By multiplying the kinetic equation by Φ and integrating over all phase space, derive the general phase-space moment equation for a collisionless plasma:

$$\frac{\partial \langle \Phi \rangle}{\partial t} - \left\langle v_x \frac{\partial \Phi}{\partial x} \right\rangle - \left\langle v_z \frac{\partial \Phi}{\partial z} \right\rangle + \frac{s}{m}\left\langle x \frac{\partial \Phi}{\partial v_x} \right\rangle - \frac{F_z}{m}\left\langle \frac{\partial \Phi}{\partial v_z} \right\rangle = 0$$

(i) Find the moment equations corresponding to $\Phi = z$ and v_z.

(ii) State why $\langle x \rangle, \langle v_x \rangle$ and $\langle F_x \rangle = -s\langle x \rangle$ should all vanish, using either physical arguments, or moment equations, or a combination of both.

(iii) Show that the phase-space moment equations for $\Phi = x^2, xv_x$, and v_x^2 are

$$\frac{d\langle x^2\rangle}{dt} = 2\langle xv_x\rangle$$
$$\frac{d\langle xv_x\rangle}{dt} = \langle v_x^2\rangle - \frac{s}{m}\langle x^2\rangle$$
$$\frac{d\langle v_x^2\rangle}{dt} = -\frac{2s}{m}\langle xv_x\rangle,$$

respectively, and solve these equations for $\langle x^2\rangle$, $\langle xv_x\rangle$ and $\langle v_x^2\rangle$ with given initial conditions $\langle x^2\rangle_0$, $\langle xv_x\rangle_0$ and $\langle v_x^2\rangle_0$, respectively.

(iv) Hence show that the *transverse emittance* (a figure of merit representing the phase-space quality of the beam),

$$\epsilon_n \equiv \sqrt{\langle x^2\rangle\langle v_x^2\rangle - \langle xv_x\rangle^2}$$

is *conserved*.

APPENDIX A

Comparison of Kinetic Theory and Quantum Mechanics

A.1 Kinetic Theory and Quantum Mechanics—A Comparison of Terms

Table A.1 Comparison of Kinetic Theory and Quantum Mechanics

	Classical Kinetic Theory	**Quantum/Atomic/ Nuclear**
Key property	Distribution function $f(\mathbf{r}, \mathbf{v}, t)$	Wave function $\Psi(\mathbf{r}, t)$
Key equation	Boltzmann's equation (1872)	Schrödinger's equation (1928)
Dimensionality	6D phase space $(\mathbf{r}, \mathbf{v})$ + time	3D space $\mathbf{r}$ + time
Measurable properties	Velocity space averages with f	Expectation values with $\lvert\Psi\rvert^2$
Eigenvalue problem	$(L - \omega)\,\Psi = 0$	$H\,\Psi = E\Psi$
Operator properties	L not Hermitian, ω complex	H Hermitian, E real
Solution techniques	Expansion, variational methods	Expansion, variational methods
Spherical harmonics	$Y_m^{(l)}(\hat{\mathbf{v}})$	$Y_m^{(l)}(\hat{\mathbf{r}})$
Basis functions	Burnett functions $\phi_m^{(\nu l)}(\mathbf{v})$	E'functions 3D harmonic oscillator
Lab $\rightarrow$ CM	Talmi transformation	Talmi transformation
Contrasting style	Often take f to be Maxwellian	Ψ is calculated, never assumed
Related analysis	Fluid equations	Virial theorem
Physics courses	Limited treatment	Standard material

APPENDIX B

Inelastic and Ionization Collision Operators for Light Particles

B.1 Inelastic Collision Operator

While the integral collision term (Equation 5.36) can be applied directly to calculation of transport properties of particles of any mass in a gas undergoing elastic and inelastic collisions, simplified representations can also be obtained for light particles. Rather than trying to expand the operator itself directly in powers of m/m_0, we employ the same technique as used in finding the elastic collision term Equation 5.23, starting with the integral of the operator multiplying by some arbitrary function of velocity $\phi(\mathbf{v})$, and then expanding that in m/m_0. The focus is on obtaining the isotropic part $\left(\frac{\partial f^{(0)}}{\partial t}\right)_{\text{col}}$ of the inelastic collision term, accurate to zero order in m/m_0. By substituting an arbitrary scalar quantity $\phi = \phi(v)$ in the general expression (Equation 5.34), and assuming the neutrals to be heavy and stationary, so that $g \approx v$, $g' \approx v'$, it follows that

$$4\pi\int_0^\infty dv\, v^2\, \phi(v)\left(\frac{\partial f^{(0)}}{\partial t}\right)_{\text{col}} = 4\pi\sum_{j,j'} n_{0j}\int_0^\infty v^2 dv\, f^{(0)}(v) \times v\sigma^{(0)}(j,j';v,v')\left[\phi(v') - \phi(v)\right], \tag{B.1}$$

where

$$\sigma^{(0)}(j,j';v,v') = \int d^2\Omega_{\mathbf{g}'}\sigma(j,j';\mathbf{v},\mathbf{v}')$$

is the integral inelastic cross section for the process, and v' and v are related through the energy conservation,

$$\epsilon + \epsilon_j = \epsilon' + \epsilon_{j'}, \tag{B.2}$$

where $\epsilon \approx \frac{1}{2}mv^2$ and $\epsilon' \approx \frac{1}{2}mv'^2$ are energies in the centre-of-mass fame. In what follows, we shall use ϵs, rather than vs, as the variables, whenever it is convenient to do so. It is clear that elastic collisions, for which $j' = j$, $\epsilon' \approx \epsilon$ and hence $v' \approx v$, $\phi(v') = \phi(v)$, do not contribute to the right hand side of Equation B.1. That is, to zero order in m/m_0, $\left(\frac{\partial f^{(0)}}{\partial t}\right)_{\text{col}}$ derives entirely from inelastic processes ($j' \neq j$), and for that reason, we write it simply as $\left(\frac{\partial f^{(0)}}{\partial t}\right)_{\text{inel}}$.

For simplicity, we take a model in which the neutral has only two states, $j = 0, 1$, both non-degenerate, with energies $\epsilon_0 = 0, \epsilon_1$, respectively.

Thus by Equation B.1, the post-collision particle energy is $\epsilon' = \epsilon - \epsilon_1$ for an inelastic collision involving excitation of the neutral from the ground to the excited state, and $\epsilon' = \epsilon + \epsilon_1$ for the inverse "superelastic" process. The corresponding cross sections are written as

$$\vec{\sigma}(\epsilon) = \sigma^{(0)}(0,\ 1; \epsilon, \epsilon - \epsilon_1)$$

and

$$\overleftarrow{\sigma}(\epsilon) = \sigma^{(0)}(1, 0; \epsilon, \epsilon + \epsilon_1),$$

respectively. Note that the microscopic reversibility condition (Equation 5.32) in this notation is

$$\epsilon\, \vec{\sigma}(\epsilon) = (\epsilon - \epsilon_1)\ \overleftarrow{\sigma}(\epsilon - \epsilon_1).$$

With the notation $\phi(v) = \phi(\epsilon)$, Equation B.1 becomes

$$\begin{aligned}
&\frac{1}{m}\sqrt{\frac{2}{m}}\int_0^\infty d\epsilon\ \epsilon^{\frac{1}{2}}\ \phi(\epsilon)\left(\frac{\partial f^{(0)}}{\partial t}\right)_{\text{inel}} \\
&= \frac{n_0}{Z}\frac{2}{m^2}\int_{\epsilon_1}^\infty d\epsilon\ \epsilon f^{(0)}(\epsilon)\vec{\sigma}(\epsilon)\left[\phi(\epsilon - \epsilon_1) - \phi(\epsilon)\right] \\
&+ n_0 \frac{e^{-\frac{\epsilon_1}{k_B T_0}}}{Z}\frac{2}{m^2}\int_0^\infty d\epsilon\ \epsilon f^{(0)}(\epsilon)\overleftarrow{\sigma}(\epsilon)\left[\phi(\epsilon + \epsilon_1) - \phi(\epsilon)\right],
\end{aligned} \tag{B.3}$$

where $Z = 1 + \exp\left(-\frac{\epsilon_1}{k_B T_0}\right)$ is the partition function.

The first integral on the right hand side of Equation B.3 has a lower limit of ϵ_1, since $\vec{\sigma}(\epsilon)$ vanishes unless $\epsilon > \epsilon_1$. In the first term, we make the substitution $\epsilon \to \epsilon + \epsilon_1$, and find

$$\int_{\epsilon_1}^\infty d\epsilon\ \epsilon\, f^{(0)}(\epsilon)\vec{\sigma}(\epsilon)\phi(\epsilon - \epsilon_1) = \int_0^\infty d\epsilon(\epsilon + \epsilon_1) f^{(0)}(\epsilon + \epsilon_1)\vec{\sigma}(\epsilon + \epsilon_1)\ \phi(\epsilon).$$

In the first term of the second integral, the substitution $\epsilon \to \epsilon - \epsilon_1$ yields

$$\int_0^\infty d\epsilon\ \epsilon\, f^{(0)}(\epsilon)\ \overleftarrow{\sigma}(\epsilon)\ \phi(\epsilon + \epsilon_1) = \int_{\epsilon_1}^\infty d\epsilon\ (\epsilon - \epsilon_1) f^{(0)}(\epsilon - \epsilon_1)\ \overleftarrow{\sigma}(\epsilon - \epsilon_1)\ \phi(\epsilon).$$

These expressions are substituted in Equation B.3, the terms gathered, and the coefficients of the arbitrary function $\phi(\epsilon)$ on the left and right hand sides equated to give

$$\left(\frac{\partial f^{(0)}}{\partial t}\right)_{\text{inel}} = \frac{n_0}{Z}\sqrt{\frac{2}{m\epsilon}}\left\{\left[(\epsilon+\epsilon_1)f^{(0)}(\epsilon+\epsilon_1)\vec{\sigma}(\epsilon+\epsilon_1) - \epsilon f^{(0)}(\epsilon)\vec{\sigma}(\epsilon)\right]\right.$$

$$\left. + e^{-\frac{\epsilon_1}{k_B T_0}}\left[(\epsilon-\epsilon_1)f^{(0)}(\epsilon-\epsilon_1)\overleftarrow{\sigma}(\epsilon-\epsilon_1) - \epsilon f^{(0)}(\epsilon)\overleftarrow{\sigma}(\epsilon)\right]\right\}. \tag{B.4}$$

This expression can readily be extended to more than two internal levels, as shown in Equation 5.39.

Including terms to first order in m/m_0 in Equation B.1 picks up contributions from elastic as well as inelastic collisions. However, since inelastic cross sections can be several orders of magnitude less than their elastic counterparts, the dominant contribution to $\left(\frac{\partial f^{(0)}}{\partial t}\right)_{\text{col}}$ to first order in m/m_0 is generally from elastic processes.

If anisotropic scattering is important, the reader is referred to Ref. [198] for expressions for the $l = 0$ and $l = 1$ collision operators.

B.2 Ionizing Collisions

B.2.1 Positron impact ionization

A process in which an atom or molecule M is ionized by positron impact,

$$e^+ + M \rightarrow M^+ + e + e^- \tag{B.5}$$

may be effectively considered as an inelastic collision. The positron loses an energy ϵ_I equal to the ionization energy of M, which is then "excited" to become an ion M^+, but positron number is conserved, and the ejected electron may or may not be of further interest. For a cold gas, the collision term for ionization by positron impact may be obtained directly from the first two terms on the right hand side of Equation B.4, with ϵ_1 replaced by ϵ_I:

$$\left(\frac{\partial f^{(0)}}{\partial t}\right)_{\text{PI}} = n_0\sqrt{\frac{2}{m\epsilon}}\left[(\epsilon+\epsilon_I)f^{(0)}(\epsilon+\epsilon_I)\sigma_{\text{PI}}(\epsilon+\epsilon_I) - \epsilon f^{(0)}(\epsilon)\sigma_{\text{PI}}(\epsilon)\right], \tag{B.6}$$

where σ_{PI} denotes the cross section for the process (Equation B.5). The above is valid, strictly speaking, only if the electron is ejected at rest. A more detailed and realistic model, allowing for non-zero ejected energy and momentum [69], gives in terms of energy and the energy-partition function, $P(\epsilon,\epsilon')$,

$$J_l^{\text{ion}}(f_l)=$$

$$n_0\sqrt{\frac{2\epsilon}{m}}\sigma^{\text{ion}}(\epsilon)\,f_l(\epsilon)-\begin{cases} n_0\sqrt{\frac{2}{m\epsilon}}\int_0^\infty \epsilon'\sigma^{\text{ion}}\left(\epsilon'\right)P(\epsilon,\epsilon')\,f_0(\epsilon')d\epsilon' & l=0 \\ 0 & l\geq 1 \end{cases}. \tag{B.7}$$

If the scattered positron leaves the collision with an exact fraction, Q, of the available energy, $\epsilon'-\epsilon_I$, where ϵ_I is the threshold energy, then the energy-partition function has the form

$$\begin{aligned} P(\epsilon,\epsilon') &= \delta\left(\epsilon-Q(\epsilon'-\epsilon_I)\right), \\ &= \frac{1}{Q}\delta\left(\epsilon'-\left(\frac{\epsilon}{Q}+\epsilon_I\right)\right), \end{aligned}$$

and the integral in Equation B.7 reduces to

$$J_l^{\text{ion}}(f_l)=\nu^{\text{ion}}(\epsilon)\,f_l(\epsilon)-\begin{cases} \frac{1}{Q}\frac{\left(\frac{\epsilon}{Q}+\epsilon_I\right)^{\frac{1}{2}}}{\epsilon^{\frac{1}{2}}}\nu^{\text{ion}}\left(\frac{\epsilon}{Q}+\epsilon_I\right)f_0\left(\frac{\epsilon}{Q}+\epsilon_I\right) & l=0, \\ 0 & l\geq 1, \end{cases} \tag{B.8}$$

where $\nu^{\text{ion}}(\epsilon)=n_0\sqrt{\frac{2\epsilon}{m}}\sigma^{\text{ion}}(\epsilon)$ is the ionization collision frequency. Equation B.8 can be considered a "modified Frost–Phelps" operator. In the case where the positron gets all of the available energy, that is, $Q=1$, Equation B.8 reduces to the standard Frost–Phelps operator, Equation B.6, as required. The reader is referred to Ref. [69] for further details.

B.2.2 Electron impact ionization

On the other hand, electron impact ionization*

$$e^- + M \rightarrow M^+ + e^- + e^- \tag{B.9}$$

does *not* conserve particle number and therefore can *not* be simply treated as a type of inelastic process, and the corresponding collision term $\left(\frac{\partial f^{(0)}}{\partial t}\right)_I$ must be obtained from first principles. If it is assumed that the gas is cold

* It is assumed for simplicity that there is only one ionization channel, but the arguments presented here can be extended in a straightforward way to the more general case.

and that the recoil of the ion M^+ can be neglected, energy conservation can be expressed as

$$\epsilon = \epsilon_I + \epsilon' + \epsilon'', \tag{B.10}$$

where $\epsilon = \frac{1}{2}mv^2$ is the initial electron energy, and $\epsilon' = \frac{1}{2}mv'^2$ and $\epsilon'' = \frac{1}{2}mv''^2$ are the energies of the two emerging electrons. The way in which the available energy can be partitioned between the two outgoing electrons is characterized by the cross section $\sigma_I(\epsilon; \epsilon', \epsilon'')$.

We can find the ionization collision term $\left(\frac{\partial f^{(0)}}{\partial t}\right)_I$ to zero order in m/m_0 using a similar strategy to that established for other types of interaction, by first formulating an expression for $4\pi\int_0^\infty dv\, v^2\phi(v)\left(\frac{\partial f^{(0)}}{\partial t}\right)_I$, where $\phi(v)$ is an arbitrary function of v. To that end, we adapt Equation B.1 to a single scattering channel with *two* outgoing particles, taking $g \approx v$, and then using energy rather than speed as the independent variable. Just as we obtained Equation B.3 for inelastic collisions, we find

$$\begin{aligned}\frac{1}{m}\sqrt{\frac{2}{m}}\int_0^\infty d\epsilon\, \epsilon^{\frac{1}{2}}\phi(\epsilon)\left(\frac{\partial f^{(0)}}{\partial t}\right)_I &= \frac{2}{m^2}n_0\int_{\epsilon_I}^\infty d\epsilon\, \epsilon f^{(0)}(\epsilon)\sigma_I(\epsilon;\epsilon',\epsilon'') \\ &\quad \times\left[\phi(\epsilon') + \phi(\epsilon'') - \phi(\epsilon)\right],\end{aligned} \tag{B.11}$$

where $\phi(\epsilon)$ is an arbitrary function of electron energy.

For purposes of illustration, assume that the energy is shared equally between the two electrons, that is, $\epsilon'' = \epsilon' = \frac{1}{2}(\epsilon - \epsilon_I)$, so that Equation B.12 becomes

$$\begin{aligned}\frac{1}{m}\sqrt{\frac{2}{m}}\int_0^\infty d\epsilon\, \epsilon^{\frac{1}{2}}\phi(\epsilon)\left(\frac{\partial f^{(0)}}{\partial t}\right)_I &= \frac{2}{m^2}n_0\int_{\epsilon_I}^\infty d\epsilon\, \epsilon f^{(0)}(\epsilon)\sigma_I(\epsilon) \\ &\quad \times\left[2\phi\left(\frac{\epsilon - \epsilon_I}{2}\right) - \phi(\epsilon)\right],\end{aligned}$$

where as a matter of notational convenience, $\sigma_I(\epsilon)$ has been written for $\sigma_I(\epsilon; \frac{\epsilon-\epsilon_I}{2}, \frac{\epsilon-\epsilon_I}{2})$. After making the substitution $\epsilon \to 2\epsilon + \epsilon_I$ in the first term in the integrand on the right hand side, and observing that $\sigma_I(\epsilon) = 0$ for $\epsilon < \epsilon_I$ in the second term, it follows that

$$\begin{aligned}&\frac{1}{m}\sqrt{\frac{2}{m}}\int_0^\infty d\epsilon\, \epsilon^{\frac{1}{2}}\phi(\epsilon)\left(\frac{\partial f^{(0)}}{\partial t}\right)_I \\ &= \frac{2}{m^2}n_0\int_0^\infty d\epsilon\left[4(2\epsilon+\epsilon_I)f^{(0)}(2\epsilon+\epsilon_I)\sigma_I(2\epsilon+\epsilon_I) - \epsilon f^{(0)}(\epsilon)\sigma_I(\epsilon)\right]\phi(\epsilon).\end{aligned}$$

Equating the coefficients of the arbitrary function $\phi(\epsilon)$ then gives

$$\left(\frac{\partial f^{(0)}}{\partial t}\right)_I = n_0\sqrt{\frac{2}{m\epsilon}}\left[4(2\epsilon+\epsilon_I)\sigma_I(2\epsilon+\epsilon_I)f^{(0)}(2\epsilon+\epsilon_I) - \epsilon\sigma_I(\epsilon)f^{(0)}(\epsilon)\right]. \tag{B.12}$$

A similar procedure can be followed for any partitioning fraction of energy between the two electrons.

APPENDIX C

The Dual Eigenvalue Problem

C.1 Green's Function for the Operator $\partial_t + L$ and the Dual Eigenvalue Problem

Let L be a linear operator acting in a space defined by the variable $x \in (a, b)$, and let t be the time. Suppose event occurs at point x_0 at time t_0 and that the subsequent evolution of the system is governed by the equation

$$\left(\partial_t + L\right) G = \delta(x - x_0)\delta(t - t_0), \tag{C.1}$$

whose solution $G = G(x, t; x_0, t_0)$ will be referred to as the Green's function. Let $\phi(x)$ be some well-behaved complex function of x satisfying sufficient integrability and differentiability conditions. If Equation C.1 is integrated with its complex conjugate $\phi^*(x)$, then there follows

$$\partial_t \left[\int_a^b \phi^* G \, dx\right] + \int_a^b (L^\dagger \phi)^* \, G \, dx = \delta(t - t_0) \, \phi^*(x), \tag{C.2}$$

where $L^\dagger$ is the adjoint of L, defined in the usual way

$$\int_a^b \phi^* L\psi \, dx = \int_a^b (L^\dagger \phi)^* \psi \, dx$$

for any functions ϕ and ψ.

We now *choose* ϕ to be one of the eigenfunctions ϕ_j of $L^\dagger$, that is

$$\left(L^\dagger - \Omega_j\right) \phi_j = 0, \tag{C.3}$$

where j is an index ordering the eigenvalues Ω_j. Hence Equation C.2 becomes

$$\left(\partial_t + \Omega_j^*\right) G_j = \delta(t - t_0)\phi_j^*(x_0), \tag{C.4}$$

where

$$G_j \equiv G_j(t; x_0, t_0) = \int_a^b \phi_j^*(x) \, G(x, t; x_0, t_0) \, dx. \tag{C.5}$$

The solution of Equation C.4 is

$$G_j(t; x_0, t_0) = \theta(t - t_0)\phi_j^*(x_0)e^{-\Omega_j^*(t-t_0)}, \tag{C.6}$$

where the step function is defined such that $\theta(t-t_0)=1$ if $t>t_0$, and 0 if $t<t_0$.

Consider now the eigenfunctions ψ_j and eigenvalues ω_j of L, which are the solutions of the problem

$$(L-\omega_j)\psi_j=0. \tag{C.7}$$

It is straightforward to show from Equations 11.28 and C.7 that the sets of eigenfunctions $\{\psi_j\}$ and $\{\phi_j\}$ are orthogonal. We may also assume normalisation, that is

$$\int_a^b \phi_j^* \psi_j \, dx = \delta_{jj'}. \tag{C.8}$$

The same analysis shows that the eigenvalues of the dual problems (Equations 11.28 and C.7) are related by

$$\omega_j = \Omega_j^*. \tag{C.9}$$

Suppose now that the set $\{\psi_j\}$ is complete on (a, b), and that we may therefore represent the Green's function in x-space as

$$G(x,t;x_0,t_0)=\sum_j A_j \, \psi_j(x).$$

Multiplying by $\phi_{j'}^*(x)$ and integrating over x and using the orthogonality property (Equation C.8) then gives

$$A_j=\int_a^b \phi_j^*(x)\, G(x,t;x_0,t_0)\, dx \equiv G_j(t;x_0,t_0)$$

and with Equation C.6, we obtain finally

$$G(x,t;x_0,t_0)=\theta(t-t_0)\sum_j \phi_j^*(x_0)e^{-\omega_j(t-t_0)}\psi_j(x). \tag{C.10}$$

This is the representation used in Chapter 8.

Finally, we make a few observations:

a. If the operator is *self-adjoint*, that is, $L^\dagger = L$, then the dual eigenvalue problems Equations 11.28 and C.7 collapse to a single equation, the eigenfunctions and eigenvalues are identical, that is, $\psi_j = \phi_j$, $\omega_j=\Omega_j$, and by Equation C.9, these eigenvalues are real. This is the situation encountered in many branches of physics, for example, quantum mechanics, where Equation C.1 corresponds to the time-dependent Schrodinger equation, and L to the hamiltonian operator (see Appendix A for further comparison). However, in kinetic theory, operators are generally not self-adjoint, eigenvalues are complex, and the dual eigenvalues problem must be solved.

b. The above can be extended to more than one dimension in a straightforward way, that is, x may be replaced by a set of phase space coordinates.

c. Although we have referred to t as time, it could in fact also represent *any* coordinate in the half plane $(0, \infty)$, if desired.

d. The solution to the more general problem,

$$\left(\partial_t + L\right) f = h(x,t),$$

where the right hand side $h(x,t)$ is a given function, can be obtained from the Green's function in the usual way, namely

$$\begin{aligned} f(x,t) &= \int_0^\infty \mathrm{d}t_0 \int_a^b \mathrm{d}x_0\, h(x_0,t_0) G(x,t;x_0,t_0) \\ &= \sum_j \psi_j(x) \left\{ \int_0^t \mathrm{d}t_0 e^{\omega_j(t-t_0)} \int_a^b \mathrm{d}x_0\, \phi_j^*(x_0)\, h(x_0,t_0) \right\}. \end{aligned} \tag{C.11}$$

Appendix D

Derivation of the Exact Expression for $\hat{n}_p(k)$

D.1 Derivation of the Exact Expression Equation 14.21

From Equations 14.17 and 14.20, we find

$$\hat{n}_p(\mathbf{k}) = -\int_{-\infty}^{0} d\sigma \frac{1}{k}\left[\left\{\nu\hat{n}_p(\mathbf{k})\exp\left(-\frac{\sigma^2}{2\alpha^2}\right) + f_0(\mathbf{k},\sigma,\mathbf{0})\right\}\right.$$
$$\left.\times\exp\left(-\frac{i\sigma(\frac{1}{2}\sigma a_{\|} - \Omega)}{k}\right)\right] \tag{D.1}$$

and substituting Equation 14.18

$$f_0(\mathbf{k},\sigma,\mathbf{0}) = n_0\exp\left\{-\frac{\sigma^2}{2(\alpha')^2}\right\} \tag{D.2}$$

into Equation D.1, there follows

$$\hat{n}_p(\mathbf{k}) = -\int_{-\infty}^{0} d\sigma \frac{1}{k}\left\{\nu\,\hat{n}_p(\mathbf{k})\exp\left(-\frac{\sigma^2}{2\beta^2}\right) + n_0\exp\left(\frac{i\sigma\Omega}{k} - \frac{\sigma^2}{2(\beta')^2}\right)\right\}, \tag{D.3}$$

where β and β' are defined by

$$\beta^{-2} \equiv \alpha^{-2} + \frac{i\mathbf{a}\cdot\mathbf{k}}{k^2} \tag{D.4}$$

and

$$\beta'^{-2} \equiv \alpha'^{-2} + \frac{i\mathbf{a}\cdot\mathbf{k}}{k^2}. \tag{D.5}$$

The integral over σ may be carried out with the help of the identity

$$\int_{-\infty}^{0} d\sigma\exp\left(\frac{i\sigma\Omega}{k} - \frac{\sigma^2}{2\beta^2}\right) = \frac{i\beta}{\sqrt{2}}Z(\zeta), \tag{D.6}$$

where the plasma dispersion function $Z(\zeta)$ and ζ are defined by Equations 14.22 and 14.23. The solution of Equation D.3 then gives Equation 14.21 immediately.

APPENDIX E

Physical Constants and Useful Formulas

E.1 Constants and Useful Relations

Fundamental constants

- Boltzmann's constant: $k = 1.381 \times 10^{-23}$ J K^{-1} $= 8.616 \times 10^{-5}$ eVK^{-1}
- Planck's constant: $h = 6.626 \times 10^{-34}$ J s; $\hbar = h/2\pi = 1.055 \times 10^{-34}$ J s
- Permittivity of free space: $\varepsilon_0 = 8.854 \times 10^{-12}$ Fm^{-1}
- Speed of light in vacuum: $c = 2.998 \times 10^8$ m s^{-1}

Electrons

- Electronic charge: $e = 1.602 \times 10^{-19}$ C
- Electron rest mass: $m = 9.109 \times 10^{-31}$ kg
- Electron rest energy: $mc^2 = 0.511$ MeV
- Electron charge to mass ratio: $e/m = 1.758 \times 10^{11}$ C kg^{-1}
- Electron de Broglie wavelength: $\lambda_{\text{dB}} = \frac{h}{\sqrt{2m\epsilon}} = \frac{1.22 \times 10^{-9}}{\sqrt{\epsilon(\text{eV})}}$ m
- Speed of a 1 eV electron: $v = \sqrt{\frac{2e}{m}} = 5.927 \times 10^5$ m s^{-1}

Other particles

- Proton rest mass: $m_p = 1.672 \times 10^{-27}$ kg
- Muon rest mass: $m_\mu = 1.883 \times 10^{-28}$ kg
- Muon life time: $\tau_\mu = 2.197 \times 10^{-6}$ s
- (Ortho) positronium life time: $\tau_{Ps} = 1.386 \times 10^{-7}$ s

Standard conditions for temperature and pressure

- 1 atm = 760 torr = 1013.25 mbar
- 0°C = 273.15 K

Number densities of background medium

- Ideal gas at standard temperature and pressure (Loschmidt's number): $n_0 \equiv n_S = 2.687 \times 10^{25}$ m^{-3}
- Liquid water at 37°C: $n_0 \equiv n_L = 3.321 \times 10^{28}$ m^{-3}

Reduced quantities

- Reduced electric field: E/n_0 (Unit: 1 Td = 1 townsend = 10^{-21} V m) $= 3.034\ \mathrm{E}/p_0$ (V cm^{-1} Torr^{-1}, at 293 K)
- Reduced magnetic field: B/n_0 (Unit: 1 Hx = 1 huxley = 10^{-27} tesla m^3) $= 3.034\ B/p_0$ (Unit: gauss Torr^{-1}, at 293K)

Frequencies

- Reduced electron collision frequency: $\nu/n_0 = \sqrt{\frac{2\epsilon}{m}}\sigma = 5.927 \times 10^{-15}\sqrt{\epsilon(\mathrm{eV})}\,\sigma(\mathring{\mathrm{A}}^2)$ s^{-1}
- Reduced electron cyclotron frequency: $\Omega_L/n_0 = eB/(n_0 m) = 1.758 \times 10^{-16}\ B/n_0(\mathrm{Hx})$ m^{3}s^{-1}

Plasma parameters

- Debye length: $\lambda_D = \sqrt{\frac{\varepsilon_0 k_B T}{ne^2}} = 69\sqrt{T\ (\mathrm{K})/n\ (\mathrm{m}^{-3})}$ m
- Electron plasma frequency: $f_P = \frac{1}{2\pi}\sqrt{\frac{ne^2}{\varepsilon_0 m}} = 9\sqrt{n\ (\mathrm{m}^{-3})}$ s^{-1}

References

1. L. Boltzmann. Weitere Studien über das Wärmegleichgewicht unter Gasmolekülen. *Wiener Berichte*, 66:275, 1872.

2. C. S. Wang Chang, G. E. Uhlenbeck. and J. de Boer. The heat conductivity and viscosity of polyatomic gases. In *Studies in Statistical Mechanics*, edited by J. de Boer and G. E. Uhlenbeck, vol. 2, pp. 241–268, New York, NY: Wiley, 1964.

3. L. S. Frost and A. V. Phelps. Rotational excitation and momentum transfer cross sections for electrons in H_2 and N_2 from transport coefficients. *Physical Review*, 127(5):1621–1633, 1962.

4. E. G. D. Cohen and W. Thirring. *The Boltzmann Equation: Theory and Applications*. Vienna: Springer, 1973.

5. M. Abria. Investigations of the enigmatic phenomenon of striations. *Annales de Chimi et de Physique*, 7:462, 1843.

6. V. I. Kolobov. Striations in rare gas plasmas. *Journal of Physics D: Applied Physics*, 39(24):R487–R506, December 2006.

7. S. C. Brown. A summary of the numerous articles published by Brown et al., on high-frequency breakdown. *Handbuch der Physik*, 21:531, 1956.

8. A. Müller. The background of Röntgen's discovery. *Nature*, 157:119, 1946.

9. L. B. Loeb. *Basic Processes of Gaseous Electronics*. CA: University of California Press, 1955.

10. L. G. H. Huxley and R. W. Crompton. *The Diffusion and Drift of Electrons in Gases*. New York, NY: Wiley, 1974.

11. V. A. Rozhansky and L. D. Tsendin. *Transport Phenomena in Partially Ionized Gases*. London: Taylor and Francis, 2001.

12. M. A. Lieberman and D. L. Lichtenberger. *Principles of Plasma Discharges and Materials Processing*, 2nd edition, New York, NY: Wiley, 2005.

13. T. Makabe and Z. Lj. Petrović. *Plasma Electronics*, 2nd edition, Boca Raton, FL: CRC Press, 2015.

14. K. Kumar. The physics of swarms and some basic questions of kinetic theory. *Physics Reports*, 112(5):319–375, 1984.

15. R. W. Crompton. Benchmark measurements of cross sections for electron collisions: Electron swarm methods. *Advances in Atomic, Molecular, and Optical Physics*, 33:97–148, 1994.

16. F. Sauli. *Gaseous Radiation Detectors: Fundamentals and Applications*. Cambridge: Cambridge University Press, 2014.

17. L. Reggiani. *Hot Electron Transport in Semiconductors*. Berlin: Springer-Verlag, 1985.

18. S. Chapman and T. G. Cowling. *The Mathematical Theory of Non-uniform Gases: An Account of the Kinetic Theory of Viscosity, Thermal Conduction and Diffusion in Gases. Cambridge Mathematical Library*. Cambridge: Cambridge University Press, 1970.

19. T. Kihara. The mathematical theory of electrical discharges in gases. B. Velocity-distribution of positive ions in a static field. *Reviews of Modern Physics*, 25:844, 1953.

20. E. A. Mason and H. W. Schamp. Mobility of gaseous ions in weak electric fields. *Annals of Physics*, 4:233, 1958.

21. B. I. Davydov. Diffusion equation with taking into account of molecular velocity. *Physikalische Zeitschrift der Sowjetunion*, 8:59, 1935.

22. H. A. Lorentz. The motion of electrons in metallic bodies. In *Proceedings of the Royal Netherlands Academy of Arts and Sciences*, 7:438–453, 1905.

23. W. P. Allis. Motions of ions and electrons. In *Handbuch der Physik*, vol. 21, ed. S. Flügge. Berlin, Heidelberg: Springer, p. 383, 1959.

24. M. Cohen and J. Lekner. Theory of hot electrons in gases, liquids, and solids. *Physical Review*, 158(2):305–309, June 1967.

25. G. H. Wannier. Motion of gaseous ions in strong electric Fields. *Bell System Technical Journal*, 32(1):170–254, 1953.

26. S. L. Lin, L. A. Viehland, and E. A. Mason. Three-temperature theory of gaseous ion transport. *Chemical Physics*, 37(3):411–424, 1979.

27. L. A. Viehland and E. A. Mason. Gaseous ion mobility in electric fields of arbitrary strength. *Annals of Physics*, 91(2):499–533, 1975.

28. L. A. Viehland and E. A. Mason. Gaseous ion mobility and diffusion in electric fields of arbitrary strength. *Annals of Physics*, 110(2):287–328, 1978.

29. S. L. Lin, R. E. Robson, and E. A. Mason. Moment theory of electron drift and diffusion in neutral gases in an electrostatic field. *Journal of Chemical Physics*, 66:435, 1979.

30. K. Kumar. The Chapman-Enskog solution of the Boltzmann equation: A reformulation in terms of irreducible tensors and matrices. *Australian Journal of Physics*, 20:205, 1967.

31. K. Kumar, H. R. Skullerud, and R. E. Robson. Kinetic theory of charged particle swarms in neutral gases. *Australian Journal of Physics*, 33:343–448, 1980.

32. L. A. Viehland. Comparison of theory and experiment for gaseous ion transport involving molecular species. *Physica Scripta*, T53:53, 1994.

33. E. A. Mason and E. W. McDaniel. *Transport Properties of Ions in Gases*. New York, NY: Wiley, 1988.

34. Z. Lj. Petrović, M. Šuvakov, Ž. Nikitović, S. Dujko, O. Šašić, J. Jovanović, G. Malović, and V. Stojanović. Kinetic phenomena in charged particle transport in gases, swarm parameters and cross section data. *Plasma Sources Science and Technology*, 16(1):S1–S12, February 2007.

35. R. D. White, R. E. Robson, S. Dujko, P. Nicoletopoulos, and B. Li. Recent advances in the application of Boltzmann equation and fluid equation methods to charged particle transport in non-equilibrium plasmas. *Journal of Physics D: Applied Physics*, 42:194001, 2009.

36. J. R. Haynes and W. Shockley. Investigation of hole injection in transistor action. *Physical Review*, 75(4):691, 1949.

37. I. M. Sokolov, J. Klafter, and A. Blumen. Fractional kinetics. *Physics Today*, 55:48–54, 2002.

38. M. Muccini. A bright future for organic field-effect transistors. *Nature Materials*, 5(8):605–613, 2006.

39. P. Peumans, S. Uchida, and S. R. Forrest. Efficient bulk heterojunction photovoltaic cells using small-molecular-weight organic thin films. *Nature*, 425(6954):158–162, September 2003.

40. B. Philippa, R. E. Robson, and R. D. White. Generalized phase-space kinetic and diffusion equations for classical and dispersive transport. *New Journal of Physics*, 16(7):073040, July 2014.

41. P. W. Stokes, B. Philippa, D. Cocks, and R. D, White. Solution of a generalised Boltzmann's equation for non-equilibrium charged particle transport via localised and delocalised states. *Physical Review E*, 93:032119, 2015.

42. J. C. Maxwell. On the dynamical theory of gases. *Philosophical Transactions of the Royal Society of London*, 157:49, 1867.

43. R. E. Robson, R. D. White, and Z. Lj. Petrović. Colloquium: Physically based fluid modeling of collisionally dominated low-temperature plasmas. *Reviews of Modern Physics*, 77:1303, 2005.

44. Z. Lj. Petrović, S. Dujko, D. Marić, G. Malović, Ž. Nikitović, O. Šašić, J. Jovanović, V. Stojanović, and M. Radmilović-Radenović. Measurement and interpretation of swarm parameters and their application in plasma modelling. *Journal of Physics D: Applied Physics*, 42(19):194002, October 2009.

45. B. Philippa, M. Stolterfoht, P. L. Burn, G. Juška, P. Meredith, R. D. White, and A. Pivrikas. The impact of hot charge carrier mobility on photocurrent losses in polymer-based solar cells. *Scientific Reports*, 4:5695, January 2014.

46. J. Franck and G. Hertz. Über Zusammenstöße zwischen Elektronen und Molekülen des Quecksilberdampfes und die Ionisierungsspannungen desselben. *Verhandlungen der Deutschen Physikalischen Gesellschaft*, 16:457, 1914.

47. J. Fletcher and P. H. Purdie. Spatial non-uniformity in discharges in low pressure helium and neon. *Australian Journal of Physics*, 40:383, 1987.

48. Y. Sakai and W. F. Schmidt. High and low mobility electrons in liquid neon. *Chemical Physics*, 164:139–152, 1992.

49. K. Huang. *Statistical Mechanics*, 2nd edition, Hoboken, NJ: Wiley, 1987.

50. K. F. Ness, Kinetic theory of charged particle swarms with applications to electrons, PhD Thesis, James Cook University, 1986.

51. K. F. Ness and R. E. Robson. Velocity distribution function and transport coefficients of electron swaiiis in gases. II. Moment equations and applications. *Physical Review A*, 34(3):2185, 1986.

52. R. E. Robson, R. Winkler, and F. Sigeneger. Multiterm spherical tensor representation of Boltzmann's equation for a nonhydrodynamic weakly ionized plasma. *Physical Review. E, Statistical, Nonlinear, and Soft Matter Physics*, 65:056410, 2002.

53. R. E. Robson, T. Mehrling, and J. Osterhoff. Phase-space moment-equation model of highly relativistic electron-beams in plasma-wakefield accelerators. *Annals of Physics*, 356:306–319, 2015.

54. H. Goldstein. *Classical Mechanics*. Boston, MA: Addison-Wesley, 1964.

55. Y. Chang and R. D. White. Linearized Boltzmann collision integral with the correct cutoff. *Physics of Plasmas*, 21:072304, 2014.

56. D. C. Montgomery and D. A. Tidman. *Plasma Kinetic Theory*. New York, NY: McGraw-Hill, New York, 1964.

57. K. Miyamoto. *Plasma Physics for Nuclear Fusion*. London: MIT Press, 1989.

58. E. A. Mason and E. W. McDaniel. *Transport Properties of Ions in Gases*. New York, NY: Wiley, 1988.

59. E. W. McDaniel. *Collision Phenomena in Ionized Gases*. New York, NY: Wiley, 1964.

60. A. S. Davydov. *Quantum Mechanics*. London: Pergamon, 1965.

61. M. J. Brunger and S. J. Buckman. Electron–molecule scattering cross-sections. I. Experimental techniques and data for diatomic molecules. *Physics Reports*, 357(3–5):215–458, January 2002.

62. M. A. Morrison. Near threshold electron-molecule scattering. *Advances in Atomic Molecular and Optical Physics*, 24:51, 1988.

63. P. L. Bhatnagar, E. P. Gross, and M. Krook. A model for collision processes in gases. I. Small amplitude processes in charged and neutral one-component systems. *Physical. Review*, 94(3):511–516, 1954.

64. R. D. White, R. E. Robson, B. Schmidt, and M. A. Morrison. Is the classical two-term approximation of electron kinetic theory satisfactory for swarms and plasmas? *Journal of Physics D: Applied Physics*, 36:3125, 2003.

65. J. Ross, J. C. Light, and K. E. Schuler. Kinetic processes in gases and plasmas. In *Kinetic Processes in Gases and Plasmas*, edited by A. R. Hochsttm. New York, NY: Academic Press, p. 281, 1969.

66. L. Waldmann. Transport phenomena in gases at moderate pressure. In *Handbuch der Physik*, S. Flügge (ed.), Berlin: Springer, 295–514, 1958.

67. R. F. Snider. Quantum mechanical modified Boltzmann equation for degenerate internal states. *Journal of Chemical Physics*, 32:1051, 1960.

68. R. F. Snider. Relaxation and transport of molecular systems in the gas phase. *International Reviews in Physical Chemistry*, 17(2):185–225, 1998.

69. G. J. Boyle, W. J. Tattersall, D. G. Cocks, S. Dujko, and R. D. White. Kinetic theory of positron-impact ionization in gases. *Physical Review A*, 91(5):1–13, 2015.

70. R. Zallen. *The Physics of Amorphous Solids*. New York, NY: John Wiley & Sons, 1983.

71. H. Bässler. Charge transport in disordered organic photoconductors a Monte Carlo simulation study. *Physica Status Solidi (B)*, 175(1):15–56, January 1993.

72. R. E. Robson and B. V. Paranjape. Interaction of plasma and lattice waves in piezoelectric semiconductors. *Physica Status Solidi*, 59:641, 1973.

73. C. N. Likos. Effective interactions in soft condensed matter physics. *Physics Report*, 348:267–439, 2001.

74. Y. Sakai. Quasifree electron transport under electric field in nonpolar simple-structured condensed matters. *Journal of Physics D: Applied Physics*, 40(24):R441–R452, December 2007.

75. G. J. Boyle, R. P. McEachran, D. Cocks, and R. D. White. Electron scattering and transport in liquid Argon. *Journal of Chemical Physics*, 142:154507, 2015.

76. L. Van Hove. Correlations in space and time and Born approximation scattering in systems of interacting particles. *Physical Review*, 95(1):249–262, 1954.

77. E. B. Wagner, F. J. Davis, and G. S. Hurst. Time of flight investigations of electron transport in some atomic and molecular gases. *Journal of Chemical Physics*, 47:3138, 1967.

78. R. D. White, R .E. Robson, and K .F. Ness. Anomalous anisotropic diffusion of electron swarms in ac electric fields. *Australian Journal of Physics*, 48(6):925–937, 1995.

79. Z. Lj. Petrović, R. W. Crompton, and G. N. Haddad. Model calculations of negative differential conductivity in gases. *Australian Journal of Physics*, 37:23–34, 1984.

80. R. E. Robson. Generalized Einstein relation and negative differential conductivity in gases. *Australian Journal of Physics*, 37:35, 1984.

81. B. Philippa, C. Vijila, R. D. White, P. Sonar, P. L. Burn, P. Meredith, and A. Pivrikas. Time-independent charge carrier mobility in a model polymer: Fullerene organic solar cell. *Organic Electronics*, 16:205–211, November 2014.

82. B. Philippa, M. Stolterfoht, R. D. White, M. Velusamy, P. L. Burn, P. Meredith, and A. Pivrikas. Molecular weight dependent bimolecular recombination in organic solar cells. *Journal of Chemical Physics*, 141:054903, 2014.

83. R. E. Robson. Physics of reacting particle swarms in gases. *Journal of Chemical Physics*, 85(8):4486, 1986.

84. R. E. Robson and K. F. Ness. Physics of reacting particle swarms. III. Effects of ionization upon transport coefficients. *Journal of Chemical Physics*, 89(8):4815, 1988.

85. P. Nicoletopoulos and R. Robson. Periodic electron structures in gases: A fluid model of the window phenomenon. *Physical Review Letters*, 100: 124502, 2008.

86. K. D. Knierim. Time-dependent moment theory of hot-atom reactions. *Journal of Chemical Physics*, 75(3):1159, 1981.

87. J. Lucas and H. T. Saelee. A comparison of a Monte Carlo simulation and the Boltzmann solution for electron swarm motion in gases. *Journal of Physics D: Applied Physics*, 8(6):640, 1975.

88. L. Verlet and J. J. Weis. Equilibrium theory of simple liquids. *Physical Review A*, 5:939, 1972.

89. W. Van Megen and P. N. Pusey. Dynamic light-scattering study of the glass transition in a colloidal suspension. *Physical Review A*, 43(10):5429–5441, 1991.

90. G. J. Boyle, R. D. White, R. E. Robson, S. Dujko, and Z. Lj. Petrović. On the approximation of transport properties in structured materials using momentum-transfer theory. *New Journal of Physics*, 14(4):045011, April 2012.

91. R. D. White and R. E. Robson. Multiterm solution of a generalized Boltzmann kinetic equation for electron and positron transport in structured and soft condensed matter. *Physical Review E*, 84(3):031125, September 2011.

92. R. E. Robson and K. F. Ness. Velocity distribution function and transport coefficients of electron swarms in gases: Spherical-harmonics decomposition of Boltzmann's equation. *Physical Review A*, 33(3):2068–2077, 1986.

93. R. E. Robson, K. F. Ness, G. E. Sneddon, and L. A. Viehland. Comment on the discrete ordinate method in the kinetic theory of gases. *Journal of Computational Physics*, 92(1):213–229, January 1991.

94. K. Kumar. Talmi transformation for unequal mass particles and related formulas. *Journal of Mathematical Physics*, 7:671, 1966.

95. R. D. White, R. E. Robson, K. F. Ness, and B. Li. Charged-particle transport in gases in electric and magnetic fields crossed at arbitrary

angles: Multiterm solution of Boltzmann's equation. *Physical. Review E*, 27(5):1249, 1999.

96. L. A. Viehland. Velocity distribution functions and transport coefficients of atomic ions in atomic gases by a Gram-Charlier approach. *Chemical Physics*, 179(1):71–92, 1994.

97. R. D. White, R. E. Robson, and K. F. Ness. Computation of electron and ion transport properties in gases. *Computer Physics Communications*, 142:349–355, 2001.

98. C. Mark. The spherical harmonics methods, II (application to problems with plane & spherical symmetry). *Atomic Energy of Canada Limited*, CRT-338, Chalk River, Ontario, 1–96, 1957.

99. J. H. Parker and J. J. Lowke. Theory of electron diffusion parallel to electric fields. I. Theory. *Physical Review*, 181(1):290–301, 1969.

100. M. M. R. Williams. *Mathematical Methods in Particle Transport Theory*. London: Butterworths, 1971.

101. F. B. Hildebrandt. *Methods in Applied Mathematics*. Upper Saddle River, NJ: Prentice-Hall, 1965.

102. R. E. Robson. Nonlinear diffusion of ions in a gas. *Australian Journal of Physics*, 28:523–531, 1975.

103. J. H. Parker. Position- and time-dependent diffusion modes for electrons in gases. *Physical Review*, 139(6A):A1792, 1965.

104. B. D. Fried and S. D. Conte. The Plasma Dispersion Function: The Hilbert Transform of the Gaussian. New York, NY: Academic Press, 1961.

105. B. Philippa, R. White, and R. Robson. Analytic solution of the fractional advection-diffusion equation for the time-of-flight experiment in a finite geometry. *Physical Review E*, 84(4):041138, October 2011.

106. R. Robson and A. Blumen. Analytically solvable model in fractional kinetic theory. *Physical Review E*, 71(6):61104, 2005.

107. H. Margeneau. Conduction and dispersion of ionized gases at high frequencies. *Physical Review*, 69:508, 1946.

108. R. E. Robson, R. D. White, and T. Makabe. Charged particle transport in harmonically varying electric fields: Foundations and phenomenology. *Annals of Physics*, 261:74113, 1997.

109. R. D. White, R. E. Robson, and K. F. Ness. Nonconservative charged-particle swarms in ac electric fields. *Physical Review E*, 60(6 Pt B):7457–7472, December 1999.

110. Z. Lj. Petrovic, Z. M. Raspopović, S. Dujko, T. Makabe, and Z. M. Raspopovic. Kinetic phenomena in electron transport in radio-frequency fields. *Applied Surface Science*, 192(1–4):1–25, May 2002.

111. R. D. White. Mass effects of light ion swarms in ac electric fields. *Physical Review E*, 64(5):56409, 2001.

112. H. R. Skullerud. Longitudinal diffusion of electrons in electrostatic fields in gases. *Journal of Physics B: Atomic and Molecular Physics*, 2(6):696–705, 1969.

113. R. D. White, Kinetic theory of charged particle swarms in a.c. fields. PhD Thesis, James Cook University, 1997.

114. R. E. Robson, R. D. White, and M. Hildebrandt. One hundred years of the Franck-Hertz experiment. *European Physical Journal D*, 68(7):188, July 2014.

115. J. Franck and G. Hertz. Die Bestätigung der Bohrschen Atomtheorie im optischen Spektrum durch Untersuchungen der unelastischen Zusammenstöße langsamer Elektronen mit Gasmolekülen. *Physikalische Zeitschrift*, 20:132, 1919.

116. J. Fletcher. Non-equilibrium in low pressure rare gas discharges. *Journal of Physics D*, 18:221, 1985.

117. R. E. Robson, B. Li, and R. D. White. Spatially periodic structures in electron swarms and the Franck-Hertz experiment. *Journal of Physics B: Atomic*, 33:507–520, 2000.

118. R. D. White, R. E. Robson, P. Nicoletopoulos, and S. Dujko. Periodic structures in the Franck–Hertz experiment with neon: Boltzmann equation and Monte-Carlo analysis. *European Physical Journal D*, 66(5):117, May 2012.

119. G. F. Hanne. What really happens in the Franck–Hertz experiment with mercury? *American Journal of Physics*, 56:696, 1988.

120. P. Magyar, I. Korolov, and Z. Donkó. Photoelectric Franck–Hertz experiment and its kinetic analysis by Monte Carlo simulation. *Physical Review E—Statistical, Nonlinear, and Soft Matter Physics*, 85(5):1–10, 2012.

121. F. Sigeneger, N. A. Dyatko, and R. Winkler. Spatial electron relaxation: Comparison of Monte Carlo and Boltzmann equation results. *Plasma Chemistry and Plasma Processing*, 23(1):103–116, 2003.

122. G. Rapior, K. Sengstock, and V. Baev. New features of the Franck-Hertz experiment. *American Journal of Physics*, 74(5):423–428, 2006.

123. B. Li, Hydrodynamic and non-hydrodynamic swarms. PhD Thesis, James Cook University, 1999.

124. M. Charlton and J. W. Humberston. *Positron Physics*. Cambridge: Cambridge University Press, 2001.

125. M. Charlton. Positron transport in gases. *Journal of Physics: Conference Series*, 162:012003, April 2009.

126. R. D. White and R. E. Robson. Positron kinetics in soft condensed matter. *Physical Review Letter*, 102(23):230602, 2009.

127. G. Garcia, Z. Lj. Petrovic, R. D. White, and St. J. Buckman. Monte carlo model of positron transport in water: Track structures based on atomic and molecular scattering data for positrons. *IEEE Transactions on Plasma Science*, 39(11):2962, 2011.

128. Z. Lj. Petrović, S. Marjanović, S. Dujko, A. Banković, G. Malović, S. Buckman, G. Garcia, R. White, and M. Brunger. On the use of Monte Carlo simulations to model transport of positrons in gases and liquids. *Applied Radiation and Isotopes*, 83:148–154, 2013.

129. R. D. White, M. J. Brunger, N. A. Garland, R. E. Robson, K. F. Ness, G. Garcia, J. de Urquijo, S. Dujko, and Z. Lj. Petrović. Electron swarm transport in THF and water mixtures. *European Physical Journal D*, 68(5):125, May 2014.

130. S. R. Cherry, J. A. Sorensen, and M. E. Phelps. *Physics in Nuclear Medicine*. Philadelphia, PA: Saunders, 2003.

131. M. Ahmadi, B. X. R. Alves, C. J. Baker, W. Bertsche, E. Butler, A. Capra, C. Carruth, C. L. Cesar, M. Charlton, S. Cohen, R. Collister, S. Eriksson, A. Evans, N. Evetts, J. Fajans, T. Friesen, M. C. Fujiwara, D. R. Gill, A. Gutierrez, J. S. Hangst, W. N. Hardy, M. E. Hayden, C. A. Isaac, A. Ishida, M. A. Johnson, S. A. Jones, S. Jonsell, L. Kurchaninov, N. Madsen, M. Mathers, D. Maxwell, J. T. K. McKenna, S. Menary, J. M. Michan, T. Momose, J. J. Munich, P. Nolan, K. Olchanski, A. Olin, P. Pusa, C. Ø. Rasmussen, F. Robicheaux, R. L. Sacramento, M. Sameed, E. Sarid, D. M. Silveira, S. Stracka, G. Stutter, C. So, T. D. Tharp, J. E. Thompson, R. I. Thompson, D. P. van der Werf, and J. S. Wurtele. Observation of the 1S–2S transition in trapped antihydrogen. *Nature*, 541:506–510, 2016.

132. W. Tattersall, R. D. White, R. E. Robson, J. P. Sullivan, and S. J. Buckman. Simulations of pulses in a buffer gas positron trap. *Journal of Physics: Conference Series*, 262:012057, January 2011.

133. W. Tattersall, L. Chiari, J. R. Machacek, E. Anderson, R. D. White, M. J. Brunger, S. J. Buckman, G. Garcia, F. Blanco, and J. P. Sullivan. Positron interactions with water–total elastic, total inelastic, and elas-

tic differential cross section measurements. *Journal of Chemical Physics*, 140(4):044320, 2014.

134. S. Marjanović, A. Banković, R. D. White, S. J. Buckman, G. Garcia, G. Malović, S. Dujko, and Z. Lj. Petrović. Chemistry induced during the thermalization and transport of positrons and secondary electrons in gases and liquids. *Plasma Sources Science and Technology*, 24:025016, 2015.

135. A. Banković, S. Dujko, R. D. White, J. P. Marler, S. J. Buckman, S. Marjanović, G. Malović, G. García, and Z. Lj. Petrović. Positron transport in water vapour. *New Journal of Physics*, 14:035003, 2012.

136. Y. S. Badyal, M. L. Saboungi, D. L. Price, S. D. Shastri, D. R. Haeffner, and A. K. Soper. Electron distribution in water. *Journal of Chemical Physics*, 112(21):9206, 2000.

137. G. J. M. Hagelaar and L. C. Pitchford. Solving the Boltzmann equation to obtain electron transport coefficients and rate coefficients for fluid models. *Plasma Sources Science and Technology*, 14(4):722–733, November 2005.

138. R. D. White, R. E. Robson, and K. F. Ness. Visualization of ion and electron velocity distribution functions in electric and magnetic fields. *Journal of Physics D: Applied Physics*, 34:2205–2210, 2001.

139. S. F. Biagi. A multiterm Boltzmann analysis of drift velocity, diffusion, gain and magnetic-field effects in argon-methane-water-vapour mixtures. *Nuclear Instruments and Methods in Physics Research Section A*, 283:716, 1989.

140. D. Loffhagen and R. Winkler. Temporal relaxation of plasma electrons acted upon by direct current electric and magnetic fields. *IEEE Transactions on Plasma Science*, 27:1262–1270, 1999.

141. R. D. White, K. F. Ness, and R. E. Robson. Velocity distribution functions for electron swarms in methane in electric and magnetic fields. *Journal of Physics D: Applied Physics*, 32:1842, 1999.

142. R. E. Robson. Approximate formulas for ion and electron transport coefficients in crossed electric and magnetic fields. *Australian Journal of Physics*, 47(2):279–304, 1994.

143. G. F. Knoll. *Radiation Detection and Measurement*. New York, NY: John Wiley & Sons, 2010.

144. W. R. Leo. *Techniques for Nuclear and Particle Physics Experiments*. Berlin: Springer-Verlag, 1994.

145. E. Rutherford and H. Geiger. An electrical method of counting the number of α-particles from radio-active substances. *Proceedings of the Royal Society A*, 81:141–161, 1908.

146. H. Geiger. Über eine einfache Methode zur Zählung von α- und β-Teilchen. *Verhandlungen der Deutschen Physikalischen Gesellschaft*, 15:534–539, 1913.

147. H. Geiger and W. Müller. Elektronenzählrohr zur Messung schwächster Aktivitäten. *Naturwissenschaften*, 16:617–618, 1928.

148. H. Geiger and W. Müller. Das Elektronenzählrohr—Wirkungsweise und Herstellung eines Zählrohres. *Physikalische Zeitschrift*, 29:839–841, 1928.

149. H. Geiger and W. Müller. Technische Bemerkungen zum Elektronenzählrohr. *Physikalische Zeitschrift*, 30:489–493, 1929.

150. H. Geiger and W. Müller. Demonstration des Elektronenzählrohrs. *Physikalische Zeitschrift*, 30:523, 1929.

151. G. Charpak. The use of multiwire proportional counters to select and localize charged particles. *Nuclear Instruments and Methods*, 62:262–268, 1968.

152. A. H. Walenta, J. Heintze, and B. Schürlein. The multiwire drift chamber - A new type of proportional wire chamber. *Nuclear Instruments and Methods*, 92:373–380, 1971.

153. J. N. Marx and D. R. Nygren. The time projection chamber. *Physics Today*, 31:46–53, 1978.

154. V. V. Parkhumchuk, Yu. N. Pestov, and N. N. Petrovykh. A spark counter with large area. *Nuclear Instruments and Methods*, 93:269–270, 1971.

155. F. Sauli and A. Sharma. Micropattern gaseous detectors. *Annual Review of Nuclear and Particle Science*, 49:341–388, 1999.

156. W. Blum, W. Riegler, and L. Rolandi. *Particle Detection with Drift Chambers*. Berlin Heidelberg: Springer-Verlag, 2008.

157. E. Nappi and V. Peskov. *Imaging Gaseous Detectors and their Applications*. Weinheim: Wiley-VCH Verlag GmbH & Co KGaA, 2013.

158. T. Francke and V. Peskov. *Innovative Applications and Developments of Micro-Pattern Gaseous Detectors*. Hershley: IGI Global, 2014.

159. C. Grupen and B. Shwartz. *Particle Detectors*. Cambridge: Cambridge University Press, 2008.

160. F. Sauli. GEM: A new concept for electron amplification in gas detectors. *Nuclear Instruments and Methods in Physics Research A*, 386:531–534, 1997.

161. L. Malter. Thin film field emission. *Physical Review*, 50:48–58, 1936.

162. H. Raether. *Electron Avalanches and Breakdown in Gases*. London: Butterworths, 1964.

163. W. D. B. Spatz. The factors influencing the plateau characteristics of self-quenching Geiger-Müller counters. *Physical Review*, 64:236–240, 1943.

164. S. S. Friedland. On the life of self-quenching counters. *Physical Review*, 74:898–901, 1948.

165. E. C. Farmer and S. C. Brown. A study of the deterioration of methane-filled Geiger-Müller counters. *Physical Review*, 74:902–905, 1948.

166. J. Va'vra. Review of wire chamber aging. *Nuclear Instruments and Methods in Physics Research A*, 252:547–563, 1986.

167. J. A. Kadyk. Wire chamber aging. *Nuclear Instruments and Methods in Physics Research A*, 300:436–479, 1991.

168. J. Va'vra. Physics and chemistry of aging-early developments. *Nuclear Instruments and Methods in Physics Research A*, 515:1–14, 2003.

169. H. Yasuda. New insights into aging phenomena from plasma chemistry. *Nuclear Instruments and Methods in Physics Research A*, 515:15–30, 2003.

170. G. Charpak, F. Sauli, and W. Duinkre. High-accuracy drift chambers and their use in strong magnetic fields. *Nuclear Instruments and Methods*, 108:413–426, 1973.

171. T. E. O. Ericson, V. W. Hughes, and D. E. Nagle. *The Meson Factories*. Berkeley, CA: University of California Press, 1991.

172. Los Alamos National Laboratory (LANL), Los Alamos, USA (http://www.lanl.gov/); Paul Scherrer Institut (PSI), Villigen, Switzerland (https://www.psi.ch/); Tri University Meson Facility (TRIUMF), Vancouver, Canada (http://www.triumf.ca/); ISIS Science & Technology Facilities Council (ISIS), Harwell Oxford, UK (http://www.isis.stfc.ac.uk/); Japan Proton Accelerator Research Complex (J-PARC), Tokai, Japan (https://j-parc.jp/)

173. D. Taqqu. Compression and extraction of stopped muons. *Physical Review Letters*, 97(19):10–13, 2006.

174. Y. Bao, A. Antognini, W. Bertl, M. Hildebrandt, K. Siang Khaw, K. Kirch, A. Papa, C. Petitjean, F. M. Piegsa, S. Ritt, K. Sedlak, A. Stoykov, and D. Taqqu. Muon cooling: Longitudinal compression. *Physical Review Letters*, 112(22):1–5, 2014.

175. V. A. Andreev, T. I. Banks, R. M. Caray, T. A. Case, S. M. Clayton, K. M. Crowe, J. Deutsch, J. Egger, S. J. Freedman, V. A. Ganzha, T. Gorringe, F. E. Gray, D. W. Hertzog, M. Hildebrandt, P. Kammel, B. Kiburg, S. Knaack, P. A. Kravtsov, A. G. Krivshich, B. Lauss, K. L. Lynch, E. M. Maev, O. E. Maev, F. Mulhauser, C. Petitjean, G. E. Petrov, R. Prieels, G. N. Schapkin, G. G. Semenchuk, M. A. Soroka, V. Tishchenko, A. A. Vasilyev, A. A. Vorobyov, M. E. Vznuzdaev, and P. Winter. Measurement of the muon capture on the proton to 1% precision and determination of the pseudoscalar coupling g_P. *Physical Review Letters*, 110:1–5, 2013.

176. J. Egger, D. Fahrni, M. Hildebrandt, A. Hofer, L. Meier, C. Petitjean, V. A. Andreev, T. I. Banks, S. M. Clayton, V. A. Ganzha, F. E. Gray, P. Kammel, B. Kiburg, P. A. Kravtsov, A. G. Krivshich, B. Lauss, E. M. Maev, O. E. Maev, G. Petrov, G. G. Semenchuk, A. A. Vasilyev, A. A. Vorobyov, M. E. Vznuzdaev, and P. Winter. A high-pressure hydrogen time projection chamber for the MuCap experiment. *European Physical Journal A*, 50(10):1–16, 2014.

177. E. A. Mason, S. L. Lin, and I. R. Gatland. Mobility and diffusion of protons and deuterons in helium-a runaway effect. *Journal of Physics B*, 12:4179, 1979.

178. S. E. Jones. Muon-catalysed fusion revisited. *Nature*, 321:127, 1989.

179. S. E. Jones. Observation of unexpected density effects in muon-catalyzed *d-t* fusion. *Physical Review Letters*, 56:588–591, 1986.

180. K. Nagamine. *Introductory Muon Science*. Cambridge: Cambridge University Press, 2007.

181. C. Petitjean. The μCF experiments at PSI–A conclusive review. *Hyperfine Interactions*, 138:191–201, 2001.

182. R. E. Robson. Physics of reacting particle swarms. II. The muon-catalyzed cold fusion cycle. *Journal of Chemical Physics*, 88(1):198, 1988.

183. K. F. Ness and R. E. Robson. Motion of muons in heavy hydrogen in an applied electrostatic field. *Physical Review A*, 39(12):6596–6599, 1989.

184. R. E. Robson and C. L. Mayocchi. Turbulent countergradient flow as a problem in kinetic theory. *AIP Conference Proceedings*, 414:255–267, 1997.

185. L. A. Viehland. Ion–atom interaction potentials and transport properties. *Computer Physics Communication*, 142:7, 2001.

186. L. A. Cottrill, A. B. Langdon, B. F. Lasinski, S. M. Lund, K. Molvig, M. Tabak, R. P. J. Town, and E. A. Williams. Kinetic and collisional effects on the linear evolution of fast ignition relevant beam instabilities. *Physics of Plasmas*, 15(8):082108, 2008.

187. J. P. England, M. T. Elford, and R. W. Crompton. A study of the vibrational excitation of H2 by measurements of the drift velocity of electrons in H2 mixtures. *Australian Journal of Physics*, 41(4):573, 1988.

188. R. W. Crompton and M. A. Morrison. Analyses of recent experimental and theoretical determinations of e-H_2 vibrational excitation cross sections: Assessing a long-standing controversy. *Australian Journal of Physics*, 46:203, 1993.

189. M. A. Morrison and W. K. Trail. Importance of bound-free correlation effects for vibrational excitation of molecules by electron impact: A sensitivity analysis. *Physical Review A*, 48(4):2874–2886, 1993.

190. S. J. Buckman, M. J. Brunger, D. S. Newman, G. Snitchler, S. Alston, D. W. Norcross, M. A. Morrison, B. C. Saha, G. Danby, and W. K. Trail. Near-threshold vibrational excitation of H_2 by electron impact: Resolution of discrepancies between experiment and theory stephen. *Physical Review Letters*, 65:3253–3256, 1991.

191. M. J. Brunger, S. J. Buckman, D. S. Newman, and D. T. Alle. Elastic scattering and rovibrational excitation of H_2 by low-energy electrons. *Journal of Physics B: Atomic, Molecular and Optical Physics*, 24(6):1435–1448, March 1991.

192. S. J. Buckman and M. J. Brunger. A critical comparison of electron scattering cross sections measured by single collision and swarm techniques. *Australian Journal of Physics*, 50:483–509, 1997.

193. M. A. Morrison, R. W. Crompton, B. C. Saha, and Z. Lj. Petrovic. Near-threshold rotational and vibrational excitation of H_2 by electron impact: Theory and Experiment. *Australian Journal of Physics*, 40:238–281, 1987.

194. A. G. Robertson, M. T. Elford, R. W. Crompton, M. A. Morrison, W. Sun, and W. K. Trail. Rotational and vibrational excitation of nitrogen by electron impact. *Australian Journal of Physics*, 50:441, 1997.

195. G. D. Morgan, Origin of Striations in Discharges. *Nature*, 172:542, 1953.

196. T. Wada and G. R. Freeman. Temperature, density, and electric-field effects on electron mobility in nitrogen vapor. *Physical Review A*, 24(2):1066–1076, 1981.

197. R. E. Robson and A. Prytz. The discrete ordinate/pseudo-spectral method: Review and application from a physicist's perspective. *Australian Journal of Physics*, 46:465–496, 1993.

198. T. Makabe and R. White. Expression for momentum-transfer scattering in inelastic collisions in electron transport in a collisional plasma. *Journal of Physics D: Applied Physics*, 48(48):485205, 2015.

Index

M

N

P

R

S

T